Carbon Capture, Storage and Use

Wilhelm Kuckshinrichs • Jürgen-Friedrich Hake
Editors

Carbon Capture, Storage and Use

Technical, Economic, Environmental and Societal Perspectives

Springer

Editors
Wilhelm Kuckshinrichs
Jürgen-Friedrich Hake
Institute of Energy and Climate
Research – Systems Analysis
and Technology Evaluation (IEK-STE)
Forschungszentrum Jülich GmbH
Jülich, Germany

Kind permission was given for the translation of German-language title "CO2-Abscheidung, -Speicherung und -Nutzung: Technische, wirtschaftliche, umweltseitige und gesellschaftliche Perspektive. Advances in Systems Analysis 2" published in 2012 by Forschungszentrum Jülich GmbH into English.

Translated by Language Services, Forschungszentrum Jülich GmbH

ISBN 978-3-319-11942-7 ISBN 978-3-319-11943-4 (eBook)
DOI 10.1007/978-3-319-11943-4
Springer Cham Heidelberg New York Dordrecht London

Library of Congress Control Number: 2014955444

Printed on acid-free paper

Springer is part of Springer Science+Business Media (www.springer.com)

Preface

In numerous national and international energy scenarios, high priority is being given to technologies for carbon capture and storage (CCS) as an option for reducing energy- and process-related CO_2 emissions. This is particularly important against the background of a globally increasing use of fossil fuels despite the growing significance of renewable energy.

Numerous core components for CCS have already reached a high level of technical maturity, even though the requirements for industrial-scale application in the energy system (first-generation processes) have revealed a need for further research and development, e.g. improved conversion efficiency, less severe environmental impacts of the use of scrubbing substances, storage that is safe in the long term, reduced investment costs, and the use of CO_2 for material production and energy generation. National and international research in this area is intensive, and some activities are already concentrating on second-generation processes.

In Germany, the transportation and storage of CO_2 in geological formations is a controversial topic. In other potential user countries in Europe and throughout the world, scepticism is growing in different sections of society with regard to the implementation of CCS technologies.

Against this background, an integrated technology study is urgently required to discuss and assess the technical, economic, environmental, and social perspectives of CCS technologies, even though a wealth of information and scientific expertise already exists on individual aspects. The Institute of Energy and Climate Research – Systems Analysis and Technology Evaluation (IEK-STE) at Forschungszentrum Jülich has therefore published this study compiled by an interdisciplinary team of engineers, economists, social scientists, and political scientists. The preliminary work for this study was completed in a series of projects funded by the Helmholtz Association and third parties. Further scientific expertise has been incorporated with contributions on using and storing CO_2 written by our colleagues at RWTH Aachen University and the German Research Centre for Geosciences in Potsdam.

The editors would like to thank the scientists for their contributions and for their unwavering motivation – right from the initial idea and the development of the framework to the actual creation and editing of the manuscripts.

Jülich, Germany
March 2014

Wilhelm Kuckshinrichs
Jürgen-Friedrich Hake

Contents

List of Figures

List of Tables

Contributors

Richard Bongartz Institute of Energy and Climate Research – Systems Analysis and Technology Evaluation (IEK-STE), Forschungszentrum Jülich GmbH, Jülich, Germany

Wolfgang Fischer Institute of Energy and Climate Research – Systems Analysis and Technology Evaluation (IEK-STE), Forschungszentrum Jülich GmbH, Jülich, Germany

Johannes Fleer Institute of Energy and Climate Research – Systems Analysis and Technology Evaluation (IEK-STE), Forschungszentrum Jülich GmbH, Jülich, Germany

Joachim Geske Institute of Energy and Climate Research – Systems Analysis and Technology Evaluation (IEK-STE), Forschungszentrum Jülich GmbH, Jülich, Germany

Jürgen-Friedrich Hake Institute of Energy and Climate Research – Systems Analysis and Technology Evaluation (IEK-STE), Forschungszentrum Jülich GmbH, Jülich, Germany

Heidi Heinrichs Institute of Energy and Climate Research – Systems Analysis and Technology Evaluation (IEK-STE), Forschungszentrum Jülich GmbH, Jülich, Germany

Thomas Kempka GFZ German Research Centre for Geosciences, Potsdam, Germany

Wilhelm Kuckshinrichs Institute of Energy and Climate Research – Systems Analysis and Technology Evaluation (IEK-STE), Forschungszentrum Jülich GmbH, Jülich, Germany

Michael Kühn GFZ German Research Centre for Geosciences, Potsdam, Germany

Walter Leitner CAT Catalytic Center and Chair of Technical Chemistry and Petrochemistry, Institute of Technical und Macromolecular Chemistry (ITMC), RWTH Aachen University, Aachen, Germany

Axel Liebscher GFZ German Research Centre for Geosciences, Potsdam, Germany

Jochen Linssen Institute of Energy and Climate Research – Systems Analysis and Technology Evaluation (IEK-STE), Forschungszentrum Jülich GmbH, Jülich, Germany

Peter Markewitz Institute of Energy and Climate Research – Systems Analysis and Technology Evaluation (IEK-STE), Forschungszentrum Jülich GmbH, Jülich, Germany

Sonja Martens GFZ German Research Centre for Geosciences, Potsdam, Germany

Dag Martinsen Institute of Energy and Climate Research – Systems Analysis and Technology Evaluation (IEK-STE), Forschungszentrum Jülich GmbH, Jülich, Germany

Josefine Marx Institute of Energy and Climate Research – Systems Analysis and Technology Evaluation (IEK-STE), Forschungszentrum Jülich GmbH, Jülich, Germany

Fabian Möller GFZ German Research Centre for Geosciences, Potsdam, Germany

Thomas E. Müller CAT Catalytic Center and Chair of Technical Chemistry and Petrochemistry, Institute of Technical und Macromolecular Chemistry (ITMC), RWTH Aachen University, Aachen, Germany

Olga Schenk School of Public and Environmental Affairs, Indiana University, Bloomington, IN, USA

Andrea Schreiber Institute of Energy and Climate Research – Systems Analysis and Technology Evaluation (IEK-STE), Forschungszentrum Jülich GmbH, Jülich, Germany

Diana Schumann Institute of Energy and Climate Research – Systems Analysis and Technology Evaluation (IEK-STE), Forschungszentrum Jülich GmbH, Jülich, Germany

Martin Streibel GFZ German Research Centre for Geosciences, Potsdam, Germany

Stefan Vögele Institute of Energy and Climate Research – Systems Analysis and Technology Evaluation (IEK-STE), Forschungszentrum Jülich GmbH, Jülich, Germany

Petra Zapp Institute of Energy and Climate Research – Systems Analysis and Technology Evaluation (IEK-STE), Forschungszentrum Jülich GmbH, Jülich, Germany

Chapter 1
Carbon Capture and Utilization as an Option for Climate Change Mitigation: Integrated Technology Assessment

Wilhelm Kuckshinrichs

Abstract Fossil-based energy conversion and energy-intensive industries are sources of a large part of global CO_2 emissions. Carbon capture and storage (CCS) technologies are regarded as important technical options to reduce worldwide CO_2 emissions. However, the discussion on the potential of CCS is highly controversial concerning four perspectives: *technology development, economic competitiveness, environmental and safety impacts,* and *social acceptance*. The following chapters focus on these aspects and analyze the potential and the possible role of CCS technologies. The study is based on methods of Integrated Technology Assessment. When regional considerations are important for evaluation, e.g. in case of social acceptance, the focus is on the German perspective.

Keywords Carbon capture and storage (CCS) • CO_2 utilization • CO_2 reduction • Assessment • Evaluation index

1.1 CCS as an Option for Climate Change Mitigation and CO_2 for Industrial Application

In order to limit the anthropogenic increase in the average global temperature by 2100 to 2 °C, the concentration of CO_2 in the atmosphere must be restricted to 450 ppmv according to the Intergovernmental Panel on Climate Change (IPCC). To achieve this target, global CO_2 emissions must be cut by 50 % by 2050 compared to levels in 1990. However, global energy consumption is growing year by year and the use of fossil energy carriers is not only continuing, but coal in particular is becoming even more important as an energy carrier globally.

In their analyses on stabilizing global CO_2 emissions, Pacala and Socolow identified strategies ('wedges') to help reduce future CO_2 emissions (Pacala and

W. Kuckshinrichs (✉)
Institute of Energy and Climate Research – Systems Analysis and Technology Evaluation (IEK-STE), Forschungszentrum Jülich GmbH, D-52425 Jülich, Germany
e-mail: w.kuckshinrichs@fz-juelich.de

W. Kuckshinrichs, J.-F. Hake (eds.), *Carbon Capture, Storage and Use*,
DOI 10.1007/978-3-319-11943-4_1

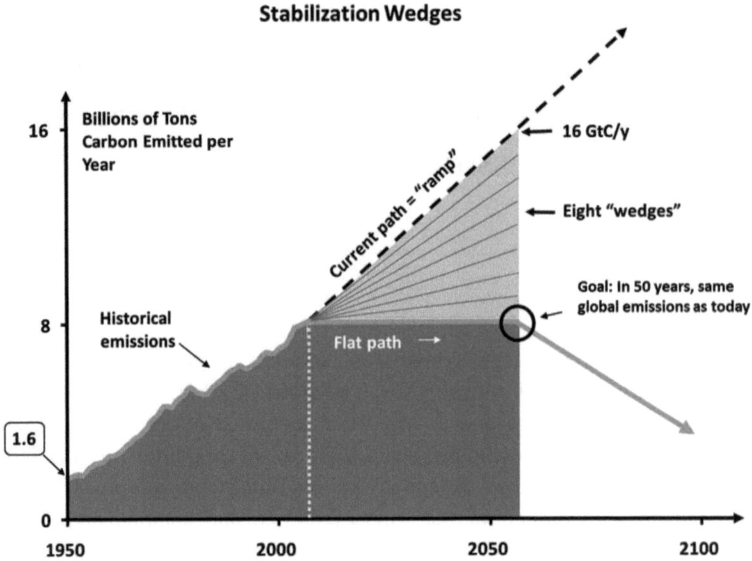

Fig. 1.1 Stabilization wedges for global CO_2 emissions (Source: Pacala and Socolow 2004; Carbon Mitigation Initiative (CMI) 2013)

Socolow 2004). A 'wedge' is a strategy or measure to reduce CO_2 emissions, which are forecast to increase in 50 years to 3.67 billion tonnes of CO_2 ($GtCO_2$) per year (= 1 GtC/a). Over 50 years, this represents a cumulative total of approx. 92 $GtCO_2$ (25 GtC). These wedges include energy efficiency, a fuel shift, nuclear energy, wind energy, solar energy, bioenergy, and natural CO_2 sinks, as well as carbon capture and storage (CCS) (Fig. 1.1).

Numerous analyses of and projections for the global energy system also emphasize the importance of CCS in strategies for reducing greenhouse gases, e.g. the Stern Report and the World Energy Outlook (IEA 2009b, 2010, 2011; Stern 2006). The IEA projects an increase in CO_2 emissions in a business-as-usual scenario from 29 $GtCO_2$ per year today to some 62 $GtCO_2$ per year by 2050 (IEA 2008). This would be accompanied by an increase in the concentration of CO_2 in the atmosphere to approx. 550 ppmv, and by a mean temperature rise of 3–4 °C. The IEA proposes two scenarios for reducing these emissions, both of which cover the period up to 2050. In the ACT Map scenario, a clear reduction in CO_2 is achieved, saving some 35 $GtCO_2$ per year by 2050 compared to the business-as-usual scenario. This would mean maintaining today's levels of CO_2 emissions in 2050, which would be equivalent to a CO_2 concentration of around 485 ppmv. The BLUE Map scenario goes even further, cutting CO_2 emissions in 2050 by 48 $GtCO_2$ per year, representing a reduction of 77 % compared to the business-as-usual scenario. This would be equivalent to a CO_2 concentration of around 445 ppmv in 2050.

In both cases, power generation would make the highest contribution of any sector and CCS would lead to the biggest reductions of any individual measure. CCS would reduce CO_2 emissions in the power sector by approx. 21 % in the ACT Map scenario and by approx. 26 % in the BLUE Map scenario. The results highlight the importance of CCS technology in the global context and show how attractive CCS is if stringent greenhouse gas reduction targets are to be achieved.

Worldwide, industrial processes are responsible for almost 30 % of CO_2 emissions (IEA 2009a), whereby some of these emissions are process-induced. CCS can therefore also help to reduce CO_2 emissions in industrial sectors (Gale 2012). The most pertinent sectors are the cement industry, the iron and steel industry, and the production of other metals, as well as industries that process crude oil.

In contrast, the current usage of CO_2 as an industrial gas amounts to approx. 20 Mt/a and as a chemical raw material around 110 Mt/a (Peters et al. 2011). The options for utilizing CO_2 in the future would mean that these two areas could contribute to a welcome, albeit limited, direct reduction in carbon dioxide emissions. The interest in utilizing carbon dioxide (CCU) stems primarily from the fact that CO_2 is a potentially recyclable material with an interesting application profile and great potential for the chemical industry. Carbon utilization would also positively affect the evaluation of strategies aiming to reduce CO_2 emissions if product-related CO_2 balances show a reduction in the emission of CO_2. In this way, the greenhouse gas carbon dioxide can be transformed on a limited scale into a raw material for the material value chain (Ausfelder and Bazzanella 2008) (see schematic in Fig. 1.2).

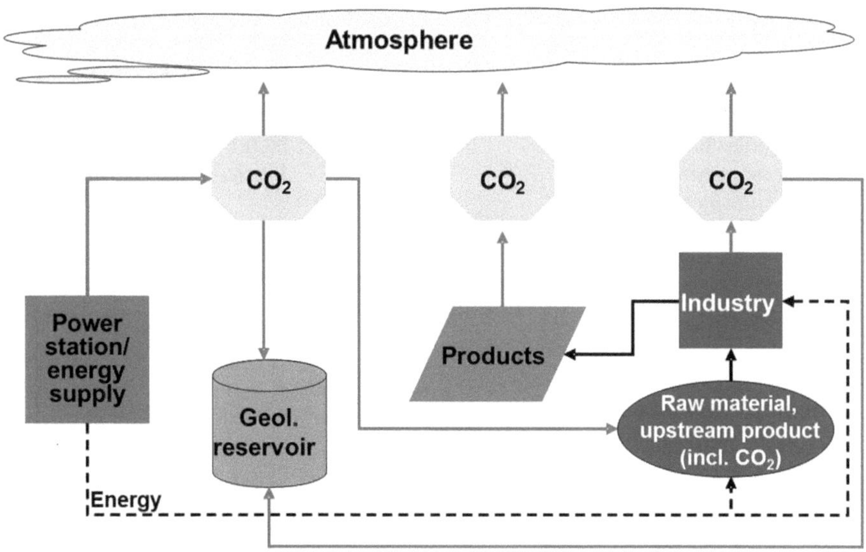

Fig. 1.2 Schematic of carbon capture and storage as well as the utilization of CO_2 as a raw material for manufacturing (Source: Kuckshinrichs et al. 2010)

1.2 Methodological Approach of an Integrated Technology Assessment for CCS and Structure of the Study

The objective of a technology evaluation is to determine the importance of a technology in relation to a set of criteria. The set of criteria selected here is rooted in the regulatory framework governing the concept of sustainable development, which has led to the need for the transformation of the energy sector in favour of sustainable technologies and systems. The principle involves investigating the development of energy technologies (and energy systems) in terms of their technical, economic, ecological, and social impacts, and thus evaluating what contribution technologies can make to the transformation of energy systems.

The range of methods for technology evaluations is very broad. They include technologically oriented methods (e.g. risk assessments), economically oriented methods (e.g. cost analyses), politically oriented methods (e.g. voting procedures), systematic considerations (e.g. cost-benefit analyses), and methods based on systems theory (e.g. scenario techniques) (Renn 2010). IEK-STE pursues a systems analysis approach here, which focuses on the interdependencies between technologies and their associated fields in the economy and in society, and is mainly based on quantitative modelling (Fig. 1.3).

This volume is a compilation of separate chapters written by a range of experts on the technological, economic, ecological and social aspects of CCS technologies. This structure allows specific aspects to be reviewed more closely on the basis of differentiated methodological approaches used to analyse possible technical

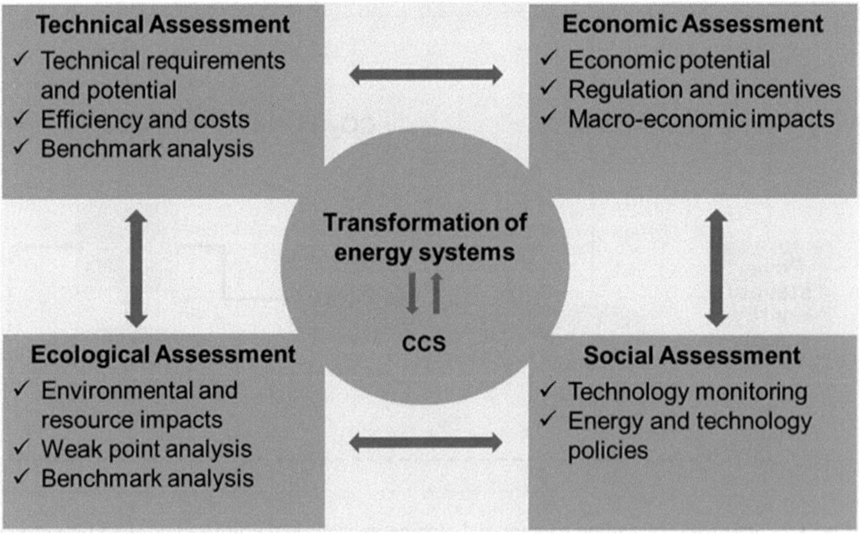

Fig. 1.3 Methodological approach of an integrated technology assessment of CCS

applications and prerequisites for application, as well as development potential, economic and social perspectives on applications in the energy sector and in industry, and also energy- and climate-policy aspects from a German and a European point of view.

1.2.1 Technical Potential, R&D Work, and Degree of Technical Maturity

Some of the technologies are characterized by a very different degree of maturity and only a few are already in commercial use. Notably, carbon capture in power plants has not yet been implemented on a commercial scale, and strategies for a broader utilization of CO_2 are still in their infancy. While some technologies are already in commercial use (e.g. enhanced oil recovery (EOR), production of urea and methanol), others are only being prepared for demonstration or are at the pilot stage (e.g. oxyfuel, production of aliphatic polycarbonates). Others again are at a very early stage of technical development (laboratory scale) or are only at the initial design phase (e.g. CO_2 membranes, artificial photosynthesis) (Fig. 1.4) (Markewitz et al. 2012).

The chapters in Part I of the study are dedicated to the technological state of the art and conceivable R&D approaches along the CCS and CCU process chain. *Markewitz and Bongartz* (Chap. 2) analyse the major development lines of first-generation carbon capture in power plants (post-combustion, pre-combustion,

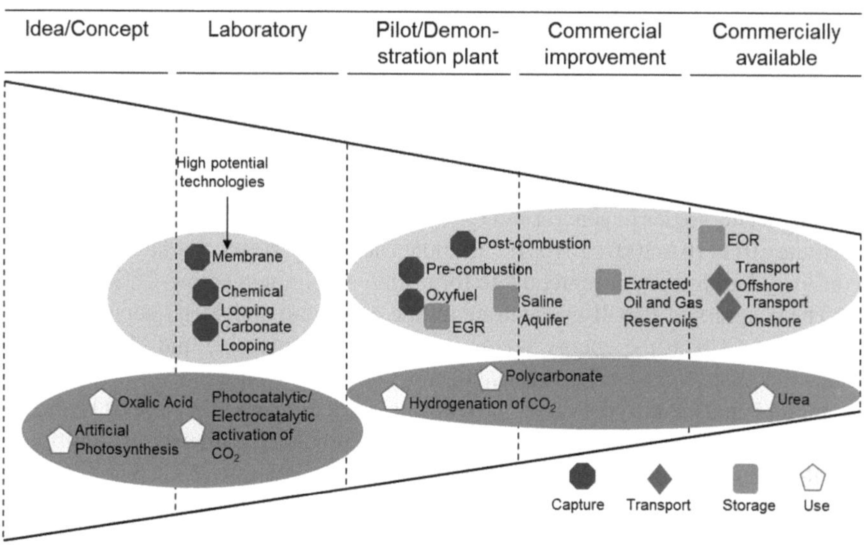

Fig. 1.4 Schematic of innovation stages for technologies for the capture, transportation, storage and utilization of CO_2 (Source: Adapted from McKinsey 2008)

oxyfuel) as well as energy-intensive and carbon-intensive industries. They also take a look at second-generation technologies such as membranes. For carbon capture systems, the most important considerations include possible improvements in efficiency, the influence of the purity of CO_2, the flexibility of system operation, and the retrofitting of coal-fired power plants. *Bongartz et al.* (Chap. 3) focus on the transportation of CO_2 and address safety issues as well as the purity of the CO_2 stream. *Müller et al.* (Chap. 4) take a look at the options and concepts for utilizing CO_2. In addition to organic-chemical usage as well as inorganic and material use, priority is given to product-related evaluation criteria such as CO_2 fixation (amount and duration), technical implementation, and total CO_2 balance. *Schreiber et al.* (Chap. 5) analyse the environmental impacts of the use of CCS technologies. Using a life cycle assessment, they create CCS process chains including upstream and downstream processes, and analyse them in their environmental impact categories. To conclude, *Kühn et al.* (Chap. 6) discuss safety issues and risks associated with the geological storage of CO_2. Here, the focus is on the underground retention of the compressed CO_2 stream and possible negative impacts on groundwater resources using the example of the Ketzin test site for carbon dioxide storage.

1.2.2 Application in Science and Industry

The use of CCS on a large scale in the energy sector and in industry can only be described within the framework of climate protection strategies. The additional costs for the implementation of CCS compared to the conventional conversion of fossil fuels into electricity are reflected in the internalization of CO_2 costs. CCS systems are characterized by high capital expenditure and long-term capital tie-up, which means that each investment decision must account for the long-term profit potential. The implications of climate, energy and technology policy decisions must be taken into consideration here, together with the development prospects of competing technologies, and the way in which society views energy and climate-friendly technologies in general and CCS in particular (ETP ZEP 2011; Global CCS Institute 2011; IEA 2007, 2010; IPCC 2005; McKinsey 2008). Social acceptance is considered an important prerequisite for testing and implementing CCS.

The chapters in Part II concentrate on the economic and social perspectives of the use of CCS in the energy sector, and in energy-intensive and CO_2-intensive industries. *Kuckshinrichs and Vögele* (Chap. 7) discuss the use of CCS in the energy sector and analyse the costs associated with electricity generation and CO_2 mitigation on the basis of technology-specific cost and process parameters. In addition, a merit-order approach is used to illustrate possible implications of CCS facilities for electricity prices and quantities of electricity, as well as the ensuing options for refinancing CCS investments. *Fleer and Kuckshinrichs* (Chap. 8) outline the costs of CCS application in energy- and CO_2-intensive industries using reference plants. *Geske* (Chap. 9) analyses the system characteristics of CCS infrastructures, and shows that the infrastructure cost function depends on the ratio of fixed to variable costs, as well as on the spatial distribution of CO_2 sources and storage facilities. With an energy system model, *Martinsen et al.* (Chap. 10) analyse cross-sector carbon mitigation strategies and their impacts on the energy and CO_2 balance. In this context, they estimate the system value should other technology lines be implemented instead of CCS. Using an acceptance analysis, *Schumann* (Chap. 11) discusses the awareness and knowledge of CCS, as well as spontaneous attitudes towards it, and how the risks and benefits of CCS are perceived in Germany. In addition, she looks at the factors that influence spontaneous attitudes towards CCS among the German population.

1.2.3 Framework for Energy and Climate Policy

Energy and industrial strategies for the development and utilization of CCS are embedded in the energy, climate and technology policy guidelines of the European Union and Germany. This is where the EU framework for the implementation of CCS (European Parliament and the Council 2009) and instruments for funding investments in demonstration projects (Europäisches Parlament und Rat 2009) come into play. The basis for German energy and climate policy is the German federal government's energy concept, which is rooted in the resolutions of 2010 and 2011 (Bundesregierung 2010, 2011), and rests upon the elements of CO_2 reduction, renewable energies, energy efficiency, and the move away from nuclear energy. To implement the European CCS Directive in national legislation, Germany has introduced a CCS law.

The chapters in Part III concentrate on aspects of energy and climate policy from a European and German perspective. *Fischer* (Chap. 12) analyses the legislative process for CCS in Germany with reference to the federal system, the parties, and social organizations in Germany. This is characterized by contradictory policies and conflicts. *Schenk and Hake* (Chap. 13) examine CCS policy in the European Union, and review political measures and challenges promoting the demonstration and commercial use of CCS. This part of the study concludes with *Hake and Schenk* (Chap. 14) analysing important international cooperation in the area of CCS and the significance of international cooperation for the implementation of CCS in Germany.

1.3 Energy and Industrial Policy Implications from a German Perspective

In the preceding chapters, the focus was on individual technical, economic, ecological, and social aspects which are important for a technology evaluation of CCS. The final chapter by *Kuckshinrichs and Markewitz* (Chap. 15) summarizes the central arguments, and draws a conclusion regarding the potential role that could be played by carbon capture and utilization within the framework of a German transformation strategy. In addition, the findings regarding prospects in Germany are presented in the European and international context.

References

Ausfelder F, Bazzanella A (2008) Verwertung und Speicherung von CO_2. Diskussionspapier der DECHEMA. DECHEMA, Frankfurt

Bundesregierung (2010) Energiekonzept für eine umweltschonende, zuverlässige und bezahlbare Energieversorgung, Berlin. http://www.bundesregierung.de/ContentArchiv/DE/Archiv17/_Anlagen/2012/02/energiekonzept-final.pdf?_blob=publicationFile&v=5

Bundesregierung (2011) Eckpunkte "Der Weg in die Energie der Zukunft", Berlin, http://www.bundesregierung.de/Content/DE/Artikel/2011/06/2011-06-06-energiewende-kabinettsbeschluss-doorpage-energiekonzept.html

Carbon Mitigation Initiative (CMI) (2013) *Stabilization wedges* [Online]. Princeton University, http://cmi.princeton.edu/wedges. (2012)

CS Institute Global (2011) Economic assessment of carbon capture and storage technologies. 2011 update. Global CCS Institute, Canberra

ETP ZEP (2011) The costs of CO_2 capture. http://www.zeroemissionsplatform.eu/library/publication/165-zep-cost-report-summary.html, 15 Aug 2012: European Technology Platform for Zero Emission Fossil Fuel Power Plants (ZEP)

Europäisches Parlament und Rat (2009) Verordnung über ein Programm zu Konjunkturbelebung durch eine finanzielle Unterstützung der Gemeinschaft zugunsten von Vorhaben im Energiebereich. Verordnung Nr. 663/2009, Brussels

European Parliament and the Council (2009) Directive on the geological storage of carbon dioxide. DIRECTIVE 2009/31/EC, Brussels

Gale J (2012) In: GHG I (ed) Challenges and opportunities of CO_2 capture and storage for the iron and steel industry. IEA GHG, London

IEA (2007) Climate policy uncertainty and investment risk. International Energy Agency IEA/OECD, Paris

IEA (2008) Energy technology perspectives – scenarios & strategies to 2050. International Energy Agency IEA/OECD, Paris

IEA (2009a) Energy technology transitions for industry. International Energy Agency IEA/OECD, Paris

IEA (2009b) World energy outlook 2009. International Energy Agency OECD/IEA, Paris

IEA (2010) Energy technology perspectives – scenarios & strategies to 2050. OECD/IEA, Paris

IEA (2011) World energy outlook 2011. International Energy Agency OECD/IEA, Paris

IPCC (2005) Carbon dioxide capture and storage – special report of the intergovernmental panel on climate change. Cambridge University Press, Cambridge

Kuckshinrichs W, Markewitz P, Linssen J, Zapp P, Peters M, Köhler B, Müller T, Leitner W (2010) Weltweite Innovationen bei der Entwicklung von CCS-Technologien und Möglichkeiten der Nutzung und des Recyclings von CO_2. Schriften des Forschungszentrums Jülich, Reihe Energie & Umwelt/Energy & Environment 60. Forschungszentrum Jülich, Jülich

Markewitz P, Kuckshinrichs W, Leitner W, Linssen J, Zapp P, Bongartz R, Schreiber A, Müller T (2012) Worldwide innovations in the development of carbon capture technologies and the utilization of CO_2. Energy Environ Sci 5:7281–7305

Mckinsey (2008) Carbon capture and storage: assessing the economics. www.mckinsey.com

Pacala S, Socolow R (2004) Stabilization wedges: solving the climate problem for the next 50 years with current technologies. Science 305:968–972
Peters M, Köhler B, Kuckshinrichs W, Leitner W, Markewitz P, Müller TE (2011) Chemical technologies for exploiting and recycling carbon dioxide into the value chain. ChemSusChem Special Issue Carbon Dioxide Recycling 4:1216–1240
Renn O (2010) Methoden und Verfahren der Technikfolgenabschätzung und der Technologiebewertung. Available: www.elib.uni-stuttgart.de/opus/volltexte/2010/5286/pdf/ren2.pdf
Stern N (2006) Stern review: the economics of climate change, London, http://webarchive.nationalarchives.gov.uk/+/http://www.hm-treasury.gov.uk/stern_review_report.htm. 21 June 2012

Part I
Technologies: Status and R&D Prospects

Chapter 2
Carbon Capture Technologies

Peter Markewitz and Richard Bongartz

Abstract This chapter focuses on carbon capture technologies that can be used in coal fired power plants and industrial processes. The three technology lines (post combustion, pre-combustion, oxyfuel) will be described in terms of state of the art, efficiency losses and advantages and disadvantages. An outlook will be given of further developments in the long term (second generation). Special attention will be paid to retrofitting options of existing coal fired power plants. An increasing share of highly volatile renewable power generation will change the flexibility requirements of coal fired power plants. Against this background flexibility options will be discussed for power plants with carbon capture technologies.

Keywords Carbon capture • Coal power plants • Industrial processes • Retrofitting • Flexibility

2.1 Introduction

According to the United Nations' Intergovernmental Panel on Climate Change (IPCC), CO_2 emissions must be reduced by at least 50 % if the expected temperature increase caused by the greenhouse effect is to be limited to around 2–3 °C. With an ambitious reduction scenario (Blue Map), the International Energy Agency (IEA) outlines what technical measures could be used to achieve a significant reduction (IEA 2010). In order to reduce emissions to 14 $GtCO_2$ by 2050, a wide range of reduction measures are essential (Fig. 2.1). The capture and storage of carbon dioxide (CCS) from large point sources is considered extremely important in this context. If CO_2 reduction targets are to be achieved, CCS technologies must reduce emissions by around 8.2 $GtCO_2$ according to the IEA. Accounting for a 19 % emissions reduction, CCS as a single measure makes the biggest contribution to reducing emissions in the IEA scenario.

This chapter is structured as follows. First, we will describe the carbon capture technologies that can be used in coal-fired power plants (Sect. 2.2). The state of the

P. Markewitz (✉) • R. Bongartz
Institute of Energy and Climate Research – Systems Analysis and Technology Evaluation (IEK-STE), Forschungszentrum Jülich GmbH, D-52425 Jülich, Germany
e-mail: p.markewitz@fz-juelich.de; r.bongartz@fz-juelich.de

W. Kuckshinrichs, J.-F. Hake (eds.), *Carbon Capture, Storage and Use*,
DOI 10.1007/978-3-319-11943-4_2

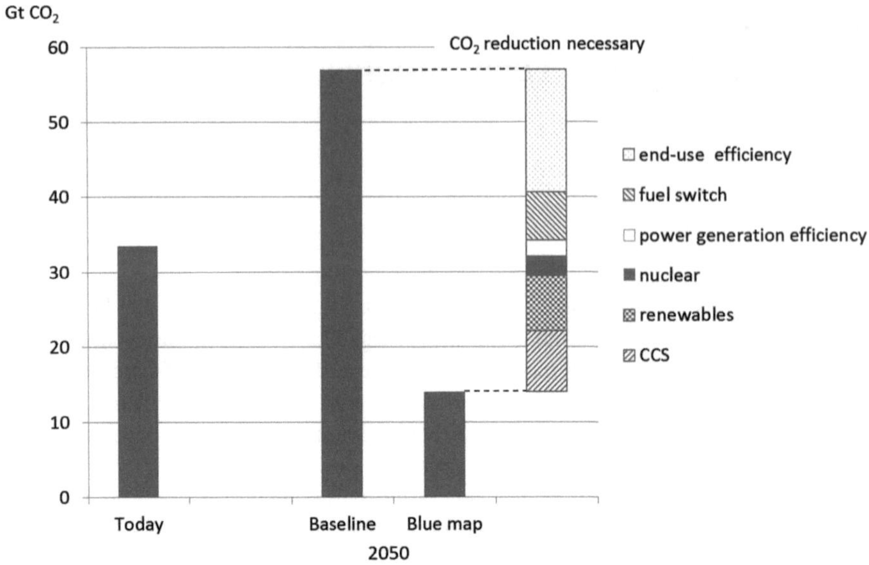

Fig. 2.1 Measures for reducing CO_2 in the International Energy Agency's Blue Map scenario (Source: IEA 2010)

art, efficiency losses and characteristics of the individual processes will be discussed. The current processes behind the three technology lines being discussed at present (post-combustion, oxyfuel, pre-combustion) will be outlined together with the 'second-generation' processes, which are still far from being ready for technical implementation but which have great development potential. Needless to say, an analysis of the operation of power plants with CCS cannot be performed in isolation, as the operating conditions must be taken into account within the context of the power generation system as a whole. For this reason, Sect. 2.3 takes a look at specific requirements (e.g. power plant operation mode, retrofitting the existing power plant fleet) that arise within a system context and are also decisive in implementing CCS technologies or CCS power plants.

As fossil-fired power plants cause a large share of global CO_2 emissions, present R&D work on carbon capture technologies concentrates on power plant application. However, the use of carbon capture processes is also conceivable in industrial processes, which likewise cause a significant share of global CO_2 emissions. Section 2.4 analyses which industrial applications and/or processes are of interest, as well as which of the capture processes currently being discussed could be used here. Carbon capture for biogas treatment, which has already been commercially implemented, will also be discussed.

2.2 Carbon Capture Technologies for Use in Coal-Fired Power Plants

Global CO_2 emissions in 2010 totalled approximately 31.3 GtCO_2 (Ziesing 2011a). With a share of 40 % of total emissions, fossil-fired power plants were the main emitters. The distribution is similar in Germany: fossil power plants produced around 42 % of total German emissions (2010: approx. 826 Mt) (Ziesing 2011b). Both the global and national CO_2 reduction targets can therefore only be achieved if CO_2 emissions from fossil-fired power plants can be significantly reduced. In addition to a transition to low-carbon fuels and an improved efficiency of power plants, carbon capture is another important reduction option. As carbon capture entails substantial efficiency losses, a high efficiency is generally desirable for the basic power plant process.

In the following, the focus will be on the carbon capture process and its integration into the power plant. For measures aiming to increase the efficiency of fossil power plants, the reader is directed to the existing literature (Markewitz et al. 2011a; Wietschel et al. 2010).

Numerous research and development projects are being conducted throughout the world on carbon capture. At present, three technical process routes are preferred (Fig. 2.2): these are post-combustion, the oxyfuel process, and pre-combustion. Other processes also exist which have a long way to go before they are ready for technical implementation but which also offer several advantages (e.g. great efficiency potential) compared to processes being developed today. They are often referred to as 'second-generation' processes. The three technology lines currently being discussed will be outlined here. The processes will be described in brief, as will the state of the art, the advantages and disadvantages, and an outlook will be given of further technical developments in the long term (second generation). The technical details of the processes will not be described here as these are provided in the extensive literature (see (Stolten and Scherer 2011; IPCC 2005; BMWi 2007; Kather et al. 2008)).

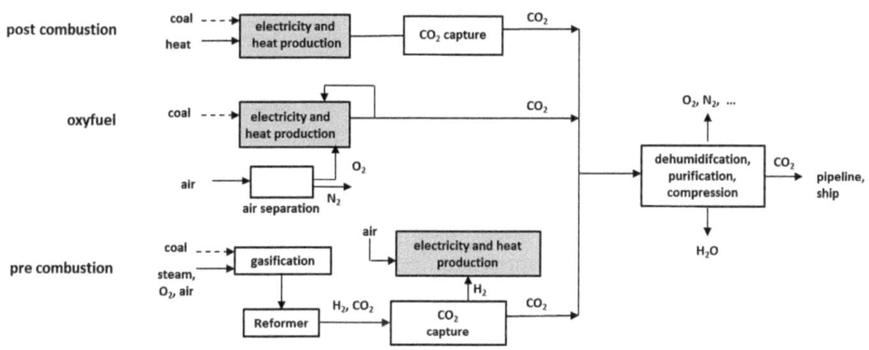

Fig. 2.2 Principles of CCS technologies

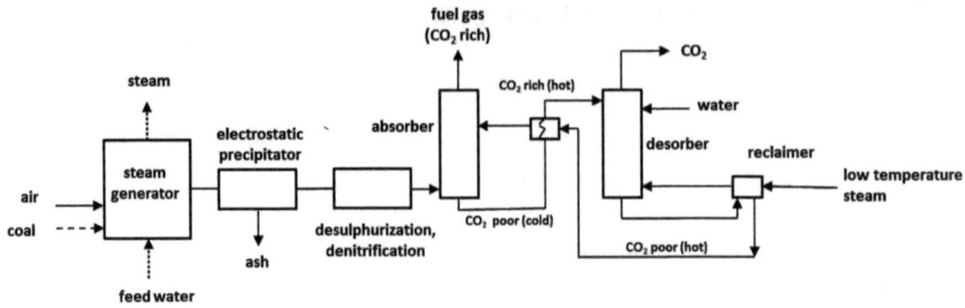

Fig. 2.3 Process flow diagram for post-combustion systems (amine scrubbing)

2.2.1 Post-combustion Processes

When carbon capture is performed after the combustion process (Fig. 2.2) including downstream flue gas purification (dedusting, desulfurization, denitrification), this is referred to as a post-combustion process. Using a suitable solvent, the carbon dioxide in the flue gas is chemically absorbed. In a second step, the CO_2 in the loaded solvent is excited by a temperature or pressure change and desorbed (Fig. 2.3). The CO_2 is then processed and compressed for transport to storage sites. The near CO_2-free solvent is then recycled and used for a new separation cycle. Suitable solvents include organic substances (e.g. alkanolamines, known as amines) and inorganic substances (e.g. alkaline earth solutions, ammonia).

2.2.1.1 State of the Art

The main aim of ongoing research and development activities is to identify suitable solvents. Considerable experience has already been obtained with alkanolamines (e.g. monoethanolamines), which are used commercially today in chemical industrial processes (e.g. ammonia production, natural gas treatment). Unfortunately, this cannot be transferred to the power plant process as the flue gas composition and the flue gas volume flow in a power plant are very different to the requirements of a chemical industrial process. A simplified process flow diagram of an amine-based post-combustion facility is shown in Fig. 2.3. CO_2 is absorbed in an absorber at a temperature between 40 and 60 °C. The amine liquid loaded with CO_2 is then fed into a desorber. The steam for regenerating the liquid (releasing CO_2) is extracted from the power plant process at a temperature of 100–140 °C and is thus no longer available for power generation. In addition to the effort required to compress the captured CO_2, this is one of the major reasons for the considerable efficiency losses of the process as a whole. An important development goal is to reduce the energy consumption for regeneration. Against this background, the use of sterically hindered or tertiary amines (aMDEA, MDEA, KS-1) is considered promising. They require considerably less desorption energy but have disadvantages with regard to the absorption process.

Basic problems include decomposition of the solvent in the presence of oxygen, as well as solvent degradation via reactions with sulfur dioxide or nitrogen oxide. Both effects mean that the solvent must be continuously replaced. Solvent decomposition caused by oxygen can be prevented by adding 'inhibitors'. Reaction with SO_2 and NO_x produces degradation products in the form of salts, which are removed at high temperatures in a reclaimer (see Fig. 2.3). Degradation can be minimized by decreasing the amount of SO_2 in the flue gas before carbon capture. It is assumed that the degradation effect could be prevented at SO_2 concentrations of 10–25 mg/m^3 (current value in coal-fired power plants: 150–200 mg/m^3) (Rao and Rubin 2002). This in turn demands the construction of much more efficient flue gas desulfurization units (see Sect. 2.3.2).

Inorganic solvents are an alternative to amine scrubbing. The use of basic alkaline and alkaline earth solutions is an interesting alternative as these are characterized by high thermal stability, resistance to oxygen, and relatively low absorption and desorption heat. The relatively low reaction rate is disadvantageous but it can be increased by including additives. The basic feasibility of this has yet to be demonstrated (BMWi 2007). A further option involves the use of ammonia. With the aid of ammonia and the addition of water, the CO_2 contained in the flue gas can be bound. The process based on the use of ammonia is characterized by a low energy demand for absorption and desorption. Furthermore, ammonia is an absorbent that can be produced cost-effectively. A variant of the ammonia-based process is the chilled ammonia process, which absorbs the CO_2 at a much lower temperature. The advantages compared to the conventional ammonia process are small volume and mass flows, and thus a lower energy demand and a lower NH_3 slip (Kozak et al. 2009).

Today, around 20 smaller post-combustion plants can be found throughout the world designed as pilot or demonstration plants. Of these, two are in operation in Germany (lignite-fired power plant Niederaussem: amine scrubbing; hard-coal-fired power plant Staudinger: amino acid salts). Another 20 post-combustion projects are currently planned worldwide (GCCS 2011). In today's test facilities, the amount of CO_2 captured ranges from 0.125 to 500 tCO_2 per day, which corresponds to an electric power of much less than 1 MW (Fahlenkamp and Dittmar 2011). In addition to optimally integrating CCS technology in the power plant process, up-scaling to larger plants will thus be a crucial development step in the future. Within the framework of a European Union research programme for the demonstration of CCS, three post-combustion projects for coal-fired power plants (Porto Tolle, Belchatow, Maasvlakte) are being funded with plant sizes equivalent to a generation capacity of around 250 MW_{el} (ZEP 2008).

2.2.1.2 Efficiency Losses

Different authors have assessed the efficiency losses associated with carbon capture as ranging from around 9–14 percentage points (including compression, liquefaction and conditioning) (IPCC 2005; Finkenrath 2010; Kather et al. 2008). The proportion of losses caused by the compression and liquefaction of carbon dioxide is between 2 and 3 percentage points. To improve efficiency, there are a number of options. These include the development of new solvents as well as an optimal thermal integration of the capture process (incl. CO_2 compression) into the power plant process. If this potential is fully exploited, according to (Kather et al. 2008), the efficiency losses for the post-combustion process should decrease in the optimal case to 9.1 percentage points (incl. compression) at a capture efficiency of 90 %.

2.2.1.3 Advantages and Disadvantages of Post-combustion Processes

An advantage of the post-combustion process is that extensive experience already exists for amine scrubbing from applications in chemical industrial processes. Furthermore, there is considerable potential for improving efficiency. Of all CCS technologies, the post-combustion process achieves the highest degree of purity for captured carbon dioxide (>99.99 %) based on what we know today (see Chap. 3). Although integrating a post-combustion process into a power plant process entails modifying the low-temperature steam part, it does not require fundamental changes in the power plant process. All necessary components are commercially available and no fundamental new developments are necessary. In addition, the post-combustion process (end-of-pipe technology) can be retrofitted in existing power plants (see Sect. 2.3.2). The investment costs, which are still quite high, are a major disadvantage (see Chap. 7). Another drawback is that it is not yet known how flexibly a power plant equipped with post-combustion technology can be operated (see Sect. 2.3.1).

2.2.1.4 Second-Generation Post-combustion Processes

Second-generation post-combustion processes include the carbonate looping technique and membrane-based processes. Carbonate looping is based on the principle of dry sorption. The main element is a dual fluidized bed reactor in which calcium oxide (CaO) is circulated continuously as the CO_2 carrier in a loop between the carbonator (absorption) and the calcinator (desorption). The process is based on a high-temperature reaction (600–700 °C) involving the reversible exothermic absorption of calcium oxide combined with the endothermic calcination of the calcium carbonate ($CaCO_3$, approx. 900 °C). In contrast to conventional scrubbing techniques, the absorption heat can be used in the power plant process, thus leading to increased efficiency. Efficiency losses are estimated by (Abanades et al. 2004) as around 7.2 percentage points (with CO_2 compression and conditioning). Ströhle et al. (2008) quote an efficiency loss of around 3 percentage points (without CO_2 compression and conditioning) at a capture rate of 70–95 %. The advantages of the process, which is currently being tested on a laboratory scale, are the high purity that can be achieved as well as its theoretical suitability for retrofitting.

Another option involves the use of membranes. Polymer and organic/inorganic hybrid membranes are the preferred membrane types for the capture of CO_2/N_2 (Reijerkerk et al. 2011). Membrane processes for the post-combustion route demand a considerable energy expenditure for flue gas compression in order to produce the required pressure difference for gas separation (cf. Göttlicher 1999). At present, membranes do not yet have the necessary characteristics (e.g. selectivities) to achieve high capture rates and purities in a single-stage membrane configuration with a reasonable energy demand. One solution that has been proposed involves a multiple-stage configuration of membrane modules and compressors/turbines. Furthermore, recirculating the retentate is one way of increasing the CO_2 concentration in the feed gas. With such concepts, efficiency losses can be limited to 7.8–9.2 percentage points depending on the respective membrane properties, and a capture efficiency of 90 % and purity of approx. 95 % can be achieved (Zhao et al. 2009, 2011).

2.2.2 *Oxyfuel Processes*

Oxyfuel is a term used to describe the combustion of carbon-containing fuels with pure oxygen, which achieves a high concentration of carbon dioxide. Compared to the flue gas of current power plants, which contains concentrations of CO_2 of around 12–15 vol. %, this figure is around 89 vol. % in oxyfuel plants (BMWi 2007). After flue gas purification and scrubbing, the flue gas is mainly comprised of a carbon dioxide/steam mixture. By condensing the steam out of the flue gas, a CO_2-rich flue gas is produced, which is then compressed ready for transport to the storage site. Oxygen for the combustion process is produced using a cryogenic air separation unit, which separates the oxygen out of the air via condensation at low temperatures (< -182 °C). Combustion with pure oxygen leads to much higher combustion temperatures than in existing power plants, and it necessitates modifications in the burner and furnace chamber due to heat- and flow-specific conditions. In order to limit the combustion temperatures, some of the CO_2-rich combustion gas is fed back into the firing chamber (Fig. 2.2). At the same time, oxygen that was not converted is fed back into the combustion process and the residual oxygen concentration in the flue gas is reduced.

2.2.2.1 State of the Art

The oxyfuel process has been tested in several test facilities worldwide, as well as in two smaller pilot plants (approx. 30 MW_{th}) in Germany (lignite-fired power plant Jänschwalde) and France (gas-fired project power plant Lacq). However, feasibility for the usual size of power plants today has not yet been demonstrated.

Cryogenic air separation units are state of the art today and are used on a large scale for industrial applications (e.g. steel industry, syngas production). Air separation demands a high energy expenditure, which causes considerable efficiency losses in an oxyfuel power plant. The challenge therefore involves optimizing the energy demand of the air separation unit and integrating it into the power plant process. By improving cryogenic air separation from a process engineering point of view, e.g. three-column process, the specific energy demand could be decreased by 20 % compared to today's plants (BMWi 2007). An important framework condition here is the required oxygen purity, which should be around 99.5 vol. % (residual components: Ar, N_2). Measures for refining conventional air separation units, however, often lead to a lower oxygen purity and a higher residual gas content. This gives rise to a need for additional gas purification, which in turn is connected with a higher energy expenditure as well as with a higher compression effort (cf. (Kather et al. 2008; Castillo 2009; Kather and Klostermann 2011)).

Combustion with pure oxygen leads to considerably lower flue gas volumes and a modified radiant heat transfer. This in turn necessitates a new design for the heat exchange surfaces, the firing chamber geometries, and the flue gas channels. In general, it is assumed that the oxygen excess is lower than for conventional firing, which could cause burn-up problems and corrosion of the combustion chamber walls.

Another problem is the infiltration of undesired air, which amounts to several percent of the total flue gas volume flow and may increase with the lifetime of the power plant. This means that the required CO_2 purity can no longer be achieved, and that additional energy is required for compression.

Several of the problems outlined here are currently being investigated in ongoing R&D work. Of particular note is the biggest oxyfuel demonstration plant in the world at the moment: a lignite-fired power plant at the Jänschwalde site with a rated thermal capacity of 30 MW. The demonstration plant went into operation in 2009, and can be operated with pure oxygen as well as with air. Tests have shown that a CO_2 purity of 99.7 % can be achieved at a capture efficiency of 90 %. Other large pilot plants (e.g. Callide Power Plant, Australia, 30 MW_{el}) are currently being planned (cf. GCCS 2011). An oxyfuel power plant is also planned for the Compostilla site in Spain as part of the EU programme for the demonstration of CCS. The envisaged design is an oxyfuel power plant with a fluidized bed reactor (phase 1: 30 MW_{th}; phase 2: 300 MW_{el}).

2.2.2.2 Efficiency Losses

The efficiency losses of an oxyfuel power plant are currently estimated to range between 8 and 11 percentage points. The operation of the air separation unit alone causes an efficiency loss of 7 percentage points. The remaining losses are due to CO_2 compression and conditioning. According to calculations performed by (Kather and Klostermann 2011), the efficiency losses are around 10 percentage points when an improved air separation unit and improved conditioning technologies are used. By optimally integrating CO_2 compression into the heat balance of the power plant, the efficiency losses can be reduced further by one percentage point. A transition to the three-column air separation process would allow total efficiency losses to be limited to around 8 percentage points. In general, it should be noted that the efficiency should always be considered in correlation with the carbon capture efficiency, the required CO_2 purity and the oxygen purity (cf. Kather and Klostermann 2011; Castillo 2011).

2.2.2.3 Advantages and Disadvantages of Cryogenic Oxyfuel Processes

As was the case for the post-combustion process, the oxyfuel process also has considerable efficiency potential. In addition, all necessary technical components are commercially available, which means that no new developments are required. As air separation units are already being built and operated as large plant units today, the oxygen volumes required for power plant operation do not pose any problems. Compared to scrubbing processes, the oxyfuel process is less complex, which is advantageous. Furthermore, there is no need to dispose of by-products. As things currently stand, the relatively high investment costs are considered a disadvantage. Whether oxyfuel processes can also be retrofitted into existing power plants is still unclear (see Sect. 2.3.2). As was the case for post-combustion processes, no conclusive statements can be made on the flexibility of the operation mode of an oxyfuel power plant (Sect. 2.3.1).

2.2.2.4 Second-Generation Oxyfuel Processes

An alternative to cryogenic air separation is chemical looping, which uses metal oxides as an oxygen carrier and where the metal oxide is circulated continuously between two reactors in a loop. In the combustion reactor, oxygen is separated from the metal oxide in an endothermic reaction, and then used for fuel combustion. The reduced metal oxide is regenerated in a second reactor using atmospheric oxygen as an oxidizing agent at temperatures of 1,200 °C. The heat flow from the oxidation reactor and the flue gas is used to generate heat (BMWi 2007). The main challenge and the focus of several ongoing research projects is the identification of suitable metal oxides with an appropriate stability and a sufficient reactivity with relevant regeneration properties (Ausfelder and Bazzanella 2008; Lyngfelt and Mattisson 2011). Epple and Ströhle (2011) assume that the efficiency losses of chemical looping are around 8 percentage points (incl. compression).

Another alternative to cryogenic air separation is to separate oxygen using oxygen-conducting membranes. Ceramics (perovskites, fluorites) are used for this purpose as they are characterized by a specific conductivity for oxygen ions and are permeable to oxygen ions at a temperature above 700 °C. A particular advantage of this process is that relatively high oxygen purities can be achieved. In addition to membrane development, thermally integrating the membrane process into a power plant process also remains challenging. Different concepts are currently being developed and tested for this (three-end process, four-end process)

(Kneer et al. 2010; Engels et al. 2010). Depending on the concept chosen and other parameters, efficiency losses range from 6 to 10 percentage points according to the literature (Kneer et al. 2010; Engels et al. 2010; Stadler et al. 2011; Castillo 2011; Beggel et al. 2011).

2.2.3 Pre-combustion Processes

In integrated gasification combined cycle (IGCC) power plants, coal or other materials (e.g. refinery residues) are partially oxidized at high temperatures and under high pressure (approx. 30 bar), and converted into a raw gas (CO, H_2 and CO_2) (Fig. 2.2). A downstream catalytic conversion uses steam as an oxidizing agent to convert the carbon monoxide into carbon dioxide and hydrogen (CO shift). The resulting syngas has a high pressure level, which allows carbon capture to be performed using physical absorption on the basis of methanol (Rectisol wash) or dimethyl ether/polyethylene glycol (Selexol scrubbing). The remaining hydrogen is subsequently used in a gas and steam process to generate power (BMWi 2007; Wietschel et al. 2010).

2.2.3.1 State of the Art

The gasification of coal and other energy carriers is state of the art, and is used throughout the world to produce syngas, which in turn is the starting material for several other industrial applications (e.g. methanol synthesis). However, gasification technology has not yet become established for power generation, and only a few IGCC power plants are commercially operated worldwide today (BMWi 2007; Wietschel et al. 2010). The efficiencies that can be currently achieved are comparable to those of conventional coal-fired power plants. Due to the high efficiency potential of coal-fired combined cycle processes and the feasibility of using established and tested processes for CO_2 separation, IGCC power plants are attracting increasing attention within the context of carbon capture.

For carbon capture, the use of Rectisol wash is preferred because it can simultaneously remove CO_2 along with the undesired accompanying substances H_2S and COS (carbonyl sulfide) from the syngas. The Rectisol process comprises several process steps, such as the absorption of sulfur-containing components, the absorption of carbon dioxide, and the desorption of the loaded solvent. In addition, the sulfur-containing accompanying substances must be removed from the solvent and treated accordingly (e.g. Claus unit). The solvent here is an absorbent with a methanol basis. Absorption occurs under pressure (30–60 bar) and at very low temperatures (−40 °C). CO_2 is separated from the loaded solvent using nitrogen and temperature changes. The CO_2 is then conditioned and compressed. The Rectisol process can generate relatively high purities of over 99 % (Kunze and Spliethoff 2010).

Syngas preparation is already common in several industrial sectors, which means that prior experience exists in this area, and the technical challenge therefore lies in implementing the basic IGCC process. Gasification technology must be improved, raw gas cooling must be refined (e.g. optimal use of waste heat, partial quench), and it must be rendered feasible to use H_2 in gas turbines (BMWi 2007; Wietschel et al. 2010). As is the case for other CCS technology lines, the optimal thermodynamic design is a crucial prerequisite for exploiting the existing efficiency potential.

Planning has begun on an IGCC power plant, which is being funded within the framework of the European Union's CCS demonstration programme. The coal-fired combined power plant (Stainforth, UK) with a capacity of 650 MW_{el} is part of the Don Valley Power Project. It is currently the only planned coal-fired IGCC plant in Europe. Further IGCC projects are planned in the USA and in China (GCCS 2011).

2.2.3.2 Efficiency Losses

The efficiency losses for an IGCC power plant with physical CO_2 scrubbing compared to an IGCC power plant with no carbon capture range from 9 to 12 percentage points (Scherer and Franz 2011; Kather et al. 2008; Göttlicher 1999; Kunze and Spliethoff 2010; Grabner et al. 2010; Rubin et al. 2007). The IGCC demonstration power plant (450 MW_{br}, full quench, 40 bar), which was originally planned for the German site Hürth, was supposed to achieve an efficiency of 34 % at a gross efficiency of 48.5 %. The transition to a partial quench would have improved the efficiency by around 1–1.5 percentage points. By exploiting the considerable efficiency potential, a gross target efficiency of 44 % was reported for a capture rate of >90 % (Renzenbrink et al. 2008).

2.2.3.3 Advantages and Disadvantages of Pre-combustion Processes

A significant advantage of the pre-combustion process is that physical scrubbing (Rectisol, Selexol) is already used commercially on a large industrial scale to purify syngas and there is thus a direct analogy to IGCC power plants. The high purities that can be achieved are also a benefit of this process. In addition, the considerable efficiency potential of the basic IGCC process must also be considered, which includes gasification, thermodynamic optimization of the entire system, and the improvement of gas turbine technology (BMWi 2007). Another key advantage of IGCC plants is the high fuel flexibility and the option of generating other products in addition to power (polygeneration). For example, it is possible to use syngas to produce other gaseous or liquid products (SNG, methanol, synthetic fuels, etc.). The high complexity of these plants compared to conventional power plants, however, is a major obstacle. A major barrier is also posed by very high investment costs for the basic IGCC power plant process, which also explains the low number of IGCC plants globally. High investment costs were also stated as the reason why some planned IGCC power plants with carbon capture were not implemented in the recent past (cf. Markewitz et al. 2012).

2.2.3.4 Second-Generation Pre-combustion Processes

Due to favourable pressure conditions, membranes can be used in the pre-combustion process as a replacement for physical scrubbing in the long term. Either H_2-selective membranes (e.g. microporous zeolite membranes, microporous sol-gel membranes, MPEC membranes) or CO_2-selective membranes (polymer membranes) can be used (Scherer and Franz 2011). The type of membrane chosen depends mainly on the IGCC concept, as this determines the prevailing framework parameters (pressure, temperature, syngas composition, etc.). Scherer and Franz (2011) reported efficiency losses ranging between 8.7 and 10.5 percentage points for the use of CO_2-selective membranes. For H_2-selective membranes, the authors reported a range of 9.1–11.1 percentage points. Compared to the efficiency losses associated with scrubbing (9–12 percentage points), membranes demonstrated only slight efficiency advantages. However, the use of catalytic high-temperature H_2 membranes, which combine H_2 separation with the CO shift reaction in one engineering process, have great potential. The advantage here is that the shift reaction can occur in a stoichiometric ratio (steam to CO). Compared to present concepts, which require a super-stoichiometric ratio, this would allow the steam demand to be reduced considerably. In contrast to a Rectisol wash, no temperature decrease is required here for the conversion reaction. This allows a gain in efficiency potential of between 0.8 and 2.9 percentage points compared to the previously mentioned membrane concepts (Scherer and Franz 2011). However, hot gas dedusting must be performed upstream of the membrane reactor.

In general, it should be noted that the membrane types currently being discussed do not yet demonstrate the necessary characteristics (e.g. selectivity, permeability, stability) for implementation in various separation concepts, and that there is a need for fundamental R&D work in the field of membrane development and fabrication.

2.3 Future Framework Conditions and Requirements for the Implementation of Power Plants with Carbon Capture

While the previous section outlined the individual carbon capture technologies and evaluated them in terms of their advantages and disadvantages, Sect. 2.3.1 will analyse CCS power plants in terms of their flexibility and the main technical requirements arising from power plant operation.

In general, it is assumed that power plants with carbon capture technologies will only be available commercially from 2020 at the earliest. This means that the installed power plant fleet in 2020 will not be equipped with CCS technology. Against this background, Sect. 2.3.2 will look at how suitable the discussed technology routes are from a technical perspective for retrofitting the installed power plant fleet and the potential that exists.

2.3.1 Flexibility of Power Plants

The liberalization of the European electricity market, which began in 1998, has meant that power plant operators have had to change track because their expected revenues now depend on market conditions on the power exchanges. This in turn has a major impact on how power plants are operated, and the demand for flexibility has increased substantially. The accelerated deployment of renewables for power generation must also be taken into account, as this will continue in future according to the plans of many countries. This large share of highly volatile power generation (wind, photovoltaic) is frequently fed into the grid preferentially in accordance with legal requirements. Covering the remaining residual load (consumer load minus volatile feed-in) necessitates controllable conventional power plant capacity with high power gradients as well as large load ranges (maximal/minimal capacity), while simultaneously accounting for a much higher number of load changes (RWE 2009).

This change in the basic conditions means that new coal-fired power plants currently being constructed will be much more flexible to operate than older existing power plants. This applies both to the power gradients and the load ranges, as shown in Table 2.1 for selected existing coal-fired power plants and new plants currently being built.

In the future, it can be assumed that not only will the number of load changes increase considerably but so too will the number of start-up and shut-down cycles (Trautmann et al. 2007; RWE 2009). According to estimates, the number of warm start-ups (>8 h downtime) and hot start-ups (<8 h downtime) will increase over the entire lifetime of a coal-fired power plant by a factor of around four and eight, respectively. Other important factors include the time required to start up and shut down a power plant. It is assumed, for example, that the usual start-up times today (cf. Table 2.2) will have to be considerably reduced (by 20–30 %) in order to meet market requirements (RWE 2009).

Table 2.1 Power gradients and load ranges for selected old and new coal-fired power plants

	Hard coal		Lignite	
	Old	New	Old	New
Maximal gradient	+/−8 MW/min	+/−27 MW/min	+/−11 MW/min	+/−30 MW/min
Maximal capacity	600 MW	800 MW	300 MW	1,000 MW
Minimal capacity	420 MW	200 MW	200 MW	500 MW

Source: RWE (2009)

Table 2.2 Start-up times of coal-fired power plants for different types of start-ups

	Start-up time
Cold start-up (after <72 h downtime)	Approx. 6 h
Warm start-up (after <48 h downtime)	Approx. 4 h
Hot start-up (after <8 h downtime)	120 min
Hot start-up (after <2 h downtime)	90 min

Source: Trautmann et al. (2007)

The requirements outlined above apply to all conventional power plant types in general and thus also to power plants with CCS technology. However, ongoing R&D activities in the area of carbon capture are currently focused on the development of CO_2 separation processes and their integration into the power plant process. Most technical and economic analyses on CCS power plants are therefore founded on a base-load mode of operation. Different publications (Chalmers 2010; Davison 2011) have therefore concluded that the flexibility of CCS power plants and the resulting technical requirements have not received enough attention and that there is a considerable need for research. Reliable estimates of temporal behaviour (start-up and shut-down cycles, load ramps) demand dynamic modelling of plants and power plant components. As almost all existing CCS power plant analyses with very few exceptions (Ziaii et al. 2009; Kvamsdal et al. 2009) have been performed with static models, a well-founded evaluation of the temporal behaviour of CCS technology routes is not yet possible. The same applies to the possibilities afforded by partial load operation, and the associated efficiency losses. Studies by Linnenberg and Kather (2009) suggest that the efficiency loss of a CCS power plant in partial load operation is greater than that of a power plant with no carbon capture in partial load operation.

The following will take a look at some options that are currently being discussed to increase the partial load flexibility of the preferred CCS technology routes in order to operate at as large a load range as possible. The discussion is based predominantly on a comprehensive literature review performed by (Chalmers 2010) on this particular topic.

2.3.1.1 Post-combustion Processes

One way of operating post-combustion power plants with a more flexible load is to bypass the carbon capture system. The low-pressure steam required for desorption could then be used for electricity generation and/or to increase the load. It should be noted, however, that the option of bypass operation must be devised when a new plant is being planned because it requires modifications in the design of the low-pressure turbines, the generator, and other components. The expected response times for load activation depend on the steam extraction concept. Economic considerations must reflect the fact that a higher load must be fed into the grid in times of high load demand and high electricity prices in order to generate higher revenues. These revenues must be balanced against the additional investments and CO_2 allowance costs for the additional CO_2 emissions.

Another possible method of increasing load is to store the CO_2-loaded solvent. This means that desorption of the solvent loaded with CO_2 is not performed for a certain period of time, and that the solvent is stored in special tanks for this interim period. The low-pressure steam originally intended for desorption can then be used for electricity generation and/or to increase load. In times when load demands and revenues are low, the power plant output can be reduced, and desorption of the stored loaded solvent can be performed using low-pressure steam.

The two concepts outlined above are still only of a purely theoretical nature, and detailed technical and economic analyses have yet to be performed. Furthermore, in addition to carbon capture technology, compression and thermal connection to the actual steam cycle as well as possible impacts on pipeline injection and operation must also be taken into account.

2.3.1.2 Oxyfuel Processes

Current concepts are designed so that the first step involves starting up an oxyfuel power plant with air and switching over to oxygen operation as soon as sufficient flame stability has been achieved. High oxygen concentrations, which could arise in the transition phase and pose a danger of explosion, are considered a potential problem. To avoid this problem, optimized burner technology is essential. The duration of the start-up process is estimated to be around 15–30 min and it depends mainly on the burner configuration (Kluger et al. 2009; Grathwohl et al. 2009).

Both start-up and shut-down cycles as well as partial load operation assume a variation in oxygen supply. The speed of this variation also depends on the flexibility of the air separation unit, for which load ramps of 1–3 % per minute have been estimated (White et al. 2009). The load change rate could be enhanced by storing oxygen in liquid or gaseous form and then using it during periods of high load change rates.

2.3.1.3 Pre-combustion Processes

Only six IGCC plants are currently in operation worldwide, which means that only limited experience exists with coal-fired combined power plants. It is therefore almost impossible to predict the load behaviour of IGCC plants with carbon capture from our point of view today. One method of varying load would be to store part of the hydrogen temporarily, and to use it as required for high load demand. This variant would have the advantage that the gasifier would not have to meet the load demands in full. As the hydrogen is used in a gas turbine, both the load flexibility and load change rate are very high. An advantage of IGCC plants is the basic feasibility of producing other products such as methanol in addition to electricity and hydrogen (polygeneration), as this could help to boost the plant flexibility. In the case of a low demand for electrical load, for example, operation could be switched to another product, reducing electricity generation and operating the coal gasifier in base load mode.

2.3.2 *Retrofitting the Existing Power Plant Fleet*

Some 41 % of global anthropogenic CO_2 emissions originate from power plants (IEA 2011). Around 73 % of these emissions come from coal-fired power plants, clearly indicating how important they are in the context of CO_2 mitigation strategies. In general, it is assumed that carbon capture technologies will be commercially available from 2020 at the earliest. The International Energy Agency (IEA) estimates that by 2020 the globally installed power plant capacity will increase from around 1,580 GW today by between approximately 350 GW (450 ppm scenario) and 700 GW (current policies scenario) (IEA 2011). Which of the existing coal-fired power plants could be retrofitted with carbon capture technologies, supplementing the predicted new builds, depends on the technical and economic prerequisites. It can be assumed that these will be relatively modern plants (cf. Finkenrath et al. 2012), namely those built between 1999 and today. Between 1999 and 2009, plants with a combined global power plant capacity of 525 GW were built (Fig. 2.4).

Since 2005, a global power plant capacity of around 272 GW has been installed. If these installed plant capacities are added to the additional new-build capacities predicted by the IEA by 2020, the potential to retrofit coal-fired power plants with carbon capture technology ranges from 875 to 1,225 GW. This emphasizes the

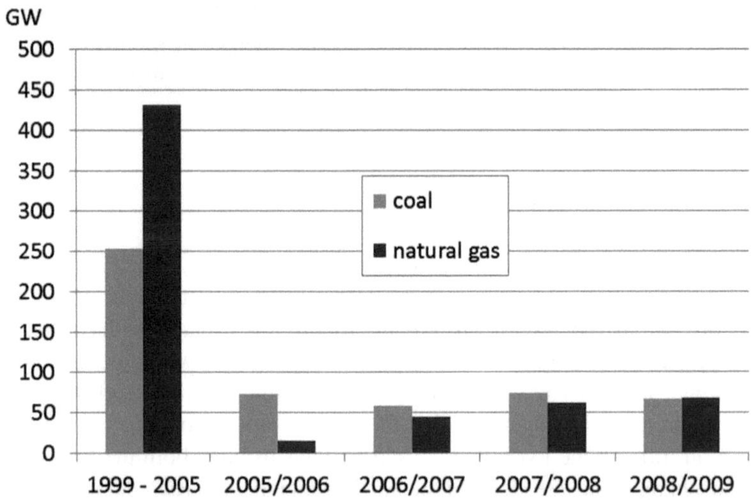

Fig. 2.4 Global expansion of coal- and gas-fired power plants (1999–2009) (Source: IEA 2000, 2006, 2008c, 2009, 2011)

growing importance of retrofitting existing power plants with carbon capture technologies.

2.3.2.1 Excursus: Germany

The installed capacity of coal-fired power plants in Germany is currently around 47 GW (lignite: 19.9 GW; hard coal: 26.8 GW). Coal-fired power plants with a capacity of 11.5 GW are currently being built and are due to go into operation by 2013 at the latest (Markewitz et al. 2011b). As a countermove, around 12 GW coal-fired power plant capacity will be taken off the grid by 2020 due to obsolescence. In the current scenario framework for the Network Development Plan approved by the Federal Network Agency for Electricity, Gas, Telecommunications, Post and Railway (Bundesnetzagentur) (Bundesnetzagentur 2011), it is assumed that coal-fired power plants will be further expanded by 2022. Depending on the scenario, additional capacity ranges between 11 and 20 GW. These capacities in particular are extremely relevant for retrofitting with carbon capture technology.

Due to the high efficiency losses associated with the use of carbon capture technologies, electricity generation drops and the net power is reduced compared to existing power plants that have not been retrofitted with CCS. These losses must be compensated by the respective operator, for example by buying additional quantities of electricity or by building new power plants. However, the limiting resource situation must be considered when building new lignite-fired power plants (Markewitz et al. 2009).

2.3.2.2 Suitability of Carbon Capture Technologies for Retrofitting

There is an ongoing discussion as to whether power plants currently being built and those planned for the near future (up to 2020) should be ‘capture ready’. TÜV Nord, for example, offers certification for this (Climate Change Standard TN-CC006). This standard comprises requirements for the technological and site-specific feasibility of retrofitting a capture unit, the availability of the necessary space, the feasibility of disposing of the captured CO_2, and/or of storing it, as well as possible impacts on plant safety and the environment. Compliance with these criteria aims to ensure the basic feasibility of a retrofit. Additional investments must also be earmarked for technical measures in advance.

For such existing power plants, the question arises as to whether component design should be planned in advance in order to simplify retrofits at a later point in time from an engineering perspective and also to improve efficiency. In other words, additional investments for this at the time of power plant construction would then be offset from the point in time of the retrofit, e.g. by a higher efficiency and shorter retrofit times (see Irons et al. 2007a). From an operator’s point of view, there are potential risks and uncertainties associated with advance investment (Irons et al. 2007a). These include future provisions of regulatory law, the future development of energy prices and CO_2 allowances, and the future organization of the electricity market. From a technical point of view, there is an inherent risk that planning for components in advance would commit operators to a technology concept that may be preferred today but by the time the plant is retrofitted could be considered antiquated.

The current discussion on retrofitting concentrates on the post-combustion process and the oxyfuel process. A special case is that of IGCC power plants: although retrofitting existing IGCC power plants with CCS is currently a topic of discussion internationally, the small number of existing and planned plants worldwide means that it is still insignificant and is therefore not discussed in the following.

2.3.2.3 Oxyfuel Processes

How suitable the oxyfuel process is for retrofitting existing conventional coal-fired power plants has yet to be clarified. The degree to which higher temperatures, altered thermal radiation conditions and flue gas compositions (e.g. via fuel gas recycling) could possibly cause high-temperature corrosion or carburization of the boiler materials and thus impair the functionality and even the lifetime of the existing steam boiler is unclear. Another challenge lies in ensuring the necessary tightness of the boiler in order to prevent air ingress (cf. Irons et al. 2007a, b). Model-based simulations of the firing and the steam boiler (cf. Tigges et al. 2009) indicate that power plant operation with air or pure oxygen is theoretically possible. In addition to providing sufficient space for the installation of new components (e.g. CO_2 compressor and conditioning unit, air separation unit), (Irons et al. 2007b) also state that the burner must be replaced, and the air and flue gas channels and peripheral units must be modified (e.g. induced draught fans). No alteration of the heat exchange surfaces is necessary according to Irons et al. (2007b). It can be assumed that the flue gas desulfurization unit must be modified in order to achieve the levels of purity required for pipeline transport.

Within the framework of the Australian research project CALLIDE, an existing coal-fired power plant (site: Callide; 120 MW_{el}) is being retrofitted with oxyfuel technology. Test operation is planned under real conditions over a period of 5 years.

2.3.2.4 Post-combustion Processes

Of all post-combustion processes, amine scrubbing is the most intensively studied with regard to retrofitting existing power plants (cf. Irons et al. 2007b; Ploumen 2006). Compared to the oxyfuel process, the main advantage of the post-combustion process is that the firing process does not have to be modified. It is also assumed that no modifications are necessary for the steam generator. As previously mentioned, too much residual SO_2 in the flue gas can cause degradation of the amine-based solvent used today. In order to prevent degradation, the flue gas should contain no more than 10 ppm (<25 mg/m^3) SO_2 (Rao and Rubin 2002). Initial tests with flue gas desulfurization at the German power plant site in Niederaussem indicate that such SO_2 volumes will most likely be achievable using a lime scrubbing (REAplus) process that is very different to present flue gas desulfurization systems (Reissner 2009). Whether existing lime scrubbing systems can be retrofitted or whether they will have to be completely replaced is still unclear.

Large amounts of low-pressure steam are necessary for desorption of the amine liquid loaded with CO_2. Thermodynamic calculations show that around 50–65 % of the total low-pressure steam in a power plant is required for this, leading to a situation where several stages of the low-pressure turbine could no longer be operated (cf. Ploumen 2006; Irons et al. 2007b). Retrofitting therefore necessitates considerable modifications of the low-pressure turbine section, the pre-heating section (heat exchanger), the condenser, and the cooling water pumps. Moreover, partial load operation poses another problem. It involves a reduced steam flow combined with a lower turbine inlet pressure in the low-pressure section. If the pressure level sinks below the pressure of the low-pressure steam required for desorption, steam must be extracted from the medium-pressure section, which then leads to additional efficiency losses (Ploumen 2006).

Compared to new power plants with carbon capture, a retrofitted existing power plant has a lower efficiency because, for example, the thermal optimization between absorption/desorption, CO_2 compression and the pre-heating section is only suboptimal.

Capture-ready measures, in contrast to retrofitting with no advance measures, can lead to substantial efficiency gains depending on the type of technical measure. According to Irons et al. (2007b), capture-ready measures for the low-pressure turbine and the flue gas desulfurization system facilitate efficiency gains of 1.4–2.3 percentage points.

Retrofitting amine scrubbing leads to an increased demand for cooling water, which is around 30 % higher than that of a conventional power plant without carbon capture (Ploumen 2006). Most of the additional water demand can be attributed to amine scrubbing components, flue gas cooling, and CO_2 compression. Flue gas cooling upstream of the absorber, in particular, leads to a clear increase in the total heat load to be dissipated. Which capture-ready measures are to be implemented for this depends on the respective cooling system. For a closed cooling loop, for example, additional space must be reserved for cooling towers. For fresh water cooling, an additional withdrawal of fresh water in accordance with the legal requirements must be possible.

2.4 Carbon Capture Processes for Industrial Applications

The development of carbon capture technologies today is generally aimed at applications in fossil-fired power plants. However, the CO_2 emissions caused by industrial processes are not without relevance. A total of around 25 % (2008: 7.4 Gt)

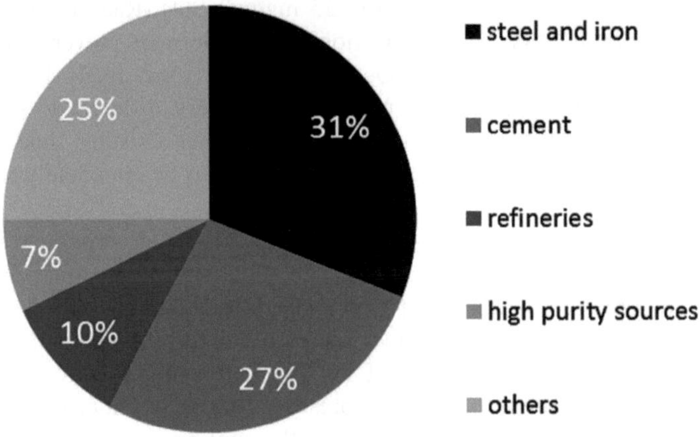

Fig. 2.5 Global share of industrial CO_2 emitters in 2008 (Source: Trudeau 2011)

of global CO_2 emissions are caused by industrial processes (Trudeau 2011). It is estimated that 75 % of these emissions (Fig. 2.5) are caused by larger point sources in the sectors iron and steel, cement, refineries and other large industrial processes (gas processing, H_2 production, ammonia production, CtL,[1] ethylene/propylene production). According to predictions by the International Energy Agency, global emissions from these sectors will increase from around 5.6 Gt today to some 10.3 Gt by 2050 (Trudeau 2011; IEA 2010).

In industry, it is important to distinguish between energy-related and process-related emissions. Process-related emissions are those CO_2 emissions where CO_2 is created as a product of a chemical reaction other than combustion. Process-related emissions are created as a result of the material conditions of the production process (e.g. pig iron production, cement and lime production) and generally cannot be avoided. Energy-related emissions are those CO_2 emissions resulting from a combustion process (e.g. process heat and power generation).

In 2010, industry was responsible for around 18.9 % of energy-related and process-related CO_2 emissions in Germany (Ziesing 2011b). This is equal to around 156 $MtCO_2$. About 98 Mt (approx. 63 %) of these emissions were generated by industrial plants involved in emissions trading. Table 2.3 provides a breakdown of plants involved in emissions trading shown according to industrial sectors.

According to this breakdown, the iron and steel sector, refineries and the cement and lime branches are the largest emitters of all industrial plants which trade emissions accounting for almost 80 %. The plant-specific emissions indicate how large the point sources are. Smelting plants are the largest industrial point sources with approx. 3.5 Mt/a, followed by coking plants (approx. 0.89 Mt/a) and refineries

[1] Coal to liquid.

Table 2.3 Number and emissions of industrial plants involved in emissions trading in Germany in 2010

	Number of plants	Emissions		Ranking Absolute emissions	∅ Emissions per plant	Ranking ∅ Emissions per plant
		$ktCO_2/a$	%		$ktCO_2/a$	
Refineries	26	22,272	22.8	1	856	3
Coking plants	4	3,568	3.7	9	892	2
Pig iron and steel production	26	5,877	6.0	5	226	6
Integrated smelting plants[a]	6	21,392	21.9	2	3,565	1
Steel processing	8	967	1.0	11	121	8
Cement clinker	39	18,577	19.0	3	476	5
Lime	69	7,757	8.0	4	112	9
Glass	85	3,701	3.8	8	43.5	12
Mineral fibres	8	356	0.4	13	44.5	11
Ceramics	134	1,327	1.4	10	9.9	14
Cellulose	5	142	0.1	14	28.4	13
Paper	122	5,715	5.9	6	46.8	10
Propylene/ethylene	8	5,169	5.3	7	646	4
Carbon black	5	725	0.7	12	145.0	7
Total	545	97,545	100		179.0	

Source: Hohlfeld et al. (2011), own calculations
[a]Blast furnace, converter, sintering and pelleting unit, hot blast stove, etc

(0.86 Mt/a). Propylene or ethylene production plants emit an average of 0.646 Mt/a and cement clinker plants around 0.48 Mt/a.[2] Although the lime branch is a considerable emitter, the large number of plants means that plant-specific emissions are much smaller. In general, it should be noted that a significant number of large industrial point sources are located in Germany.

Whether and which carbon capture process could help to reduce these emissions will be analysed in the following using examples of important industrial processes. A direct analogy to or transferability of the CO_2 process routes discussed previously for use in power plants does not exist. We will therefore discuss which capture processes could be useful for industrial applications. A special case is the oxyfuel process, which does not directly capture CO_2. The oxyfuel process aims at a high CO_2 concentration in the combustion off-gases by using pure oxygen for the combustion process. Whether this technique can be transferred to industrial processes will also be discussed in the following.

[2] For comparison, a hard-coal-fired power plant (500 MW, 4500 full load hours) emits around 1.7 $MtCO_2$ per year.

Treating biogas for injection into the natural gas grid is becoming increasingly more common. Biogas treatment comprises several technical process steps and includes the separation of CO_2 that arises during the fermentation process. Different carbon capture processes are already used commercially for this today. Although biogas treatment is not considered an industrial process, it is of significant relevance. It will therefore be dealt with here separately.

2.4.1 Steel and Iron Production

Around 1.5 Gt crude steel were produced globally in 2011. Of this, around 70 % was produced with the emissions-relevant oxygen steel process and around 30 % by the electric steel process (Wirtschaftsvereinigung Stahl 2011). In 2008, global emissions from the pig iron and steel sector amounted to around 2.3 $GtCO_2$ (cf. Fig. 2.5). In 2011, crude steel production in Germany was approx. 44.2 Mt, and 68 % of this was produced using the oxygen steel process. The emissions by the German pig iron and steel sector (excluding coking plants) amounted to some 30 $MtCO_2$ in 2010 (Table 2.3).

An integrated smelting plant for producing oxygen steel comprises several plant components, which are usually emission sources. The largest point source is the blast furnace, which accounts for around two-thirds of the total emissions of an integrated smelting plant (Fleer 2011). The CO_2 formed during the blast furnace process (energy- and process-related) is part of what is known as the top gas, which is also used for other purposes (e.g. hot blast stove, power plant). The proportion of carbon dioxide contained in the top gas depends primarily on the fuels and reducing agent used. The ranges for a typical composition are given in Table 2.4.

There are numerous options for reducing CO_2 emissions. The spectrum ranges from improved heat recovery, new coal-based processes (direct reduction of melt and iron) to the substitution of fuels (e.g. natural-gas-based direct reduction) (Fleer 2011). Carbon capture is another option.

Considerations focus on the use of CCS technologies along the blast furnace converter section. A very promising option is top gas recycling, where the CO_2 is captured using chemical or physical scrubbing, and then fed back into the blast furnace process as a reducing agent. By blowing oxygen into the blast furnace, the nitrogen concentration in the top gas is reduced and the properties of the top gas are improved (Fleer 2011; UNIDO 2010).

Table 2.4 Composition of top gas

Components	Concentration vol. %
N_2	50–55
CO	20–28
CO_2	17–25
H_2	1–5

Source: Hohlfeld et al. (2011)

Another option involves converting the carbon monoxide in the top gas into CO_2 via a shift reaction and then using physical scrubbing to separate the CO_2 contained in the syngas. The remaining H_2/N_2 mixture could be utilized with the aid of a gas turbine process (Gielen 2003).

As previously mentioned, direct reduction processes are another option for steel production. Direct reduction occurs with a reduction gas based on natural gas. The reduced iron is subsequently smelted with an electric arc furnace. One advantage of the process is that coking plants would no longer be needed, which would considerably cut CO_2 emissions. The CO_2 emissions resulting from the reduction process could then be separated using conventional CCS techniques. A key obstacle to the use of direct reduction processes is the poor economic efficiency and the limited spectrum of starting materials (only high-quality iron ore) compared to conventional oxygen steel processes (Fleer 2011; UNIDO 2010; IEA 2010).

It is also possible to integrate carbon capture processes into a melt reduction process (e.g. Hlsarna smelting technology) as coal gasification with pure oxygen leads to a relatively high CO_2 concentration in the off-gas. However, this process is still in the R&D phase, which means that the classic oxygen steel process will not be replaced any time in the near future (IEA 2010).

To summarize, there are a number of options for deploying carbon capture processes for crude steel production. According to UNIDO (2010), the range of chemical and physical scrubbing processes also includes pressure swing adsorption and cryogenic processes. The combination of different processes is also believed to be possible.

2.4.2 Cement and Clinker Production

Cement is a hydraulic binder used to produce building materials (e.g. mortar, concrete). It contains clinker (approx. 70 %), which is produced in a clinker burning process, and calcium sulfate (gypsum). During cement clinker production, direct CO_2 emissions are released (approx. one-third) through rotary kiln operation. Around two-thirds of the emissions are process-related and are released in the calciner. Possibilities for reducing CO_2 emissions include optimizing the processes, fuel substitution for kiln operation, and reducing the clinker volume in cement. Another conceivable variant is the use of carbon capture processes, whereby the CO_2 in the off-gas (approx. 14–33 vol. % CO_2) is separated downstream of the clinker kiln with the aid of chemical scrubbing (IEA 2008a; Barker 2010). In contrast to a power plant, the regeneration of the loaded solvent occurs in a separate steam process. Another possibility involves operating the calciner with pure oxygen in analogy to the oxyfuel process. This concentrates the CO_2 in the off-gas, which is then cleaned and compressed. One problem associated with this process is the higher CO_2 partial pressure, which could impact negatively on the calcination reaction (Fleer 2011; IEA 2008a). Compared to today's plants, the calciner would have to be redesigned (De Coninck and Mikunda 2010).

2.4.3 *Refineries*

A refinery transforms crude oil into marketable products (e.g. petrol, fuel oil, kerosene, lubricants, etc.). In contrast to other industrial plants (e.g. blast furnace, cement), several refinery processes produce a large number of products. Thermal energy is required for the main process groups, namely distillation, conversion (cracking, coking, reforming), and the post-treatment and refining of products. Accordingly, the CO_2 emissions produced in a refinery are distributed among different sources (Table 2.5). In some refineries, the necessary process steam is produced in power plants where CCS technologies can be used, as discussed in the previous sections. In addition to an improved efficiency of the furnaces, improved thermal integration, better process control and CHP utilization, the use of carbon capture processes (integrated in the refinery process) has also been discussed as a further mitigation measure for the refinery process (De Coninck and Mikunda 2010).

In principle, all of the CCS technologies currently being discussed could be implemented here. For furnaces and steam generators, both chemical absorption scrubbing and physical absorption techniques are possible. The latter is often discussed in combination with a gasification of petroleum coke. A potential problem is the relatively low CO_2 concentration in the off-gas flow (Table 2.5), which could be solved by increasing the concentration using the oxyfuel process (Fleer 2011).

Approximately 5–20 % of CO_2 emissions result from the production of hydrogen, which is required for diverse cracking processes, and is currently produced using the steam reforming of natural gas or the gasification of heavy refinery residues (e.g. flexicoking) (De Coninck and Mikunda 2010). The deployment of CCS technologies (e.g. physical scrubbing) would also be feasible here.

In summary, different carbon capture systems could be implemented in different processes in a refinery. This applies in particular to off-gas flows with a high CO_2 content. In addition to the lack of economic efficiency, technically integrating these

Table 2.5 CO_2 emission sources in a typical refinery

Emission source	Description	Share of CO_2 emissions (%)	CO_2 concentration (vol.) in the off-gas flow (%)
Process furnaces	Heat generation via combustion of fossil energy carriers for distillation columns and reactors	30–60	8–10
Steam generators	Process steam generation via combustion of fossil energy carriers	20–50	4–15
Catalytic crackers	Burn-up of petroleum coke	20–50	10–20
Hydrogen production	Reforming of hydrocarbons to H_2 and CO_2	5–20	20–99

Source: Fleer (2011), Brown (2010), Van Straelen et al. (2009), own data

processes into the refinery process also poses a problem. Furthermore, the need for additional equipment could lead to space problems. We are not aware of any demonstration project at the moment testing the application of carbon capture processes in refineries.

2.4.4 Ammonia Synthesis

For the production of ammonia, hydrogen is required, which is produced in today's plants via the steam reforming of natural gas (share: 85 %) or via the gasification of solid energy carriers (share: 15 %) (IPCC 2005). The resulting syngas (CO and H_2) is then converted in another step to CO_2 and H_2. Following this, chemical absorption is used in most plants to separate the CO_2 (IEA 2008b). Around two-thirds of the total CO_2 is used in many ammonia synthesis plants for the further production of uric acid (end product: fertilizers) and thus does not enter the atmosphere directly (see contribution on CO_2 utilization). Some of the excess CO_2 is also further utilized (e.g. enhanced oil recovery). As long as the captured CO_2 is not intended for further use, it has to be prepared for storage and compressed. Ammonia synthesis including chemical scrubbing is an established process and is used throughout the world on a large industrial scale. It is estimated that worldwide around 180 Mt of unused CO_2 could be stored. For Germany, this figure is estimated to be around 3 Mt (McKinsey 2007).

2.4.5 Ethylene Oxide Production

Ethylene oxide is an important raw material that is required for the fabrication of a variety of products (e.g. ethylene glycol, organic insulating materials, textile fibres). The starting point for the production of ethylene oxide is ethylene (C_2H_4), which is usually fabricated today in refineries using naphtha steam cracking. Ethylene oxide is produced by adding oxygen and using suitable catalysts (e.g. silver) at temperatures of around 200–300 °C and a pressure of 10 bar. In addition to the main reaction, a number of side reactions occur in which carbon dioxide is formed (via over-oxidation) and must be separated in a subsequent process. In 2008, around 19 Mt of ethylene oxide were produced worldwide. In Germany, production is estimated to be almost 1 Mt. However, it is unclear how much CO_2 is produced. De Coninck and Mikunda (2010) estimates that for every unit of ethylene oxide around 0.3 units of carbon dioxide are produced. The amounts of CO_2 assumed by the IPCC (2005), on the other hand, are lower by a factor of four. Ausfelder and Bazzanella (2008) estimate that the annual global CO_2 emissions caused by ethylene oxide production amount to some 5.1 Mt.

2.4.6 Excursus: Carbon Capture During Biogas Treatment

In 2011, there were some 83 biogas treatment and feed-in plants in Germany, and their biomethane feed-in capacity totalled around 460 million m^3 (DBFZ 2012). Around two-thirds of the plants have a plant-specific treatment capacity of 350–700 Nm^3/h (DBFZ 2012). According to the German federal government's plans, biogas injection is to be increased by 2020 to 6 billion m^3 by 2020.

Before biogas can be fed into the low-pressure gas grid, diverse requirements must be met. These are set down in the relevant regulations (DVGW worksheets G280, G685). This includes compliance with the combustion parameters, a set CO_2 content in the conditioned biogas (max. 6 %), a set oxygen content (max. 3 %), and a maximal permissible water content. In order to achieve as high a methane content as possible and adhere to the necessary limits, technical processing that also captures the CO_2 contained in the raw biogas is essential. The CO_2 contents in the untreated biogas range between 25 and 55 % depending on origin (Table 2.6), and is thus much higher than the CO_2 contained in flue gases from coal-fired power plants (see Section 2.2).

According to DBFZ (2012), different processes are used to capture CO_2 in Germany. Around 25 % of all plants are equipped with pressure swing adsorption, 29 % with pressurized water scrubbing (chemical absorption), 31 % with amine scrubbing, and 10 % with physical scrubbing (e.g. Genosorb®)[3]. While amine scrubbing and physical scrubbing for biogas treatment are still only at the prototype stage, pressure swing adsorption and pressurized water scrubbing are established commercial and technically mature processes.

Although amine scrubbing and physical scrubbing achieve the highest methane purities with extremely low methane losses, the regeneration of the loaded scrubbing liquid necessitates process heat, which is not available at every plant site. Amine scrubbing also requires upstream desulfurization of the biogas. Another drawback is that the degraded scrubbing liquid must be continuously replaced, which has a negative effect on economic viability. The use of pressurized water

Table 2.6 Biogas composition

	Share (ranges)
CH_4	45–70 %
CO_2	25–55 %
Air	0.05–5 %
H_2S	10–30,000 mg/m^3
Organic sulfur	<0.1–30 mg/m^3
Ammonia	0.01–2.5 mg/m^3
Steam (at 25 °C and 1 bar)	Approx. 3 %

Source: Urban et al. (2012)

[3] Absorbent is tetraethylene glycol dimethyl ether.

scrubbing has a number of key advantages. Even though the CO_2 loading capacity of water is lower than that of amine-based liquids and larger quantities of scrubbing liquid are required, leading to higher pumping efforts, the amount of water needed can be reduced by increased pressure (higher loading capacity). Water also has the potential to dissolve other acidic components (e.g. H_2S), which does away with the need for an additional purification step. The regeneration of the loaded scrubbing water is simple and does not require process steam. It is done by relieving the pressure and stripping the scrubbing water with air (Urban et al. 2012). A key advantage over amine-based scrubbing is the almost unlimited availability and inexpensiveness of water. In addition, the waste heat can be used, for example for the fermentation process. The specific capacity of plants today ranges from 300 to 1,250 m^3/h. All scrubbing processes (chemical and physical scrubbing) are connected to a subsequent gas drying unit, which is not required for pressure swing adsorption. Pressure swing adsorption compresses the raw biogas (4–7 bar) before cooling it down to below 40 °C, which considerably improves the adsorption properties. The gas is then fed through an adsorber where a carbon molecular sieve retains the carbon dioxide. The adsorbents are regenerated using a vacuum pump. The advantages of the process include high stability and the inexpensive availability of the carbon molecular sieve. Another advantage is that pressure swing adsorption needs no additional process heat. As with pressurized water scrubbing, the waste heat can be used for the fermentation process. Disadvantages include pre-treatment of the raw gas, which is necessary for high H_2S contents, and the relatively low purities (>96 %), which however still meet the relevant regulations (DBFZ 2012).

All of these deliberations show that a variety of industrial processes are theoretically suitable for separating carbon dioxide. The emissions are – depending on the industrial process – either energy-related and/or process-related. Whether or not a capture technology is suitable for use depends on the individual constraints associated with the respective industrial process. The large heterogeneity of industrial processes means that different carbon capture technologies are conceivable, as shown in Table 2.7.

Table 2.7 Carbon capture processes for selected industrial applications

	Chemical absorption	CO_2-upgraded oxyfuel	Pressure swing adsorption	Cryogenic	Physical absorption
Cement	×	×			×
Ammonia synthesis	×				
Natural gas treatment	×			×	×
Biogas treatment	×		×		×
Iron and steel	×	×	×		×
Refineries	×	×			×

The discussion on biogas treatment confirms that the use of carbon capture processes is not necessarily confined to large point sources, but that these processes can also be technically implemented in smaller plants – and to some extent in a cost-effective manner today. In addition, the range of techniques currently used for biogas treatment also confirms that site-specific criteria (e.g. unavailability of process heat) play a decisive role in determining which carbon capture process is ultimately implemented.

2.5 Summary and Conclusions

Over 40 % of global carbon dioxide emissions are caused by fossil-fired power plants. According to estimates by the International Energy Agency (IEA), power generation from fossil-fired power plants will continue to increase considerably in the future. In order to limit the temperature rise expected because of the greenhouse effect, carbon dioxide emissions must be drastically reduced. A key role here is played by reducing CO_2 emissions from fossil-fired power plants. Capturing carbon dioxide is an important mitigation measure in this context and is the subject of numerous research and development projects throughout the world. At present, three technology lines are favoured: post-combustion, oxyfuel, and pre-combustion. Both post-combustion and oxyfuel processes are currently being tested in smaller demonstration plants. It is not yet clear whether the transition to commercial power plant block sizes (i.e. up-scaling) will be accomplished in one stage. The plant capacity of the demonstration power plants funded by the European Union suggests that up-scaling to commercial plant sizes will probably occur in two stages.

All three technology lines have considerable potential for increasing efficiency, which can be improved depending on the process using different measures. The thermodynamic integration of the carbon capture process is particularly challenging for all three technology lines. In the long term, there are a number of interesting options that could replace physical and chemical scrubbing, which is preferred today. Alternatives include the use of membranes and carbonate looping. For the oxyfuel process, the cryogenic air separation process could be improved (three-column process) and the transition to other oxygen production processes (use of membranes, chemical looping) is also possible.

The mode of operation of a power plant will also have to meet much higher demands in the future (e.g. higher load ramps, larger load ranges, higher number of start-up and shut-down cycles). Whether power plants with carbon capture will be able to fulfil these requirements is a question that will have to be answered in the future. Due to the considerable efficiency losses caused by carbon capture, a basic power plant process with the highest possible efficiencies is generally assumed. A significant efficiency increase can only be achieved in coal-fired power plants by increasing live steam parameters, which has adverse effects on the flexibility. Increasing the flexibility of coal-fired power plants with and without carbon capture is one of the main challenges for the future from a technical point of view.

It is generally assumed that CCS technology will be commercially available from 2020 at the earliest. Against the background of planned fossil-fired power plants worldwide, retrofitting with carbon capture technologies will play a particularly important role. Based on information available today, post-combustion processes appear to be the most promising technology line for retrofitting. A big advantage compared to other technology lines is that the modification of the power plant process would not involve too much effort.

According to estimates by the International Energy Agency, carbon dioxide emissions from industrial processes (iron/steel, cement production, etc.) will increase substantially in future. This often involves large point sources. For iron and steel production, refineries or cement production, there is a range of different options for implementing various carbon capture processes. In the long term, considerable potential is also seen here in Germany to reduce emissions by some 38 $MtCO_2$ by 2030 (McKinsey 2007). Envisaged applications include blast furnaces, ammonia synthesis, and clinker production. At the same time, it should be noted that these technologies are still not economically viable.

The motivation for carbon capture in biogas treatment is rather different compared to the previously discussed power plants and industrial processes. However, biogas treatment is a good example showing how carbon capture is also possible and indeed viable for smaller CO_2 sources depending on the local conditions. The processes currently in use (pressurized water scrubbing, pressure swing adsorption) are technically mature and are particularly suitable for capturing small amounts of CO_2.

References

Abanades JC, Anthony JC, Alvarez D, Lu DY, Salvador C (2004) Capture of CO_2 from combustion gases in a fluidized bed of CaO. J Am Inst Chem Eng (AICHE) 50:1614–1622

Ausfelder F, Bazzanella A (2008) Verwertung und Speicherung von CO_2. Diskussionspapier, DECHEMA, www.dechema.de/studien-path-1.html

Barker D (2010) Global technology roadmap for CCS in industry – cement sector. Conference Global Technology Roadmap for CCS in Industry. Abu Dhabi, 30.6.-1-7.2010: United Nations Industrial Development Organization

Beggel F, Nauels N, Modigell M (2011) CO_2 separation via the oxyfuel process with O_2-transport membranes in coal power plants. In: Scherer V, Stolten D (eds) Efficient carbon capture for coal power plants. Wiley-VCH Verlag, Weinheim

BMWI (2007) Leuchtturm COORETEC – Der Weg zum zukunftsfähigen Kraftwerk mit fossilen Brennstoffen. Bundesministerium für Wirtschaft und Technologie (BMWi), Berlin

Brown J (2010) Global technology roadmap for industry – refineries. UNIDO Workshop Global technology roadmap for industry. Abu Dhabi, 30 Juni–1 Juli 2010: United Nations Industrial Development Organisation

Bundesnetzagentur (2011) Genehmigung des Szenariorahmens für die Netzenwicklungsplanung gemäß §12a Abs. 3 EnWG. Bundesnetzagentur, Bonn

Castillo R (2009) Technical evaluation of CO_2 compression and purification in CCS power plants. 4th international conference on clean coal technologies, Dresden

Castillo R (2011) Thermodynamic analysis of a hard coal oxyfuel power plant with high temperature three-end membrane for air separation. Appl Energy 88:1480–1493

Chalmers H (2010) Flexible operation of coal-fired power plant with CO_2 capture. In: Centre ICC (ed) IEA Clean Coal Centre Report CCC/160, www.iea-coal.org.uk

Davison J (2011) Flexible CCS plants – a key to near-zero emission electricity systems. Energy Procedia 4:2548–2555

DBFZ (2012) Monitoring zur Wirkung des Erneuerbare-Energien-Gesetz (EEG) auf die Entwicklung von Biomasse – Endbericht zur EEG Periode 2009 bis 2011. Studie im Auftrag des Bundesministeriums für Umwelt und Reaktorsicherheit, Förderkennzeichen: 03MAP138. www.dbfz.de

De Coninck H, Mikunda T (2010) Global Technology Roadmap for CCS in Industry – background paper. ECN/United Nations Industrial Development Organization

Engels S, Beggel F, Modigell M, Stadler H (2010) Simulation of a membrane unit for oxyfuel power plants under consideration of realistic BSCF membrane properties. J Membr Sci 359:93–101

Epple B, Ströhle J (2011) Chemical looping in power plants. In: Scherer V, Stolten D (eds) Efficient carbon capture for coal power plants. Wiley-VCH Verlag, Weinheim

Fahlenkamp H, Dittmar M (2011) CO_2 removal in coal power plants via post-combustion with absorbents. In: Scherer V, Stolten D (eds) Efficient carbon capture for coal power plants. Wiley-VCH Verlag GmbH, Weinheim

Finkenrath M (2010) Cost and performance of carbon dioxide capture from power Generation. In: IEA (ed) Paris, www.iea.org

Finkenrath M, Smith J, Volk D (2012) Analysis of the globally installed coal-fired power plant fleet. Information paper, Internationale Energieagentur (IEA)

Fleer J (2011) Technische und ökonomische Analyse von CCS für CO_2-intensive Industrieprozesse. Diplomarbeit Ruhr-Universität Bochum (LEE), STE Student Research Report 09/2011, Forschungszentrum Jülich (IEK-STE)

GCCS (2011) The global status of CCS 2011. Global CCS Institute (GCCS)

Gielen D (2003) CO_2 removal in the iron and steel industry. Energy Convers Manag 44:1027–1037

Göttlicher G (1999) Energetik der Kohlendioxidrückhaltung in Kraftwerken. VDI Verlag, Düsseldorf

Grabner M, von Morstein O, Rappold D, Gunster W, Beysel G, Meyer B (2010) Constructability study on a German reference IGCC power plant with and without CO(2)-capture for hard coal and lignite. Energy Convers Manag 51:2179–2187

Grathwohl S, Lemp O, Schnell U, Maier J, Scheffknecht G, Kluger F (2009) Highly flexible burner concept for oxyfuel combustion. 1st international oxyfuel combustion conference 7–11 September 2009, Cottbus, Germany

Hohlfeld A, Katthöfer V, Kühleis C, Olaniyon A, Thorne C, Weiß J (2011) Kohlendioxidemissionen der emissionshandelspflichtigen stationären Anlagen im Jahr 2010 in Deutschland. Deutsche Emissionshandelsstelle

IEA (2000) World energy outlook 2000. International Energy Agency OECD/IEA, www.iea.org

IEA (2006) World energy outlook 2006. International Energy Agency OECD/IEA, www.iea.org

IEA (2008a) CO_2 capture in the cement industry. IEA Greenhouse Gas R&D Programme, Report number 2008/3

IEA (2008b) Energy technology perspectives – scenarios & strategies to 2050. International Energy Agency OECD/IEA, Paris, www.iea.org

IEA (2008c) World energy outlook 2008. International Energy Agency OECD/IEA, www.iea.org

IEA (2009) World energy outlook 2009. International Energy Agency OECD/IEA, www.iea.org

IEA (ed) (2010) Energy technology perspectives – scenarios & strategies to 2050. International Energy Agency OECD/IEA, Paris, www.iea.org
IEA (2011) World energy outlook 2011. International Energy Agency OECD/IEA, www.iea.org
IPCC (2005) Carbon dioxide capture and storage – special report of the intergovernmental panel on climate change. Cambridge University Press, Cambridge, UK
Irons R, Davison J, Sekkapaan G, Gibbins J (2007a) Implications of building 'capture ready' power plants. 3rd international conference on clean coal technologies for our future, Cagliari, Sardinia
Irons R, Sekkappan G, Panesar R, Gibbins J, Lucquiaud M (2007b) CO_2 capture ready plants. Technical Study Report Nr. 2007/4, IEA Greenhouse Gas R&D Programme
Kather A, Klostermann M (2011) CO_2 capture via the oxyfuel process with cryogenic air. In: Scherer V, Stolten D (eds) Efficient carbon capture for coal power plants. Wiley-VCH Verlag, Weinheim
Kather A, Rafailidis S, Hermsdorf C, Klostermann M, Maschman A, Mieske K, Oexmann J, Pfaff I, Rohloff J (2008) Research and development needs for clean coal deployment. IEA Clean Coal Centre, London
Kluger F, Mönckert P, Krohmer B, Stramatelopoulos G, Jacoby J, Burchardt U (2009) Oxyfuel pulverized coal steam generator development – 30 MW_{th} pilot steam generator commissioning and testing. 1st international oxyfuel combustion conference, 7–11 September 2009, Cottbus, Germany
Kneer R, Toporov D, Forster M, Christ D, Broeckmann C, Pfaff E, Zwick M, Engels S, Modigell M (2010) OXYCOAL-AC: towards an integrated coal-fired power plant process with ion transport membrane-based oxygen supply. Energy Environ Sci 3:198–207
Kozak F, Petig A, Morris E, Rhudy R, Thimsen D (2009) Chilled ammonia process for CO_2 capture. Energy Procedia 1:1419–1426
Kunze C, Spliethoff H (2010) Modelling of an IGCC plant with carbon capture for 2020. Fuel Process Technol 91:934–941
Kvamsdal HM, Jakobsen JP, Hoff KA (2009) Dynamic modeling and simulation of a CO_2 absorber column for post-combustion CO_2 capture. Chem Eng Process Process Intensif 48:135–144
Linnenberg S, Kather A (2009) Evaluation of an integrated post-combustion CO_2 capture process for varying loads in a coal-fired power plant using monoethanolamine. 4th international conference on clean coal technologies, 18–21 May 2009, Dresden
Lyngfelt A, Mattisson T (2011) Chemical looping materials for CO_2-separation. In: Scherer V, Stolten D (eds) Efficient carbon capture for coal power plants. Wiley-VCH Verlag, Weinheim
Markewitz P, Schreiber A, Zapp P, Vögele S (2009) Kohlekraftwerke mit CO_2-Abscheidung: Strategien, Rahmenbedingungen und umweltseitige Auswirkungen. Zeitschrift für Energiewirtschaft 33:31–41
Markewitz P, Bongartz R, Birnbaum U, Linssen J, Vögele S (2011a) Energy technologies 2050: R&D priority setting of coal fired power plants in Germany. Conference on clean coal technologies CCT 2011. IEA Clean Coal Centre, Zaragoza
Markewitz P, Kuckshinrichs W, Hake JF, Fischer W, Bongartz R, Martinsen D, Pesch T, Vögele S (2011b) Transformation des Stromerzeugungssystems mit forciertem Ausstieg aus der Kernenergie – Ein Beitrag zur Diskussion nachhaltiger Energiesysteme nach dem Reaktorunfall in Fukushima. STE Research Report 06/2011, Forschungszentrum Jülich
Markewitz P, Kuckshinrichs W, Leitner W, Müller TE, Linssen J, Zapp P (2012) Worldwide innovations in the development of carbon capture technologies and the utilization of CO_2. Energy Environ Sci 5:7281–7305
Mckinsey (2007) Kosten und Potenziale der Vermeidung von Treibhausgasemissionen in Deutschland – Sektorperspektive Industrie. In: Klimaschutz SIAVBI-WFD (ed) Studie im Auftrag von BDI initiativ-Wirtschaft für den Klimaschutz. Studie im Auftrag von BDI initiativ-Wirtschaft für den Klimaschutz ed.: Mc Kinsey & Company, Inc

Ploumen P (2006) Retrofit of CO_2 capture at coal-fired power plants in the Netherlands. In: IEA-GHG (ed) 8th International conference on greenhouse gas control technologies, 19th–21st June 2006. IEA GHG, Trondheim

Rao AB, Rubin ES (2002) A technical, economic, and environmental assessment of amine-based CO_2 capture technology for power plant greenhouse gas control. Environ Sci Technol 36:4467–4475

Reijerkerk S, Nijmeijer K, Potreck J, Simons K, Wessling M (2011) Polymer membranes for CO_2-separation. In: Scherer V, Stolten D (eds) Efficient carbon capture for coal power plants. Wiley-VCH Verlag, Weinheim

Reissner H (2009) Flue gas desulpherization as a key requirement of PCC. International RWE symposium on post combustion capture. Neuss 30.-31.07.2009

Renzenbrink W, Evers J, Keller D, Wolf KJ, Apel W (2008) RWE's 450 MW IGCC/CCS project – status and outlook. Energy Procedia 1:615–622

Rubin ES, Chen C, Rao AB (2007) Cost and performance of fossil fuel power plants with CO_2 capture and storage. Energy Policy 35:4444–4454

RWE (2009) The need for smart megawatts – power generation in Europe. Facts and trends. RWE AG, Essen

Scherer V, Franz J (2011) CO_2 separation via pre-combustion utilizing membranes in coal power plants. In: Scherer V, Stolten D (eds) Efficient carbon capture for coal power plants. Wiley-VCH Verlag, Weinheim

Stadler H, Beggel F, Habermehl M, Persigehl B, Kneer R (2011) Oxyfuel coal combustion by efficient integration of oxygen transport membranes. Int J Greenh Gas Control 5:7–15

Stolten D, Scherer V (eds) (2011) Efficient carbon capture for coal power plants. Wiley-VCH Verlag, Weinheim

Ströhle J, Galloy A, Epple P (2008) Feasibility study on the carbonate looping process for post-combustion CO_2 capture from coal-fired power plants. Energy Procedia 1:1313–1320

Tigges KD, Klauke F, Bergins C, Busekrus K, Niesbach J, Ehmann M, Kuhr C, Hoffmeister F, Vollmer B, Buddenberg T, Wu S, Kukoski A (2009) Conversion of existing coal-fired power plants to oxyfuel combustion: Case study with experimental results and CFD-simulations. Energy Procedia 1:549–556

Trautmann G, Döring M, Heim M, Schneider PG (2007) Optimierung des dynamischen Verhaltens kohlefebeuerter Kraftwerke. VDI Bericht Nr. 180

Trudeau N (2011) Technology roadmap: carbon capture and storage in industrial applications. Challenges and opportunities of CO_2 capture & storage in the iron and steel industry. Steel Institute VDEh Germany, Düsseldorf

UNIDO (2010) Carbon capture and storage in industrial applications: technology synthesis report. Working paper of the United Nations Industrial Development Organisation (UNIDO) – November 2010, www.unido.org

Urban W, Lohmann H, Girod K (2012) BMBF Verbundprojekt "Biogaseinspeisung". In: Forschung BFBU (ed) Fraunhofer- Institut UMSICHT, BASE Technologies GmbH

van Straelen J, Geuzebroek N, Goodchild G, Protopapas L, Mahony L (2009) CO_2 capture for refineries – a practical approach. Energy Procedia 1:179–185

White V, Armstrong P, Fogash K (2009) Oxygen supply for oxyfuel CO_2 capture. 1st international oxyfuel conference, 7–9 September 2009, Cottbus, Germany

Wietschel M, Arens M, Dötsch C, Herkel S, Krewitt W, Markewitz P, Möst D, Scheufen M (eds) (2010) Energietechnologien 2050 – Schwerpunkte für Forschung und Entwicklung: Technologiebericht. Fraunhofer Verlag, Karlruhe

Wirtschaftsvereinigung Stahl (2011) Stahlmarkt Deutschland. Available: http://www.stahl-online.de/

ZEP (2008) EU demonstration programme for CO_2 capture and storage (CCS) – ZEP's proposal. In: European Technology Platform for Zero Emission Fossil Fuel Power Plants (ZEP) (ed) Brussels, http://www.zero-emissionplatform.eu/website/docs/ETP%20ZEP/EU%20Demonstration%20Programme%20for%20CCS%20-%20ZEP's%20Proposal.pdf, 15 Aug 2012

Zhao L, Menzer R, Riensche E, Blum L, Stolten D (2009) Concepts and investment cost analyses of multi-stage membrane systems used in post-combustion processes. Energy Procedia 1:269–278

Zhao L, Riensche E, Blum L, Stolten D (2011) How gas separation membrane competes with chemical absorption in postcombustion capture. Energy Procedia 4:629–636

Ziaii S, COHEN S, Rochelle GT, Edgar TF, Weber ME (2009) Dynamic operation of amine scrubbing in response to electricity demand and pricing. Energy Procedia 1:4047–4053

Ziesing J (2011a) Der erwartete Rückschlag für den Klimaschutz: Weltweite CO_2-Emissionen 2010 kräftig gestiegen. Energiewirtschaftliche Tagesfragen 61:67–78

Ziesing J (2011b) Kräftiger Anstieg der CO_2-Emissionen in Deutschland. Energiewirtschaftliche Tagesfragen 61:61–68

Chapter 3
CO_2 Transportation

Richard Bongartz, Jochen Linssen, and Peter Markewitz

Abstract This chapter discusses the technical options for transporting CO_2. The current situation as well as the potential risks of transporting CO_2 in pipelines will be outlined. In addition, the purity requirements for the CO_2 to be transported will be discussed.

Pipelines are particularly interesting for transporting large amounts of CO_2 over long distances. At present, CO_2 pipeline grids throughout the world have a total length of more than 4,000 km (USA: 2,600 km). The main causes of damage to CO_2 pipelines are pressure relief valves, sealing problems/faulty welds, and damage caused by corrosion. CO_2 purity is particularly relevant because it can be interpreted as indirect protection against corrosion. The impurities contained in captured CO_2 streams from power plants are very different to those contained in the volumes of CO_2 currently transported in the USA (Enhanced Oil Recovery). The quality requirements for pipeline transportation of CO_2 in the USA can therefore only be transferred to the transportation of CO_2 mixtures produced in power plant processes to a limited extent.

The release of large amounts of CO_2 can pose local risks to humans and the environment. Risk assessments were performed based on probabilistic approaches within the context of different studies. The available studies were used to qualitatively evaluate the categorized transportation risks (e.g. valve leak, pipeline leak, pipeline rupture) in terms of frequency and range of critical CO_2 concentrations. The results clearly show that most of the risks associated with transportation of CO_2 are negligible or only very low.

Keywords CO_2 transportation • CO_2 pipeline • CO_2 purity • Risk assessment

---

R. Bongartz • J. Linssen (✉) • P. Markewitz
Institute of Energy and Climate Research – Systems Analysis and Technology Evaluation (IEK-STE), Forschungszentrum Jülich GmbH, D-52425 Jülich, Germany
e-mail: r.bongartz@fz-juelich.de; j.linssen@fz-juelich.de; p.markewitz@fz-juelich.de

W. Kuckshinrichs, J.-F. Hake (eds.), *Carbon Capture, Storage and Use*,
DOI 10.1007/978-3-319-11943-4_3

3.1 Introduction

CO_2 captured from power plants and large-scale industrial facilities cannot generally be stored on site the capture plant. Suitable storage options can be located up to several thousand kilometres away from the CO_2 emission source. The choice of transportation technology essentially depends on economic criteria, as well as on other framework conditions (e.g. geology, existing infrastructures). An alternative option, which is attractive from a long-term perspective, is to relocate emission sources closer to suitable storage sites, thus obviating the need for CO_2 transportation. Against this background, coal gasification in close proximity to storage sites may also be an option, which would require the construction of a hydrogen infrastructure (cf. Baufumé et al. 2011). Such a hydrogen infrastructure could also be used when the CO_2 storage capacities have been exhausted.

This chapter discusses the technical options for transporting CO_2. The current situation as well as the potential risks of transporting CO_2 in pipelines will be outlined. In addition, the purity requirements for the CO_2 to be transported will also be discussed.

3.2 Current Situation

For a cost-effective transportation of the CO_2 captured from power plants or industry, the only viable option for logistical and economic reasons is to transport it at high densities. Figure 3.1 shows the phase diagram for pure CO_2 with the melting-point pressure and vapour-pressure curve.

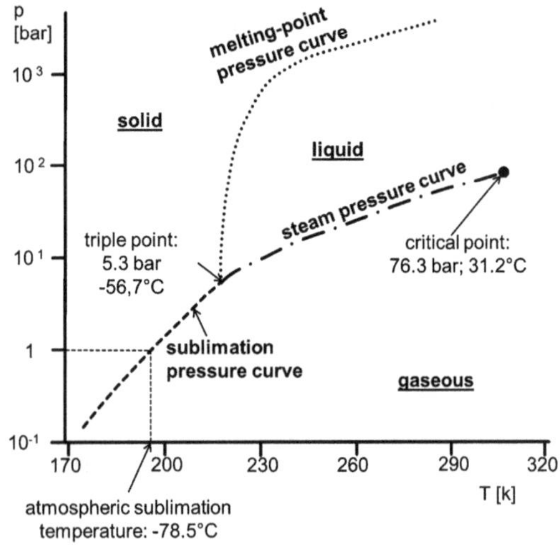

Fig. 3.1 Phase diagram for CO_2 as a pure substance

As it is not generally possible to actively cool transport pipelines or pressure tank trucks, both the operating pressure of the pipeline as well as the ambient temperature is decisive for the aggregate state of the carbon dioxide during transportation. Above the critical temperature of 304.2 K (31.2 °C), CO_2 is in a supercritical state, and its density and flow behaviour are similar to those of a liquid. CO_2 sublimes under ambient pressure at a temperature of 194.5 K (−78.5 °C). In the supercritical state, CO_2 has a high dissolving power for several hydrophobic substances, and it is already used as a solvent in technical applications, for instance in the food industry or in petrochemistry (for the usage of CO_2, see Chap. 4). Under standardized conditions, approximately 0.9 l of gaseous CO_2 dissolves in 1 l of water to form carbonic acid.

The transportation of CO_2 by pipeline and ship has already been implemented on a large industrial scale. For small quantities (a few kt CO_2 per year) or short distances, pressure tank trucks are also a possible transportation option. The criteria for selecting suitable transportation technology not only include costs and capacities, but also geographical conditions, safety issues, the type of CO_2 storage, and the flexibility of the means of transportation.

Current established practice is to transport large quantities of CO_2 – chiefly for enhanced oil recovery (EOR) – primarily by pipeline. The state of development of CO_2 pipeline transportation can therefore be considered to be commercial.

Although CO_2 pipelines do exist throughout the world today, the area covered and their interconnections are not comparable with natural gas or crude oil pipeline grids. The total length of the pipeline grid is estimated to be more than 3,000 km, and is located predominantly in the USA and Canada. The transport capacity today is slightly under 4 Mt CO_2 per year. In comparison, there are over 536,000 km of pipelines for transporting natural gas in the USA only. Table 3.1 contains a selection of CO_2 pipelines operated worldwide.

Studies on CO_2 pipeline design (e.g. Odenberger and Svensson 2003, IEA 2008) based on technical criteria and material design enable the costs of new pipeline construction projects to be evaluated (see Chap. 9).

Potential options of reducing the costs of CO_2 pipelines can be the use of improved materials and new installation techniques. As a significant proportion of the costs are associated with planning measures, rights of way etc., it may be possible to cut costs further by selecting optimized pipeline routes, for instance along existing infrastructure corridors.

In this context, the long-term nature of pipeline construction should be mentioned, the duration of which is chiefly determined by planning processes. When constructing a CO_2 transport grid, therefore, long lead times must be allowed for in order to avoid any delays in establishing a generation of CCS power plants.

In relation to CO_2 sources and sinks and their long-term availability, it should be noted that as pipeline transportation is location-dependent, it does not have the same flexibility as transportation by ship. Although the large-scale transportation of carbon dioxide by ship is technically feasible, there is still a lack of notable operating experience (Aspelund et al. 2006). Today, CO_2 is mainly transported in small ships or barges. Analyses of the technical design and competitiveness of ships

Table 3.1 Technical characteristics of selected CO_2 pipelines

Pipeline	Location	Operator	Transport capacity Mt/a	Length km	Diameter mm	Pressure bar	Put into operation Year
Weyburn	USA/CAN	North Dakota Gasification	5	328	305–356	152	2000
Cortez	USA	Kinder Morgan	19.3	803	762	186	1984
Sheep Mountain	USA	BP Amoco	ND	296	508	ND	1983
Sheep Mountain North	USA	BP Amoco	ND	360	610	132	1983
Bravo	USA	Kinder Morgan	7.3	350	508	165	1984
Central Basin	USA	Kinder Morgan	20	278	400–650	170	1985
Bati Raman	Turkey	Turkish Petroleum	1.1	90	ND	170	1983
Canyon Reef Barriers	USA	Kinder Morgan	4.4	352	400	140	1972
Val Verde	USA	Petro Source	2.5	130	250	ND	1986
Bairoil	USA	ND	8.3	180	ND	ND	1986
Sleipner	North Sea	Statoil	1	160	ND	ND	1996
Snohvit	North Sea	Statoil	0.7	153	200	100	2006

Source: IPCC (2005a) and Seevam et al. 2007)
ND no data

for CO_2 transportation (e.g. IPCC 2005b) assume that transportation by ship will be commercially advantageous from a daily transportation capacity of 20,000 t CO_2 and a distance of 700 km. With larger quantities of CO_2, this break-even point will shift towards greater transport distances. If CO_2 captured on a continuous basis from power plants is to be transported on a discontinuous basis by ship, an intermittent storage option is necessary.

In the year 2010 a 'CO_2 hub' was planned for the port of Rotterdam in the Netherlands. At this hub, CO_2 delivered from different sources via different transportation systems (pipeline or barge) should be collected and processed. It should then be transported onwards in liquid form to offshore storage sites in the North Sea by ship or via an offshore pipeline (Schreurs 2009). The advantages of this 'collection concept' are the flexibility of the system and the high capacity utilization of the offshore facilities. A comprehensive economic evaluation of the concept has yet to be conducted.

3.3 Purity Level and Quality Criteria

CO_2 captured from power plants contains constituents of other substances, which are commonly referred to as impurities. The captured CO_2 is conditioned for transportation and then compressed. Conditioning involves purification of the CO_2 using distillation and/or flash techniques. This removes impurities (H_2O, SO_2, etc.) which may damage pipelines. Furthermore, a higher proportion of impurities increases the amount of compression work necessary (Kuckshinrichs et al. 2012).

Aside from CO_2 transportation, CO_2 storage and certain types of CO_2 utilization also make certain demands on CO_2 purity. Detailed limits for permissible amounts of impurities to meet the requirements for transportation, storage, and utilization have yet to be set. The European Union's CCS Directive states that the gas mixture to be stored must consist mainly of CO_2, and that the concentration of impurities must not adversely affect the integrity of the storage facility or the transportation infrastructure. The Directive does not lay down concrete limits for impurities.

Equally as important as setting limits for an absolute purity level is a clear statement on the composition of impurities which ultimately lead to undesired side effects (e.g. corrosion). As the pipeline materials used are well-known, and, moreover, as many years of experience in CO_2 transportation have been gathered, it is possible to specify the technical requirements for the material based on impurities. Research is still being conducted on the effects of impurities on CO_2 storage (interactions with the geological storage environment and peripheral elements of storage sites, in particular cement).

The composition of impurities in the CO_2 stream captured depends on both the power plant process and the carbon capture process used. Table 3.2 contains a list of typical impurities, broken down according to capture technology and energy carrier. The differences between the individual processes are significant. While in the oxyfuel process, the proportion of argon and nitrogen is dominant due to air penetration, the methane and hydrogen impurities in the pre-combustion process are the result of the actual gasification and reforming process.

Table 3.2 Typical impurities in flue gases

Type of power plant	Components	Coal (vol.%)	Natural gas (vol.%)
Post-combustion	SO_2	<0.01	<0.01
	NO_x	<0.01	<0.01
	$N_2/Ar/O_2$	0.01	0.01
Pre-combustion	H_2S	0.01–0.6	<0.01
	H_2	0.8–2.0	1
	CO	0.03–0.4	0.04
	CH_4	0.01	2
Oxyfuel	SO_2	0.5	<0.01
	NO_X	0.01	<0.01
	$N_2/Ar/O_2$	3.7	4.1

Source: IEA (2004) and Seevam et al. (2008)

Table 3.3 shows the composition of the gases transported by a selection of American CO_2 pipelines. Some of the CO_2 from anthropogenic sources originates from syngas production, which notably explains the H_2S content, as well as the proportion of light hydrocarbons. When compared with the typical impurities in Table 3.2, it is clear that the composition of impurities in the CO_2 that has to be transported is different in the case of CCS. The flue gases from power plants contain SO_2, NO_x, Ar, and O_2, which can influence pipeline design and material selection.

The US company Kinder Morgan, which builds and operates CO_2 pipelines in the USA, has set standards for CO_2 transportation and acceptable ranges for impurities (see Table 3.4). The criteria are not always defined based on transportation. For example, a minimum purity of 95 % CO_2 is required when CO_2 is used for EOR in order to ensure solubility with crude oil. The solubility pressure is also influenced by impurities. Sulphur, nitrogen and nitrogen oxides have a negative effect, while H_2S and light hydrocarbons have a positive effect. This explains the relatively high limit for hydrocarbons. Contamination with water, oxygen, hydrogen sulphide, and glycol is limited for reasons related to the material, corrosion and safety (Seevam et al. 2008).

The standards set in the USA – predominantly for EOR – can only be applied to a certain extent to the transportation of CO_2 mixtures produced in power plant processes. As part of the EU project DYNAMIS,[1] standards were developed for CO_2 pipelines intended for transporting CO_2 mixtures captured from pre-combustion or post-combustion power plants. Table 3.5 shows technical, safety-related, and health-related criteria, as well as the proposed standards for pipeline operation.

The reasons for limiting the individual substances are also shown in Table 3.5. A higher water content in the CO_2 mixture, for example, is disadvantageous for pipeline transportation because it could cause corrosion and the formation of hydrates. When solid hydrates accumulate, there is a danger of pipes clogging and the pipeline peripheral elements becoming damaged (e.g. valves, etc.).

There is a danger of corrosion when acids form in combination with CO_2 or other components (e.g. SO_2, H_2S). The formation of (free) water is particularly disadvantageous because it considerably increases the danger of corrosion. Too high a hydrogen content is generally undesirable because it reduces the energy efficiency of the process as a whole by increasing the energy requirements for compression. In addition, hydrogen involves a risk that the properties of the pipeline material (carbon steel) will be negatively affected by hydrogen embrittlement. Oxygen is just as undesirable an impurity, because in combination with water, it can accelerate oxidizing reactions that could lead to corrosion damage. It should be noted that the described effects can occur to different extents depending on the pressure, temperature and type of impurities (Doctor 2000; Gale and Davison 2004; Visser et al. 2008).

[1] www.dynamis-hypogen.com

Table 3.3 Average gas compositions of existing pipelines

	CO_2	CH_4	N_2	H_2S	C_2H_n	CO	O_2	Source
Canyon Reef	95 %	5 %	<0.5 %	100 ppm	–	–	–	Anthropogenic
Central Basin	98.5 %	0.2 %	1.3 %	–	–	–	–	Natural
Sheep Mountain	96.8 %	1.7 %	0.9 %	–	0.6 %	–	–	Natural
Bravo Dome	99.7 %	–	0.3 %	–	–	–	–	Natural
Weyburn	96 %	0.7 %	<300 ppm	0.9 %	2.3 %	0.1 %	<50 ppm	Anthropogenic

Source: Seevam et al. (2008)

Table 3.4 Quality requirements for CO_2 pipelines in the USA

Components	Concentration	Minimum/maximum	Criterion
CO_2	95 vol.%	Minimum	Solubility pressure
N_2	4 vol.%	Maximum	Solubility pressure
C_mH_n	5 vol.%	Maximum	Solubility pressure
H_2O	480 mg/m^3	Maximum	Corrosion
O_2	10 ppm	Maximum	Corrosion
H_2S	10–200 ppm	Maximum	Safety
Glycol	0.04 ppmv	Maximum	Operation
Temperature	50 °C	Maximum	Material

Source: Visser et al. (2008)

Table 3.5 Recommended quality of CO_2 streams for pipeline transportation

Gas	Limit	Reason
H_2O	500 ppm	Technical: below the solubility limit of H_2O in CO_2
H_2S	200 ppm	Health- and safety-related aspect
CO	2000 ppm	Health- and safety-related aspect
O_2	Aquifer <4 vol.%, EOR 100–1,000 ppm	Set technically based on EOR because of the lack of practical experience with the impacts of O_2 in the underground
CH_4	Aquifer <4 vol.%, EOR 2 vol.%	Proposed by the ENCA project
N_2	Proposed by the ENCA project	Proposed by the ENCA project
Ar	<4 vol.% (all condensable gases)	Proposed by the ENCA project
H_2	<4 vol.% (all condensable gases)	Further reduction of hydrogen is desired (efficiency of the capture process)
SO_x	100 ppm	Health- and safety-related aspect
NO_x	100 ppm	Health- and safety-related aspect
CO_2	>95.5 vol.%	Main component

Source: Visser et al. (2008)

3.4 Risks of CO_2 Transportation

3.4.1 Dangers of CO_2

CO_2 is heavier than air and can therefore accumulate near the ground in high concentrations. The effects of increased carbon dioxide concentrations on humans are outlined in Table 3.6. Other potential effects include the acidification of drinking water reserves and negative impacts on flora and fauna.

Table 3.6 Health effects of increased CO_2 concentrations

Concentration	Effect
0.05 %	Natural concentration in air
0.5 %	MAC for daily exposure of 8 h/day
4 %	Concentration in exhaled breath
5 %	Headaches, dizziness, unconsciousness
8 %	Unconsciousness, death after 30–60 min
>17 %	Unconsciousness and death within 1 min

Source: Seliger et al. (2009) and DNV 2008)

3.4.2 Hazard Potential

Depending on their dimensions and lengths, CO_2 pipelines can contain thousands of tonnes of highly compressed CO_2. Automatic shut-off devices at intervals of e.g. 8 km can considerably limit potential leaks. A pipeline with a diameter of 400 mm, through which around 3 Mt CO_2 is transported every year (equivalent to emissions from a 700 MW coal-fired power plant), contains around 850 t CO_2 in every 8 km section (for a 1,000 mm pipeline, this would be around 5,300 t). Should the pipeline rupture and the section affected be automatically shut off, this amount alone would escape from the pipeline and pose a hazard potential for the affected area.

3.4.3 Operational Experience

CO_2 pipelines (mainly for EOR) have been installed since the 1980s. Pipeline operation, which has been under way in some instances for more than 20 years, is governed by regulations which also entail monitoring. Based on experience to date, it can be said that transporting carbon dioxide in pipelines fulfils the specified safety requirements. The damage statistics of American pipelines were compared for different transport media (CO_2, natural gas, liquids) by (Gale and Davison 2004) (see Table 3.7). In contrast to natural gas pipelines, no severe accidents have occurred with CO_2 according to this study. The monetary damage of the ten accidents between 1986 and 2001 was also comparatively low. It should, however, be noted that this CO_2 pipelines are far away from heavily populated areas, which had a positive effect on the evaluation of damage. The causes of damage to the CO_2 pipelines are defective pressure relief valves, sealing problems/faulty welds, and corrosion. The evaluated causes of damage to natural gas pipelines were mainly due to corrosion and external damage (e.g. excavation work).

For the CO_2 pipeline grid in the USA (2,600 km), based on operational experience for 1994–2006 in the case of failures caused by 'ruptures', (DOE 2007) determined a frequency of occurrence of $5.92 \cdot 10^{-5}$ per pipeline kilometre and year, and for those caused by 'leaks' a rate of $1.18 \cdot 10^{-4}$/(km a).

Table 3.7 Pipeline damage in the USA

	High-pressure natural gas (1986–2001)	Hazardous fluids (1986–2001)	CO_2 (1990–2001)
Number of accidents	1,287	3,035	10
Number of fatalities	58	36	0
Number of casualties	217	249	0
Material damage	$ 285 million	$ 764 million	$ 0.5 million
Accidents per 1,000 km and year	0.17	0.82	0.32
Material damage per 1,000 km and year	$ 37,000	$ 205,400	$ 15,200

Source: Gale and Davison (2004)

Statistics from the European Gas Pipeline Incident Data Group (EGIG 2008) for the period 1970–2007 indicate a failure frequency for European natural gas pipelines of 0.37 per 1,000 km and year ($3.7 \cdot 10^{-5}$/km a). For the last 5 years, this value is 0.14 ($1.4 \cdot 10^{-5}$/km a). The failures are divided into three leak categories (<2 cm diameter/>2 cm/rupture). In the statistics, pipelines with diameters of 125–425 mm were most common. With a failure frequency of around 0.58/(1,000 km a), pipelines of the lowest dimensions ($\leq$125 mm) contributed disproportionately to the overall result.

The main source of failure with a share of almost 50 % is external interference (e.g. excavation, ploughing, work on pipelines), followed by construction defects/material failures (16.5 %), internal and external corrosion (15.4 %), and ground movement (7.3 %). Depth of cover has an important impact on the failure frequency in the case of external influences. For a depth of cover of <80 cm, it is approx. 0.7/(1,000 km and year), for a depth of cover of >100 cm, it is less than 0.1/(1,000 km and year).

The evaluation of operational experience to date for European natural gas pipelines for all causes of failure, periods of time, and pipeline dimensions resulted in a frequency for ruptures of $5.0 \cdot 10^{-5}$, of $1.3 \cdot 10^{-4}$ for larger leaks (>2 cm), and of $1.9 \cdot 10^{-4}$ per pipeline kilometre and year for very small leaks (<2 cm). These failure rates agree largely with those values in DOE (2007) and McGillivray and Wilday (2009).

Operating experience are difficult to transfer from American pipelines to the European situation, because the conditions of approval and the operating conditions are very different. This can be shown using natural gas pipelines as an example. A comparison with the damage and accident statistics of the German Association for Gas and Water shows that the safety of German pipelines is much higher.

3.4.4 *Measures Minimizing Risks*

In order to prevent damage to CO_2 pipelines or to limit the impact of damage, a range of technical and administrative measures are possible:

- CO_2 purity requirements (Sect. 3.3)
- Continuous monitoring of gas purity
- Monitoring of leaks
- Maintenance measures
- Shut-off valves
- Crack stopper to limit crack propagation
- Monitoring of pipeline corridors
- Safety zones

Most of these measures are standard in the case of natural gas pipelines They are described in more detail in Bongartz et al. (2010).

As mentioned in this section, a considerable proportion of pipeline damage is due to corrosion. Particular attention should therefore be paid to CO_2 purity requirements.

3.4.5 *Evaluation of Transportation Risks*

IPCC (2005b) outlines the basic risk of CO_2 pipeline transportation, and emphasizes the increased risk when pipelines are laid in densely populated areas. Furthermore, existing natural gas pipelines are used as an example to draw analogies to CO_2 pipelines.

Over the past few years, there has been an increase in studies (Koornneef et al. 2009; IEA 2009; Eldevik et al. 2009; McGillivray and Wilday 2009) concerned with the assessment of risks associated with the transportation of CO_2. This emphasizes the current relevance of this topic. The issues investigated range from the analysis of potential causes of failures and measures for preventing damage to quantitative risk assessment (QRA), in which different damage scenarios are assumed and analysed. QRAs estimate potential risks in the direct vicinity of a particular location where a defined accident has occurred using dispersion models and assuming different parameters (e.g. meteorology, crack/puncture, etc.). Koornneef et al. (2009) ascertained that although a large number of QRAs have been performed, there is no consensus with regard to assessing the risks of CO_2 transportation.

In some European countries (e.g. UK, NL), risk-based assessment procedures are used in the area of urban and rural development. This necessitates the definition of thresholds for risks, which for example in the United Kingdom is (10^{-5}/a – 3×10^{-7}/a) for individual risk (DNV 2005). In Germany, instead of a risk-based assessment approach, a deterministic assessment approach is used ('major

Table 3.8 Relationship between duration of exposure and CO_2 concentration

Duration of exposure/concentration in air for % fatalities		
	1–5 % fatalities (SLOT)	50 % fatalities (SLOD)
60 min	63,000 ppm	84,000 ppm
30 min	69,000 ppm	92,000 ppm
5 min	86,000 ppm	115,000 ppm
1 min	105,000 ppm	140,000 ppm

Source: Harper (2010)

accidents despite precautions') in line with legislation of the major accidents ordinance (Uth 2009). However, the required deterministic or probabilistic distances proposed in the European Seveso II Directive governing the location of hazardous industrial facilities do not apply to pipelines.

Quantitative risk assessments are used by the British Health and Safety Executive (HSE) as the basis for defining risk zones (inner, middle and outer zones), which in turn provide the basis for the approval and construction of high-risk facilities or for the approval of construction projects in the vicinity of high-risk facilities. Dangerous toxic loads (DTLs) are used to assess the level of harm and thus the level of risk of toxic substances. According to DNV (2005), a DTL is considered to exist when a large number of people would be injured and require medical attention, when individuals would be severely injured and require hospitalization, or when a number of fatalities would result (e.g. 1 %).

There are still no standardized exposure thresholds or uniform dose-response relationships for toxicological exposure to CO_2. Table 3.8 shows the correlation between duration of exposure and concentration in air for the specified level of toxicity (SLOT) and the significant likelihood of death (SLOD) according to McGillivray and Wilday (2009). The former applies to a mortality rate of 1–5 % and the latter to a mortality rate of 50 % of the population affected.

In McGillivray and Wilday (2009), HSE's probabilistic approach is used to estimate and compare the risks of CO_2 and natural gas pipelines. Figure 3.2 shows the different accident scenarios modelled for three failure categories for CO_2 pipelines.

3.4.6 *Estimation of Risk Zones*

In order to determine the risk of the CO_2 release scenarios considered (see Fig. 3.2), McGillivray and Wilday (2009) used consequence modelling to determined hazard zones for the two levels of toxicity. Table 3.9 shows the results of these calculations for the SLOT. The determined hazard zones provide the input parameters for calculating the frequency of occurrence of a hazardous dose of CO_2 at a certain distance from the pipeline.

The results of such risk assessments are used in the United Kingdom to define risk zones (inner, middle, and outer zones), which in turn provide the basis for the

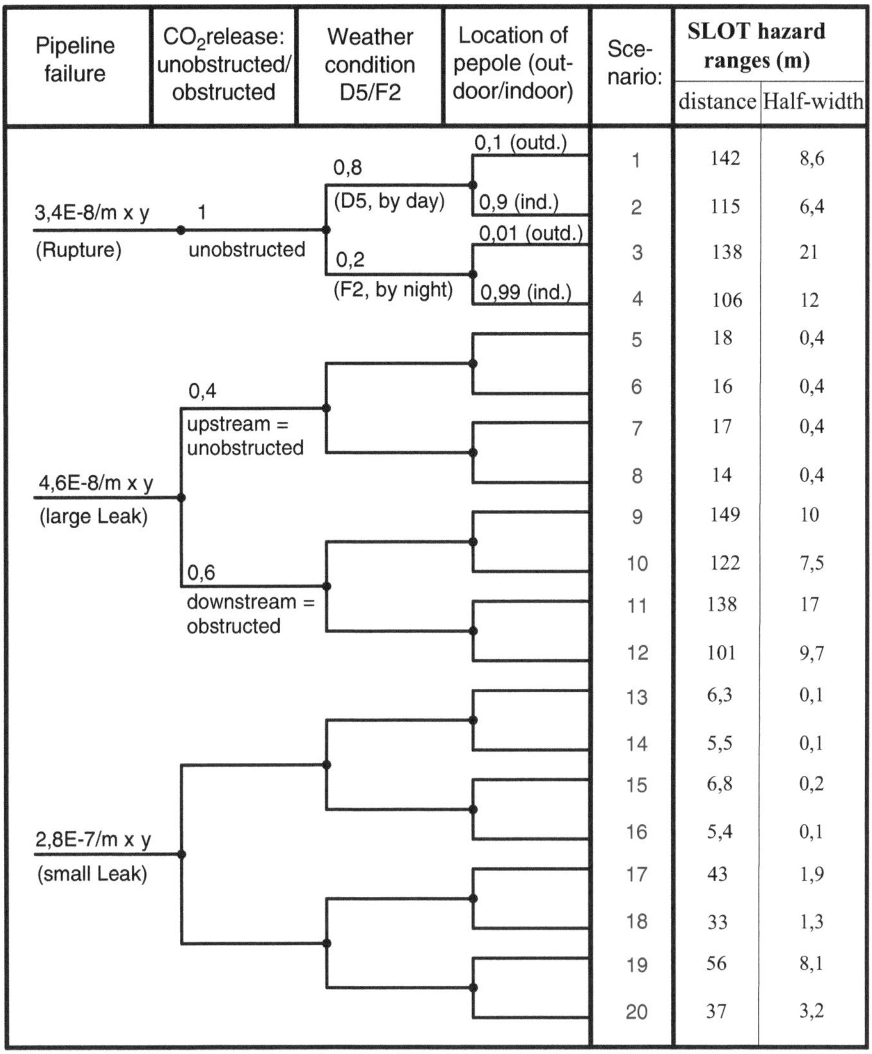

Fig. 3.2 Accident scenarios for CO_2 pipeline failures (Source: Adapted from McGillivray and Wilday 2009)

approval of the construction of a facility (here pipelines) or for the approval of construction projects in the vicinity of high-risk facilities.

The SLOT calculations showed that the threshold limits for implementing an inner risk zone were not reached (1×10^{-5}/a) in this case. The level of toxicity fell below the threshold value for the middle zone (1×10^{-6}/a) after 20 m, and below that of the outer zone (3×10^{-7}/a) after 70 m.

Table 3.9 shows the most important results of the risk assessments by DOE (2007) for CO_2 pipelines (diameters of 325–490 mm with lengths of up to 100 km).

Table 3.9 Assessment of CO_2 pipeline risks by DOE

Pipeline failure[a] (size of leak, release duration)	Frequency per km × a	Hazard range (70,000 ppmv CO_2)
Rupture (0.189 m^2, minutes)	5.92×10^{-5}	66 m
Leak (0.0019 m^2, hours)	1.18×10^{-4}	35 m

Source: Adapted from DOE (2007)
[a]CO_2 inventory: 1,290 t (8 km pipeline sections, diameters of 490 mm)

Their operating pressure was assumed to be 150 bar, and sectional shut-offs were assumed at intervals of 8 km. The pipeline sections contain 723–1,290 t CO_2. In the case of ruptures, the leak rate fluctuates because of the different pipeline cross-sections between 3,500 and 7,950 kg/s; for leaks of 0.0019 m^2 it is only 81.4 kg/s. This results in release durations of 162 s for ruptures, and of 7,000 s (723 t CO_2) and 15,800 s (1,290 t CO_2) for leaks.

The zones with life-threatening CO_2 concentrations of 70,000 ppmv, which were determined using dispersion calculations, stretch up to distances of 66 m in the case of CO_2 pipeline ruptures and up to 35 m for leaks. The distances are maximal values that apply for unfavourable weather conditions (F2 weather).

3.4.7 Categorization of Technical Risks

The technical risks of transporting CO_2 primarily concern the failure of the containment function of pipelines, which leads to the inadvertent release of CO_2 into the environment, and can endanger humans, flora, and fauna in the vicinity surrounding the site of the leak. CO_2 release can range from minor leaks to sudden massive releases, and can be caused by maintenance errors, leaks or ruptures. The extent of damage depends on how high the CO_2 concentration is and its dispersion. Concentrations of 7 % are classified as life-threatening by (DOE 2007), while other sources quote concentrations of 7–10 vol.%. The size of the endangered zone depends on the amount of CO_2 released, the release rate, and the weather conditions. The largest zones arise when CO_2 is released very quickly after ruptures and during stable weather conditions with low wind speeds (F2 weather).

The spectrum of potential CO_2 pipeline failures can be broken down into the following six accident categories depending on the cause of failure, leak rates, and frequency of occurrence:

1. Leakage at valves or compression units
2. Very small pipeline leaks (e.g. diameter <2 cm)
3. Small pipeline leaks (e.g. diameter <5 cm)
4. Large pipeline leaks (diameter >15 cm)
5. Pipeline ruptures (pipeline cross-section)
6. Pipeline leaks with failure of the automatic shut-off

Table 3.10 Categorization of transportation risks

Pipeline accident category	Frequency per year	Zone of critical CO_2 concentration
P1: Valve leak	Moderate	Very small
P2: Very small leak (<2 cm)	Moderate	Small
P3: Small leak (>5 cm)	Moderate	Moderate
P4: Large leak (>15 cm)	Low	Large
P5: Rupture (1 F rupture)	Low	Large
P6: Rupture + failure of shut-off	Very low	Large

Using the results in McGillivray and Wilday (2009) and DOE (2007), the six accident categories were subjected to an initial qualitative evaluation (Table 3.10). The frequency with which the accident categories occur and the associated ranges of critical CO_2 concentrations (approx. 7 %) were assigned to one of four frequency classes (high, moderate, low, very low) and one of four hazard class zones (large, moderate, small, very small).

The frequency classes correspond to the following division of how often accidents occur: ≥1/year of operation, ≥1/service life, <1/service life, <<1/service life. For the hazard classes, the ranges are divided as follows: >120 m, 40–120 m, <40 m and <<40 m.

During pipeline service life, it is expected that accident category 1 (leakage at valves or compression units) will occur a number of times. However, critical CO_2 concentrations are only expected in the immediate vicinity of the site of the leak. The risk for category 1 can therefore be classified as negligible.

In the case of accident category 2 (very small pipeline leaks), only 'small' hazard radii (of a few metres) are expected. Such accidents are expected to occur around once every 10 years of operation. The risk for category 2 can therefore be classified as minor.

Leaks in category 3 (small leaks) under unfavourable conditions (e.g. weather) can lead to CO_2 concentrations of approx. 7 % being detected tens of metres away from the site of the leak. In DOE (2007), they were determined at up to 35 m, and in McGillivray and Wilday (2009) up to 56 m. The frequency of these leaks is comparable to that of very small leaks. The risk associated with category 3 is slightly higher than that of category 2.

Similar risks are assumed for category 4 (large leaks). In McGillivray and Wilday (2009), relevant CO_2 concentrations were determined at distances of up to some 150 m. Under favourable conditions, the distances are much smaller (see Fig. 3.2). However, category 4 is expected to occur less than once per pipeline service life.

Relatively high risks are expected for accident category 5 (pipeline rupture) because even under favourable accident conditions, CO_2 concentrations with an impact on health can occur at relatively 'large' distances from the site of the leak. In McGillivray and Wilday (2009), distances of some 100 m are quoted (see Fig. 3.2),

Frequency class: → Hazard class: ↓	4: Very low (<< 1 x per pipeline service life)	3: Low (< 1 x per pipeline service life)	2: Moderate (≥ 1 x per pipeline service life)	1: High (≥ 1 x per pipeline year of operation)
1: High (large zone of critical CO_2 concentration)	Risk field 13: 'negligible' **[P6]**	Risk field 9: 'moderate risk' **[P4, P5]**	Risk field 5: 'high risk'	Risk field 1: 'high risk'
2: Moderate (medium-sized zone of critical CO_2 concentration)	Risk field 14: 'negligible'	Risk field 10: 'low risk'	Risk field 6: 'moderate risk' **[P3]**	Risk field 2: 'high risk'
3: Low (small zone of critical CO_2 concentration)	Risk field 15: 'negligible'	Risk field 11: 'negligible'	Risk field 7: 'low risk' **[P2]**	Risk field 3: 'moderate risk'
4: Very low (very small zone of critical CO_2 concentration)	Risk field 16: 'negligible'	Risk field 12: 'negligible'	Risk field 8: 'negligible' **[P1]**	Risk field 4: 'low risk'

Fig. 3.3 Risk matrix for evaluating risks associated with CO_2 pipelines

while DOE (2007) estimates these distances to be max. 66 m (Table 3.9). The frequency with which category 5 occurs is comparable to that of category 4.

In the case of accident category 6, the pipeline ruptures and the shut-off valve fails to isolate the section from neighbouring pipeline sections. In such a case, twice the amount of CO_2 can escape, thus leading to an expansion of the hazard zone. The frequency with which such a combined accident occurs, however, is 'very low'. Category 6 is therefore not considered high-risk.

Using the risk matrix shown in Fig. 3.3, a general classification of risks can be performed for risks associated with the transportation of CO_2. The risk matrix links the previously specified frequency and hazard classes in a risk field matrix classifying the six defined accident categories. As shown in the figure, the classification and evaluation did not allot any of the six accident categories to the critical risk fields 1, 2 or 5. The accident categories P3, P4 and P5 are allotted to fields with a justifiable risk ('moderate risk'). They may have to be assessed in more detail. The category P2 (very small leaks) is considered low-risk, while category P1 (valve leaks) and P6 (combination of pipeline rupture and failure of shut-off valve) are considered to pose a negligible risk. The latter is considered to be low-risk because it occurs so rarely.

3.4.8 Uncertainties in the Assessment

Existing risk assessments of CO_2 pipeline transportation differ in their results and their conclusions. This is mainly due to the existence of a series of uncertainties and gaps in knowledge, particularly in relation to the dispersion behaviour of CO_2 and the modelling of the release of supercritical CO_2. In addition, there is a lack of standardized exposure thresholds and uniform dose-response relationships. According to Koornneef et al. (2009), these uncertainties and gaps in knowledge have a great influence on the accuracy of risk assessment and are the cause of discrepancies in the results of analyses performed to date.

3.5 Summary and Conclusions

Pipelines are particularly interesting as a means of transporting large amounts of CO_2 over long distances. At present, CO_2 pipeline grids throughout the world have a total length of more than 4,000 km. The transportation of carbon dioxide is state of the art.

The release of large amounts of CO_2 can pose local risks to humans and the environment. As CO_2 is heavier than air under ambient conditions, it can collect in sinks, for example, and at very high concentrations (7–10 vol.%) it can pose a life-threatening danger. Comparisons of natural gas and CO_2 pipelines demonstrate that the frequency of failures is similar. However, the extent of damage caused by natural gas pipelines is much severer. The main causes of damage to CO_2 pipelines are defective pressure relief valves, sealing problems/faulty welds, and damage caused by corrosion. The spectrum of measures minimizing the risks ranges from leak monitoring, shut-off valves, crack stoppers, and the monitoring of pipelines up to the definition of safety distances. Gas purities are particularly relevant because they can be interpreted as indirect protection against corrosion. The quality requirements for pipeline transportation of CO_2 in the USA are partially motivated by the transport itself. However, they also consider the further use of CO_2, e.g. to ensure a high solubility with crude oil (enhanced oil recovery). The standards in the USA can therefore only be transferred to the European situation to a limited extent. In addition, the impurities contained in captured CO_2 streams from power plants are very different to the impurities contained in the volumes of CO_2 currently transported in the USA.

Risk assessments were performed based on probabilistic approaches within the context of different studies. In many cases, different accident scenarios were analysed in which factors such as weather conditions and leak sizes/leak rates were varied. The available studies were used to qualitatively evaluate the categorized transportation risks (e.g. valve leak, pipeline leak, pipeline rupture) in terms of frequency and range of critical CO_2 concentrations. This was done with the aid of a risk matrix (frequency classes, hazard classes), which was used to roughly classify the risks associated with transportation. The results clearly show that most of the risks associated with transportation of CO_2 are negligible or only very low.

References

Aspelund A, Mølnvik MJ, Koeijer GD (2006) Ship transport of CO_2: technical solutions and analysis of costs, energy utilization, exergy efficiency and CO_2 emissions. Chem Eng Res Des 84:847–855

Baufumé S, Hake J-F, Linssen J, Markewitz P (2011) Carbon capture and storage: a possible bridge to a future hydrogen infrastructure for Germany? Int J Hydrogen Energy 36:8809–8821

Bongartz R, Markewitz P, Zapp P (2010) Prozesskette von CCS-Technologien: Technische Risiken und CO_2-Emissionen. STE-research report 04/2010. http://www.fz-juelich.de/iek/iek-ste/DE/Forschung/Publikationen/Research%20Reports%202010/Research%20Reports%202010.html

DNV (2005) Praxis bei der Ermittlung von Betrieben nach der Seveso-II-Richtlinie in Europa und entsprechenden Betrieben in Nordamerika. Available: http://www.dnv.com. Accessed 14 Oktober 2005

DNV (2008) Mapping of potential HSE issues related to large-scale capture, transport and storage of CO_2. DNV Report no. 2008-1993. Det Norske Veritas, www.ptil.no/getfile.php/PDF/Ptil%20CCS%202008.pdf

Doctor FD (2000) Transporting carbon dioxide recovered from fossil energy cycles. In: 5th international conference on greenhouse gas control technologies, Cairns, Australia

DOE (2007) Final risk assessment report for the FutureGen project environment impact statement [Online]. Department of Energy. Available: http://www.doe.gov/sites/prod/files/EIS-0394-DEIS-RiskAssessmentReport-2007.pdf. Accessed Oct 2007

EGIG (2008) Gas pipeline incidents. Available: http://www.egig.nl/downloads/7_report_EGIG.pdf

Eldevik F, Graver B, Torbergsen LE, Saugerud OT (2009) Development of a guideline for safe, reliable and cost efficient transmission of CO_2 in pipelines. Energy Procedia 1(2009):1579–1585

Gale J, Davison J (2004) Transmission of CO_2 – safety and economic considerations. Energy 29:1319–1328

Harper P (2010) The inclusion of CO_2 as a hazardous substance in the Seveso Directive, HSE paper to EC review of the Seveso II Directive [Online]. Health and Safety Executive. Available: http://www.hse.gov.uk/seveso/co2-hazardous-substance.pdf

IEA (2004) Greenhouse Gas R&D Programme: impurities on CO_2 capture, transport and storage. Report no. PH4/32. OECD/IEA, London. www.ieaghg.org/docs/General_Docs/Reports/Ph4-32%20Impurities.pdf

IEA (2008) Energy technology perspectives – scenarios & strategies to 2050. OECD/IEA, Paris

IEA (2009) Greenhouse Gas R&D Programme: safety in carbon dioxide capture, transport and storage. Report no. 2009/6. www.ieaghg.org/docs/General_Docs/Reports/2009-6.pdf

IPCC (2005a) Carbon dioxide capture and storage – special report of the intergovernmental panel on climate change. Cambridge University Press, Cambridge

IPCC (2005b) Special report on carbon dioxide capture and storage. http://www.ipcc.ch/activity/srccs/index.htm. IPCC, 15 Aug 2012

Koornneef J, Spruijt M, Molag M, Ramierez A, Faaij A, Turkenburg W (2009) Uncertainties in risk assessment of CO_2 pipelines. Energy Procedia 1(2009):1587–1594

Kuckshinrichs W, Markewitz P, Leitner H, Linssen J, Zapp P, Bongartz R, Schreiber A, Müller T (2012) Worldwide innovations in the development of carbon capture technologies and the utilisation of CO_2. Energy Environ Sci 5:7281–7305

McGillivray A, Wilday J (2009) Comparison of risk from carbon dioxide and natural gas pipelines. Health and Safety Executive. Research report 749. Health and Safety Executive. www.hse.gov.uk/research/rrpdf/rr749.pdf

Odenberger M, Svensson R (2003) Transportation systems for CO_2 – application to carbon sequestration. Technical report no. T2003-273, Chalmers University of Technology. www.entek.chalmers.se/~klon/msc

Schreurs H (2009) CO_2 capture, transport and storage in Rotterdam Rijnmond – State of the art 2009. CSLF Ministerial Meeting, London, UK

Seevam PN, Race JM, Downie MJ (2007) Carbon dioxide pipelines for sequestration in the UK: an engineering gap analysis. J Pipeline Eng 6:133–146

Seevam P, Race JM, Downie MJ (2008) The next generation of CO_2 pipelines. CCS Summer School, Vancouver Island

Seliger U, Keutel K, Koch M (2009) Entwicklung eines toxischen Gasgemisches in Räumen durch Restglut im Holzkohlengrill. Jahresbericht 2006/2007/2008. IdF Sachsen-Anhalt. http://www.idf.sachsen-anhalt.de/fileadmin/Bibliothek/Politik_und_Verwaltung/MI/IDF/jb06-08/JB06-08_SeligerKeutelKoch.pdf

Uth H-J (2009) Störfallvorsorge durch Raumplanung. Technische Überwachung, Heft 1–2:18–22

Visser E, Hendricks C, Barrio M, Mölnvik M, Koeijer G, Liljemark S, Austegard A, Gallo Y (2008) DYNAMIS CO_2 quality recommendations. Int J Greenh Gas Control 2:478–484

Chapter 4
Opportunities for Utilizing and Recycling CO_2

Thomas E. Müller, Walter Leitner, Peter Markewitz, and Wilhelm Kuckshinrichs

Abstract Complementing Carbon Capture and Storage (CCS), the utilization of carbon dioxide (CCU) as chemical feedstock and versatile processing fluid is attracting rapidly growing interest in science and industry. The chemical exploitation of carbon dioxide aims to generate value by producing polymeric and inorganic materials, fine chemicals and other products in which large amounts of carbon are fixated for an extended period of time. Provided that the reaction of the CO_2 molecule is enabled by the use of appropriate catalysts and process conditions and that the overall carbon footprint of the CO_2-based process chain is competitive with conventional chemical production, carbon dioxide can be a promising carbon source with practically unlimited availability.

Keywords CO_2 as raw material • Energy balance • Evaluation criteria • Carbon footprint • Substitution potential • Polymer building block • Fuels • Fine chemicals • Physical use

4.1 Motivation and Background

Material utilization and recycling of CO_2 could make a welcome yet limited contribution for reducing carbon dioxide emissions. Already today, the material use of CO_2 is common practice, whereby the CO_2 mainly stems as a by-product of various processes of the chemical industry. Utilizing CO_2 from power plant exhaust gases for physical or chemical applications has not yet been realized on an industrial scale. Thus, a future possible capture of the huge amounts of carbon dioxide

T.E. Müller (✉) • W. Leitner
CAT Catalytic Center and Chair of Technical Chemistry and Petrochemistry, Institute of Technical und Macromolecular Chemistry (ITMC), RWTH Aachen University, Worringerweg 2, D-52074 Aachen, Germany
e-mail: thomas.mueller@catalyticcenter.rwth-aachen.de; leitner@itmc.rwth-aachen.de

P. Markewitz • W. Kuckshinrichs
Institute of Energy and Climate Research – Systems Analysis and Technology Evaluation (IEK-STE), Forschungszentrum Jülich GmbH, D-52425 Jülich, Germany
e-mail: p.markewitz@fz-juzelich.de; w.kuckshinrichs@fz-juelich.de

W. Kuckshinrichs, J.-F. Hake (eds.), *Carbon Capture, Storage and Use*,
DOI 10.1007/978-3-319-11943-4_4

resulting from power plants has increasingly fueled current discussions of CO_2 -utilization to complement CO_2-storage (Markewitz et al. 2012).

The interest in using carbon dioxide (Carbon Capture and Recycling: CCR) is triggered by the fact that CO_2 is a potentially valuable raw material that has an interesting application profile and may enhance value-chains for the chemical industry. Its economic exploitation can positively affect strategies for reducing CO_2-emissions by means of capture, transport and storage of CO_2 (CCS). Consequently, the greenhouse gas CO_2 may become a raw material for the material value chain.

This paper mainly focusses on the organochemical opportunities to harness carbon dioxide on a large industrial scale. Supplementing already existing literature (Hölscher et al. 2012; Aresta and Dibenedetto 2007; Ausfelder and Bazzanella 2008; Sakakura et al. 2007; Song 2006; Yu et al. 2008), this current paper aims to discuss in detail the industrial applicability. The possibilities for technical realization have been comprehensively described in Peters et al. (Peters et al. 2011). Since the incorporation of CO_2 in inorganic materials (O'Connor et al. 2000), the production of fuels from CO_2, as well as the use of CO_2 as a raw material for biotechnological syntheses (*e.g.*, by applying microbes (Wang and Lan 2010)) has been extensively discussed in the literature (Ausfelder and Bazzanella 2008; Lambertz and Ewers 2009), such topics will not be handled in detail here.

4.2 Evaluation Framework and Criteria

4.2.1 Potential for the Material Utilization and Recycling of CO_2

Worldwide, almost 130 million tonnes CO_2 are currently used per year. Roughly 110 million tonnes thereof are used as a raw material and the remaining 20 million tonnes are applied as industrial gas (Fig. 4.1).

By far, the largest established product made from CO_2 is urea (146 million tonnes) with a CO_2-utilization of 107 million tonnes (International Fertilizer Industry Association 2009) followed by methanol (30 million tonnes) for which approx. 2 million tonnes CO_2 are used. Up to now, the remainder of the carbon in the methanol manufactured stems from conventional synthesis gas (CO/H_2). Furthermore, CO_2 is utilized in the production of cyclic carbonates (0.08 million tonnes) with a CO_2-consumption of 0.04 million tonnes and of salicylic acid (0.07 million tonnes) with a CO_2-consumption of 0.03 million tonnes (Dittmeyer et al. 2005).

Twenty million tonnes CO_2 are applied as industrial gas, whereby this involves mainly physical use of CO_2, for example, as an inert gas, extractant, or in the beverage industry (Ausfelder and Bazzanella 2008).

The largest hurdle for industrially realizing a material utilization of CO_2 is its low energy level (Fig. 4.2).

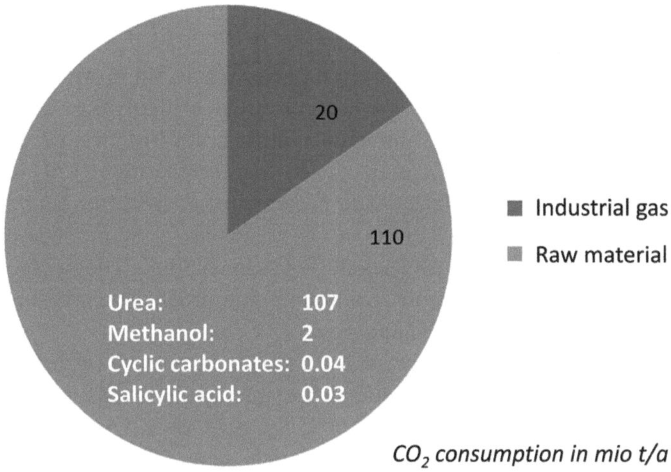

Fig. 4.1 Worldwide utilization of CO_2 as raw material in chemical industry

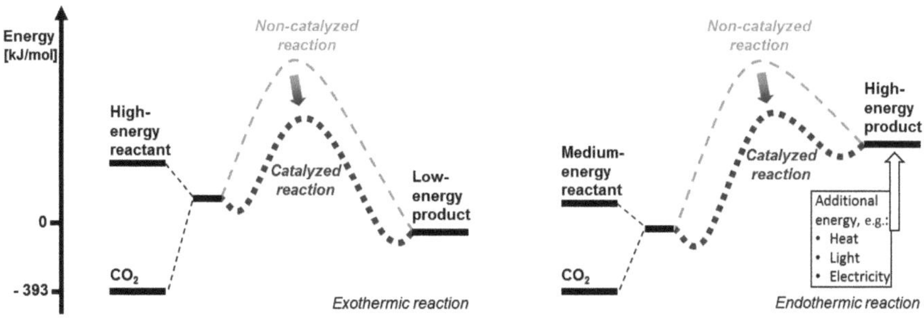

Fig. 4.2 Energy balance of chemical reactions for the fixation of CO_2 (*e.g.*, *via* energy-rich reactants (*left*) or by utilizing external energy carriers (*right*)) (Source: Markewitz et al. 2012)

As a result, energy has to be applied in the form of light, electrical energy or heat (Markewitz et al. 2012). Alternatively, high-energy reaction partners can be used for the chemical reaction such as hydrogen, strained cyclic molecules or unsaturated compounds. In addition, binding to the catalyst may sufficiently activate CO_2 to allow for the subsequent reaction steps (Elmas et al. 2013). Target molecules, *e.g.*, organic carbonates, preferentially lie on a lower energy level than the starting compounds (Sakakura et al. 2007; Peters et al. 2011). However, endothermic reactions can be driven by additional energy input, which can be supplied, *e.g.*, as heat, light or electricity.

4.2.2 Sources and Purity of CO_2

Besides the capture of carbon dioxide from power plants, there are several various sources from which CO_2 is made available. For instance, highly pure CO_2 is yielded in the chemical industry. During ammonia synthesis, for instance, ca. 120 million tonnes CO_2 are separated as a by-product each year. Furthermore, CO_2 is separated from gas streams during ethylene oxide production, in refineries and during the purification of natural gas. Likewise, CO_2 results as a by-product during the production of synthesis gas as well as during fermentation. The degree of purity of the carbon dioxide, for example, from ammonia production and fermentation, suffices for practically all synthesis purposes.

With respect to CO_2 from power plants, it may be necessary to clean the carbon dioxide gas before its material use due to possible contaminants in the stack gas. Such exhaust gas contaminants include O_2, N_2, H_2O, H_2S, CO, CO_2, SO_2, NO_x, heavy metals or also hydrocarbons (Kuckshinrichs et al. 2009).

Particular criteria need to be heeded in the cleaning of exhaust gases. From a chemical point of view, the stability of the catalysts applied in down-stream processes plays an especially important role and has to be controlled. Modern catalysts are sensitive to catalyst toxins. For example, in methanol synthesis, the H_2S-concentration in the CO_2 from combustion processes needs to be reduced to below 0.1 ppm. For other syntheses, however, the level of CO_2-purity established for pipeline transport seems to be sufficient.

The type of contamination of the CO_2 is decisive for the later application areas of the products. The application areas may be sensitive to certain contaminants (*e.g.*, heavy metals in urea used as fertilizer), or there are high purity requirements arising from legal licensing regulations as in the production of active ingredients in pharmaceuticals (*e.g.*, salicylic acid). During classic chemical syntheses, there are mostly subsequent clean-up steps after the incorporation of CO_2.

The purity of the CO_2-stream is less important for CO_2-utilization in the petroleum and natural gas industry. Far more important, however, is the degree of CO_2-purity in other physical applications. Since carbon dioxide is often used there without further purification steps such as in the food and beverage industry, contaminants in CO_2, with the exception of inert gases, have to be avoided as best as possible. From an ecological standpoint, the additional energy expenditure for cleaning CO_2 likewise plays a decisive role.

4.2.3 Evaluation Criteria for CO_2-Utilization

Aside from the necessary purity of carbon dioxide, there are other factors that are important in the evaluation of a possible material utilization of CO_2. In particular, these factors are the CO_2-fixation period and –amount, energy- and CO_2-balances as well as value creation (Fig. 4.3).

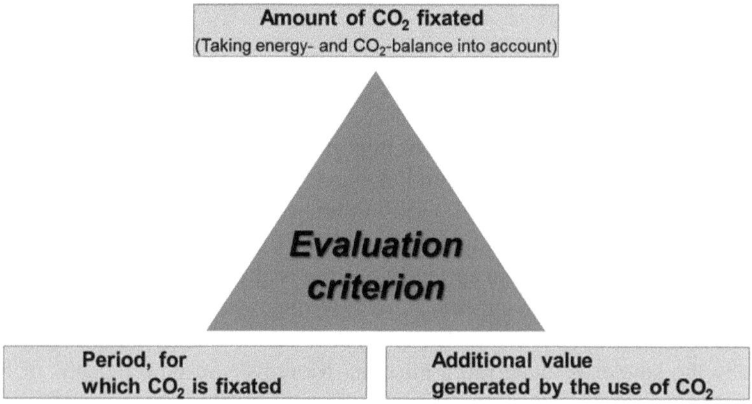

Fig. 4.3 Evaluation criteria for the material utilization of CO_2

Even though it is difficult to compile complete energy- and CO_2-balances for processes and products, estimating the fixation quantity is necessary and methodical *via* the Life Cycle Assessment (LCA). Especially important here are the system limitations in order to ensure the comparability with other balances but also to include possible substitution of CO_2 from other non-CCS-sources.

The CO_2-fixation period strongly depends on the later use of the respective product. For example, should CO_2 be incorporated in urea for producing fertilizers, large amounts of CO_2 can indeed be fixated; however, these quantities are immediately released into the environment after the application. In comparison, even though polymers can fixate less CO_2, the fixation may last over years or even decades. Similarly, only small amounts of carbon dioxide can be incorporated in fine chemicals. Y*et al*so here the fixation, depending on the application, can last for years.

Regarding CO_2-fixation, the physical utilization of carbon dioxide is least attractive due to its mostly fast release into the environment. Nonetheless, the amounts of physically exploitable CO_2 may be large.

The added value generated by manufacturing a product as well as during its application represents a measure for the economic relevance of CO_2-utilization. Yet it is extremely difficult to estimate the added value of CO_2-utilization on the basis of individual products due to the lacking predictability of how individual product markets will develop in the future.

4.3 Organochemical Utilization of CO_2

There are many possible uses of CO_2 as a C1-building block (Liu et al. 2012). The following deals with CO_2-utilization for urea, methanol, salicylic acid, *p*-hydroxybenzoic acid, formic acid, cyclic carbonates, dimethylcarbonate, polymers, polymer building blocks and for fine chemicals.

4.3.1 *Applications*

4.3.1.1 Urea

Today, urea is quantitatively the most important product of the chemical industry that is produced from CO_2 as C1-building block. In the annual production of around 146 million tonnes urea, about 107 million tonnes CO_2 are materially used (International Fertilizer Industry Association 2009). At present, the largest amounts of urea are applied as fertilizer in agriculture. Urea is also used for producing thawing aids. In the pharmaceutical industry, urea is frequently applied as a moisturizer. In the cigarette industry, urea is added to tobacco for increased nicotine absorption, whereby the smoke is made more alkaline. Moreover, in power plants or trucks, urea is added to the combustion gases so as to reduce the nitrogen oxide content of the exhaust gases *via* a subsequent Selective Catalytic Reduction (SCR).

Urea is likewise used for synthesizing fine chemicals. Applications include the production of a urea-hydrogen peroxide complex (UHP), used as a bleaching agent for cellulose, textiles and paper as well as for bleaching of teeth (Wielicka et al. 2003). The nitrate of urea serves as a starting product for nitro-urea, used as a mild nitrating reagent (Almog et al. 2006). Because it is easy to produce, urea nitrate itself is used as an explosive. Another application is for making phthalocyanine pigments (*e.g*., Heliogen®) by reaction of urea and phthalic anhydride in the presence of a central ion, typically copper (Sawada 1990). Often urea also serves as a reagent in the synthesis of heterocycles. With respect to these aforementioned applications of urea and its secondary products, the CO_2-fixation times are relatively short, as CO_2 is mostly released again immediately after the use.

Amino plastics, as downstream products of urea, show a higher CO_2-fixation potential. These include urea-formaldehyde resins (UF-resins), which make up ca. 80 % of the amide resins and melamine-formaldehyde resins (MF-resins), which almost completely cover the rest (Hischier et al. 2005). The main fields of application of the amino plastics are as binding agents for wood materials (plywood, lightweight boards), glues, varnishes (acid-containing and baking enamels), textile auxiliary agents and educts for duromer foams. Compared to UF-resins, MF-resins are more prized in all applications. In general, amino plastics have a very high CO_2-fixation potential as the materials are used frequently in fields of application where the CO_2-fixation can last for decades. Urea-formaldehyde resins (UF-resins) are formed by reacting urea with formaldehyde, a secondary product of methanol. As neither the synthesis of urea nor the preparation of UF-resins necessitates the use of a catalyst, the purity of CO_2 required for these applications is classified as low.

Melamine, the most important downstream product of urea, is needed for manufacturing melamine-formaldehyde resins (MF-resins). Melamine is produced from urea and ammonia under high pressure and high temperature, which is why melamine is often produced nearby a urea production site. Analogous to UF-resins,

MF-resins are produced from melamine and formaldehyde, possibly in the presence of alcohols.

Another polymeric secondary product of urea is cellulose carbamate, produced by treating base-activated cellulose with urea, whereby a fraction of the hydroxyl groups are converted to carbamate groups (—O—CO—NH_2) under release of ammonia. Cellulose carbamate is soluble in diluted sodium hydroxide and can subsequently be processed to fibers or foils. This is an environmentally friendly alternative to the xanthogenate process, in which the hydroxyl groups of cellulose are esterified by carbon disulfide. New large-scale application areas can be exploited for cellulose carbamate, *e.g.*, to manufacture foams for insulation or soundproofing, whereby a considerable amount of CO_2 can be fixated.

4.3.1.2 Methanol

Currently, there is an estimated annual methanol demand of about 30 million tonnes. More than 70 % of the produced methanol is applied in chemical synthesis (Hischier et al. 2005). For many of the methanol applications, there is no CO_2-fixation extending beyond use. This applies, in particular, for the use as biofuel or as biodiesel (as methyl ester) even when these applications consume large amounts of methanol. Nevertheless, liquid fuels represent an interesting transportable and storable energy form, that – in the case of storage – leads to a longer fixation periods.

Regarding fuels, there are multiple application areas for methanol. For instance, methanol itself or in the dehydrated form as dimethylether can be used as a fuel for combustion machines. Likewise established is the conversion of methanol to methyl tert.-butyl ether (MTBE; production 2000: 7.8 million tonnes) used as an anti-knocking additive in gasoline (Hischier et al. 2005). Furthermore, there is the transesterification with methanol to produce plant oil methyl esters that can be used as biodiesel.

It is also possible to convert methanol to hydrocarbon mixtures (Centi et al. 2011). Well-known procedures are the so-called MTG-process (methanol to gasoline) and indirectly *via* the Fischer-Tropsch process, where hydrocarbons (alkanes, alkenes and alkanols) are obtained from synthesis gas (Dorner et al. 2010). Thus, a CO_2-fixating access to petroleum chemistry is feasible with synthesis gas or methanol from renewable resources.

Another important secondary product is methyl methacrylate (MMA; production 2000: 0.8 million tonnes) (Hischier et al. 2005). Hereby, CO_2 can be technically fixated in the form of polymethylmethacrylate (PMMA). Nonetheless, the amount of fixated carbon dioxide is low, because the methyl group to be substituted makes up only 17 % of the polymer.

A longer term CO_2-fixation occurs *via* formaldehyde (production 2000: 9.6 million tonnes), another important downstream product of methanol that is produced in large amounts in the chemical industry (Hischier et al. 2005).

Formaldehyde is synthesized from methanol by air oxidation in an exothermic reaction, thereby requiring no substantial additional energy input with respect to methanol.

For the methanol synthesis, however, the applied CO_2 needs to be highly pure with respect to the content of H_2S ($H_2S < 0.1$ ppm) as the applied catalysts are highly sensitive. Among others, formaldehyde is used as a starting material for producing amide resins such as urea-formaldehyde (UF)-, melamine-formaldehyde (MF)- as well as phenol-formaldehyde (PF)- resins. As varnishes, wood working materials, composites or foams, the resulting duromers have a long lifespan over years or even decades.

The formaldehyde-based duromers open up new application opportunities within the scope of CO_2-fixation. Hereby, the controversial formaldehyde emissions no longer play a significant role due to the addition of particular additives. One modern example is photo-catalytic paint suitable for indoors and outdoors. These paints are spiked with titanium dioxide in the form of photo-catalytically active anatas nanocrystals so as to reduce the band gap of 3.2 eV (corresponding to near UV-light at 390 nm), thereby enabling artificial light or light filtered through glass panes to suffice for the oxidization of formaldehyde.

Forcing the production of known polymers might contribute towards CO_2-fixation. Amino plastics (UF- und MF-resins) are technically available duromers whose carbon fraction might be built completely out of fixated CO_2 (formaldehyde *via* methanol; urea directly from CO_2).

A re-substitution might take place there, where petrochemical-based thermoplasts had substantially displaced duromers during the past 50 years. The extra costs might be lower than by introducing new substances. If about 10 % of the worldwide used plastics were to be constructed from technically fixated CO_2, the potential would amount to ca. 20 million tonnes/year (Bazzanella et al. 2010; Müller 2008).

With respect to wood-amino plastic working materials, new application areas, *e.g.*, for heat insulation in buildings, can be exploited. Furthermore, composites can be principally made of amino plastics along with fibers from sustainable resources (jute, flax, hemp, wood pulp). Turning plant fibers into usable composites with technical thermoplasts requires special treatment, whereby the surface of the fibers and the polymer matrix are modified in such a way that chemical bonds are formed at the interface. For amino plastics, the hydroxymethyl groups on the nitrogen readily react with the OH-functionalized, mostly cellulose-based natural fibers. Thus, amide resins are attractive polymer systems that might be produced without the use of carbon from fossil energy resources.

An already establish technical thermoplast is polyoxymethylene (polyformaldehyde, POM), produced by the cationic polymerization of trioxane, the cyclic trimer of formaldehyde. Upon using suitable hydrogen sources, POM might be completely constructed from technically fixated CO_2. Non-reinforced POM has the highest Young's modulus of elasticity among the existing large-scale thermoplasts without liquid-crystalline properties. Toughness can be improved *via* co-monomers such as dioxolane or by external impact-strength modifiers such as thermoplastic polyurethanes (TPU).

Already today, POM is an established material, of which about 0.5 million tonnes are produced annually. Through the use of POM, polyolefins might be substituted, which generally goes hand in hand with a rise in quality and thus, with a higher lifespan of the consumption article. This would be equivalent to longer fixation times.

Acetic acid is another secondary product that can be manufactured from methanol and carbon monoxide. Around 2.7 million tonnes acetic acid was produced in 2000 (Hischier et al. 2005). The applied CO can likewise stem from renewable resources. Moreover, vinyl acetate, used as a co-monomer, can be produced from acetic acid, ethylene and atmospheric oxygen with catalysis by palladium compounds. In this way, CO_2 can also be fixated in products derived from acetic acid, such as polyethylene vinyl acetate (EVA) and similar polymers.

4.3.1.3 Salicylic Acid and p-Hydroxybenzoic Acid

Some aromatic carbonic acids, such as salicylic acid, *p*-hydroxybenzoic acid and hydroxynaphthoic acid are readily produced from CO_2 and the corresponding aromatics *via* the Kolbe-Schmitt reaction.

Salicylic acid is used for producing acetyl salicylic acid sold under the trade name Aspirin®. Nevertheless, the period of CO_2-fixation is rather low, since CO_2 is immediately released after application.

p-Hydroxybenzoic acid is used mainly for producing its ester, parabene, applied as a preservative. A CO_2-fixating application is the production of liquid-crystal (LC) polyesters, primarily applied as high-performance material in the electronic industry. The main component here is often *p*-hydroxybenzoic acid, sometimes together with hydroxynaphthoic acid.

Salicylic acid and *p*-hydroxybenzoic acid may also be used as co-monomers in standard polyesters such as polyethylene terephthalate (PET) or polybutyleneterephthalate (PBT). Thereby, the corresponding acetoxybenzoic acids are used as reactive components.

Also in bisphenol-A-polycarbonates, both building blocks, but in particular, *p*-hydroxybenzoic acid, may be incorporated. Transesterification with diphenylcarbonate results in a polyester carbonate. Based on an annual polycarbonate production of over one million tonnes, the fraction of fixated CO_2 might be considerably increased above the 10 % stemming from the carbonate unit. Since the Kolbe-Schmitt synthesis takes place with relatively large amounts of insensitive alkali hydroxides as catalyst, the requirements concerning the purity of the applied CO_2 are low (pipeline-acceptable CO_2 seems to suffice).

4.3.1.4 Formic Acid

Carbon dioxide can be reduced to formic acid or formates. The current worldwide market for formic acid is about 0.6 million t/a (Ausfelder and Bazzanella 2008).

Formic acid and its salts are used as preservatives and de-icing agents. The acid is used as a decalcifying agent, mordant and as a solution adhesive for polyamides.

Formates are good carbonylating agents, which can be employed, *e.g.*, for the catalytic conversion of phenol to methyl phenyl carbonate (Yalfani et al. 2013). Dimethylformamide (DMF) and formamide can be produced from methyl formate. The former serves as a solvent, *e.g.*, for the spinning of polyacrylonitrile, while the latter acts as an intermediate in the production of cyanic acid or as a stabilizer of single-stranded nucleic acids.

Sodium oxalate can be fabricated by heating sodium formate. Thus, oxalic acid and its esters are secondary products of formic acid.

More innovative seems to be using formic acid as a reducing agent, *e.g.*, in the reduction of unsaturated compounds (Lange and Leitner 2002; Leitner et al. 1993) or in the Leukart-Wallach reaction, whereby formic acid serves as a hydrogen equivalent. Similarly, formic acid can be used as a reducing agent in welding processes. Formic acid can also be used as fuel in fuel cells, either directly (formira fuel cell of Tekion) or after cleavage to CO_2 and hydrogen. Hereby, CO_2 acts as a reversible transport vehicle for hydrogen, because it allows itself to be hydrated to formic acid (Leitner 1995) and it can release hydrogen again at another site (Laurenczy 2011; Ritter 2007; Hull et al. 2012; Boddien et al. 2010, 2011). Note, however, that storage density is limited on a weight basis.

4.3.1.5 Cyclic Carbonates

Cyclic carbonates, such as ethylene carbonate and propylene carbonate, can be manufactured from CO_2 and epoxides or by dehydration of a mixture of CO_2 and diols. An indirect access *via* oxidative carbonylation of diols is also possible (Doro et al. 2011; Gürtler et al. 2011b, 2012), whereby the CO can be derived from CO_2.

Ethylene carbonate is used as a solvent, *e.g.*, in the spinning of polyacrylonitrile, electrolyte in lithium batteries and as intermediate for synthesizing heterocyclic compounds. Propylene carbonate has similar application areas, whereby it is also used as a plasticizer or in binding agents for core sand in the foundry industry.

One ethylene carbonate application that has not been used yet on a large scale is the trans-esterification with methanol to dimethyl carbonate and further with phenol to diphenylcarbonate (Müller et al. 2012b). This would raise the perspective of binding CO_2 for years to decades in the carbonate unit of the polymeric material bisphenol-A-polycarbonate (PC). In addition, the cyclic carbonates allow themselves to be directly polymerized *via* ring opening (Müller et al. 2012a).

The fraction of CO_2 fixated in bisphenol-A-polycarbonate is limited to 11 % because of the small fraction, by weight, of the incorporated CO-group in the total polymer. Alternatively, the fixation of CO_2 in polycarbonate materials is principally feasible *via* the reaction of urea with methanol to dimethylcarbonate and subsequently with phenol to diphenylcarbonate or directly with phenol to diphenylcarbonate.

4.3.1.6 Dimethyl Carbonate

Dimethyl carbonate is most frequently used in chemical synthesis as a methylating agent (Tundo et al. 2004). Hereby the bound CO_2 is released again. Yet polymer applications lead to CO_2-fixation. Thus, aliphatic diols can be reacted with dimethyl carbonate to form oligomeric carbonate diols which are used as hydroxy-functional telechelic structures for producing polyurethanes (Hofacker 2006). The already mentioned transesterification with phenol leads to diphenylcarbonate with can be further transesterified with bisphenol-A to form polycarbonate (Duex et al. 2009).

Furthermore, dimethyl carbonate as well as diphenylcarbonate can be used to convert di-amines into the bis-carbamates, which can then be eliminated to the diisocyanates (Garcia Deleon et al. 2002). In this way, fixated CO_2 can be introduced into polyurethane chemistry. Thus, hexamethylenediisocyanate (HDI) can be constructed in principle from up to 33 % fixated CO_2. Alternatively, CO_2 can be applied for making phosgene, which would open access to the classical polyurethane chemistry. Note that isocyanates also react with CO_2 to form oxadiazinetriones (Katmatani and Fujita 1986).

4.3.1.7 Polymers (Copolymerization of Reactive Monomers with CO_2)

CO_2 can be reacted with reactive monomers such as epoxides to form non-alternating copolymers by the use of suitable catalysts, such as the double metal cyanate (DMC) catalyst (Müller et al. 2013b, d; Gürtler et al. 2013). Polyethercarbonates obtained *via* non-alternating copolymerization of CO_2 and epoxides (*e.g.*, Dienes et al. 2012) often exhibit glass-transition temperatures of below 0 °C. Polyethercarbonates with terminal hydroxyl groups are used for producing polyurethanes (Gürtler et al. 2011a, b; Gao et al. 2012; Hofmann et al. 2012; Müller et al. 2013a, c) and, thus, open the way for CO_2 into the classical polyurethane markets, *e.g.*, for building materials, construction foams and insulating materials (Langanke et al. 2014). The advantage of this application is the possible fixation of CO_2 over a long period (years or decades). Due to the low glass-transition temperature, polyethercarbonates are also usable as soft segments for thermoplastic elastomers.

By using suitable catalysts, CO_2 can also be reacted with epoxides to form alternating copolymers (Darensbourg et al. 2010; Elmas et al. 2012). Currently, aliphatic polycarbonates are mainly used as pore-forming agents in the ceramic industry due to their residue-free degradation behavior. Thermoplastics represent potential application areas for polyalkylene carbonates obtained by the alternating copolymerization of CO_2 and epoxides (Kember et al. 2011). Amorphous thermoplastics require a glass-transition temperature[1] of over 100 °C, whereas semi-

[1] The glass-transition temperature (Tg) is the temperature at which amorphous or semi-crystalline polymers transfer from a solid state to the liquid state. Since this temperature (T_g) is specific for every type of plastics, one can differentiate plastics on hand of their particular glass-transition temperature.

crystalline thermoplastics exhibit a crystallite melting point of over 100 °C. Exceptions are niche applications in the biomedical field or in the field of biodegradable plastics. As alternative strategy, the use of thermoplastic polycarbonates as blend partners or for the manufacture of composite materials is being explored. Some particular polycarbonates, such as polycyclohexylene carbonate or polylimonene oxide carbonate have sufficiently high glass-transition temperatures, although they are relatively brittle materials.

4.3.1.8 Further Polymer Building Blocks

Multifunctional cyclic carbonates can serve as building blocks for polymers. Using CO_2, cyclic carbonates can be produced from epoxidized soy oil (Doll and Erhan 2005). These cyclic carbonates, in turn, react with bifunctional primary or secondary amines, with formation of urethane groups, to yield a duromer network. This results in a symbiosis between the themes of CO_2-fixation and renewable resources. Interesting is that in the reaction of cyclic carbonates with amines, urethane groups can be formed without the involvement of an isocyanate.

p-Hydroxybenzoic acid (PHB) and, to a limited degree, also salicylic acid can be applied as co-monomers in polyesters or polycarbonate (vide infra). Liquid crystalline polyester (LPC) also can be constructed to 50–80 % from PHB.

Potentially interesting reactions are the direct carboxylations of olefins or of non-activated aromatics to acrylic acids or aromatic diacids (Graham et al. 2007). The direct carboxylation of ethylene has been skeptically evaluated in the literature, whereas the direct carboxylation of naphthalene at low yield has been described in patent literature (Brownscombe et al. 2001).

Carbon dioxide can be incorporated in low amounts in carboxy-terminated telechelic compounds, in which anionically initiated polymerizations, *e.g.*, of butadiene or styrene are terminated with CO_2. Furthermore, it is possible to produce aliphatic oligocarbonates from dimethyl carbonate and diols.

An indirect fixation is realized by copolymerization of alkenes (ethylene) and carbon monoxide (Müller et al. 2013e–g), whereby the carbon monoxide can be obtained in turn from CO_2.

4.3.1.9 Pharmaceuticals and Fine Chemicals

The reaction of Grignard reagents or of other metal alkylenes with CO_2 has been a tested method since 1906 for producing carbonic acids on a laboratory scale. This reaction, however, is ill suited for producing chemicals on a large industrial scale. Nevertheless, a few syntheses of active ingredients exploit this method, *e.g.*, in the production of biotin (Koop et al. 2001).

Another example is the production of isotope-marked carbonic acids for metabolite studies or for diagnostic procedures (Wirz and Kueng 1983; Yajima et al. 1996). Here, $^{11}CO_2$ or $^{14}CO_2$ is reacted with Grignard reagents such as heptyl magnesium chloride or benzyloxymethylmagnesium chloride.

Urea is a typical starting product of heterocycle syntheses and appears relatively frequently as an intermediate in the production of active ingredients. The Biginelli reaction is applied for fabricating the pharmacologically active dihydropyrimidinone (Kappe 2000). Here, it deals with a tricomponent reaction of urea, benzaldehydes and acetoacetates that runs under mild conditions. The reaction can also be automated and miniaturized on solid phases, whereby through combination reactions, several molecules can be produced from which those with the required properties can be selected. Also enantioselective variations of the reaction have been developed.

The Bucherer reaction enables the production of hydantoins from cyanohydrins and ammonium carbonate or from α-aminonitriles, intermediates of the amino acid synthesis according to Strecker, and CO_2 (Bucherer and Steiner 1934). The N, N-dibromo- und N,N-bishydroxymethyl-derivatives of 5,5-dimethylhydantoin are used as pesticides. An analogous reaction is the synthesis of quinazolines from anthranilic acid nitrile and CO_2 often with the use of 1,8-diazabicyclo(5.4.0)undec-7-ene (DBU) as catalyst (Kikelj 2004; Mizuno 2002).

4.3.2 Outlook

Particularly promising application areas of urea are its downstream products such as urea-formaldehyde resins (UF-resins) and melamine-formaldehyde resins (MF-resins). These products already show a CO_2-fixation potential of several million tonnes (upon switching the synthesis gas to CO_2 and biogas as raw material sources), and this potential might be raised to ca. 10 million tonnes by expanding the areas of use. Depending on the application, CO_2 could be fixated for years or even decades. Moreover, cellulose carbamate is likewise interesting, because it can be used for foams or insulation and, thus, likewise shows a high CO_2-fixation potential.

Amino plastics (amide resins) are especially interesting for future applications, because already now they are the largest group of technically available polymer materials whose carbon fraction can be completely constructed from fixated CO_2 (formaldehyde *via* methanol, urea directly from CO_2). In particular, composites made of amino plastics with fibers from renewable resources seem to be very promising. The CO_2-fixation potential is very high due to the long-term applications. Thus, amide resins make up attractive polymer systems that can be fabricated without using carbon from fossil sources. Regarding its CO_2-fixation potential, POM may become one of the most important thermoplastics.

Also promising is the use of salicylic acid and *p*-hydroxybenzoic acid as co-monomers in bis-phenol-A-polycarbonates and polyesters. LC-polyesters might make a smaller contribution as high-performance- and high-modular polymers.

In the future, formic acid might be used in special applications for operating fuel cells. After complete technological development, a fixation of the carbon dioxide is feasible, since the CO_2 could be principally recycled.

Also promising is the fixation of CO_2 in the carbonate unit of the polymer material, polycarbonate, because the CO_2 could be bound for years or decades. Bisphenol-A-polycarbonate is most important from a quantitative perspective. In the production, cyclic carbonates (ethylene carbonate) can be used indirectly as carriers of the carbonate unit.

Interesting is also the use of dimethyl carbonate as an active carbonic acid derivative in polymer applications such as in the production of polycarbonates and polyurethanes.

Upon using polyethercarbonates with terminal hydroxyl groups, polyurethane applications have excellent prospects due to their high CO_2-fixation period. In order to apply aliphatic polycarbonates as technical thermoplastics, further preliminary work has to be done, as the temperature of usage and the decomposition temperature are too low.

The degree of purity of the needed CO_2 depends on the catalysts applied. Zinc glutarate used as catalyst in the copolymerization of CO_2 with epoxides is relatively insensitive to SO_2 und H_2S-impurities of $<1{,}000$ ppm in CO_2. Up to now, there are no literature data regarding the maximally permissible level of impurities in CO_2 for double metal cyanide (DMC)- and salen-catalysts.

Regarding polymers constructed on the basis of CO_2, primarily polyurethanes and polyesters play a large role due to their long CO_2-fixation period in typical fields of their application.

4.4 Inorganic Substances

4.4.1 Calcite

An important inorganic product from CO_2 is calcite, $CaCO_3$, which is produced in defined grades of purity and particle sizes (Fakeev 2004; Tamura and Tsuge 2006). Calcite is precipitated with CO_2 from calcium hydroxide suspensions or calcium salts in the presence of ammonia. Calcite is used as filler for plastics, varnishes and pharmaceutical preparations, as a glass raw material in pure form for optic glass or for lumiphores. No CO_2-fixation takes place, because the calcium oxide needed in the synthesis has been produced initially by lime burning.

4.4.2 Hydrotalcite

Hydrotalcite is produced from magnesium oxide, CO_2 and sodium aluminate, whereby a part of the CO_2 remains in the product (Yon et al. 2002). In the field of medicine, hydrotalcite is used as a neutralizing agent for gastric acid. Other applications are in catalysis (aldol reactions, transesterification of polyesters, as carrier material in methane-reforming). Hydrotalcite can also be applied for the reversible adsorption of CO_2.

4.4.3 Other Application Areas

Carbon dioxide is often used as a precipitating agent for isolating metals (Sanuki et al. 2000; Xu and Zhu 2005). Thus, Nd(III)carbonate is precipitated from a solution of the versatate in hydrocarbons or $SrCO_3$ is precipitated from a solution of the respective chloride in isoamyl alcohol. Carbon dioxide can also serve as a neutralizing agent for wastewater, *e.g.*, from the steel industry, or for setting the pH-value (paper industry) (Choi et al. 2002; Leigraf et al. 2002). Furthermore, the reaction of CO_2 with magnesium- or calcium silicates can contribute towards its long-term fixation (Zevenhoven et al. 2006; Zevenhoven and Fagerlund 2010).

4.5 Physical Utilization

4.5.1 Enhanced Oil Recovery/Enhanced Gas Recovery

Since the 1980s, CO_2 has been used for improving the extraction of petroleum and natural gas (EOR: Enhanced Oil Recovery; EGR: Enhanced Gas Recovery). By the high-pressure injection of CO_2 into oil fields, up to 15 % more petroleum can be yielded – a fact which makes the utilization of carbon dioxide economically attractive (Davison et al. 2001).

Since EOR using carbon dioxide has been practiced for many years, much experience has been gained with these procedures which have reached state-of-the-art. However, in the field of EGR, researchers cannot currently rely on experiences based on demonstration projects. Few practical experiences have been acquired about the mixing processes in the gas reservoirs of injected CO_2/natural gas and about the necessity of a gas separation for processing the natural gas (Radgen et al. 2006). Yet theoretical laboratory tests currently allow the conclusion to be drawn that a mixing of natural gas and carbon dioxide can be avoided by suitable injection procedures (*e.g.*, variously deep injection and extraction borings).

Up to now, EOR/EGR focused on maximizing the amount of recovered hydrocarbons. Yet for a long time, proving the safe sequestration of CO_2 in the reservoirs was not considered important. Thus, a fitting of the procedures to new framework conditions is absolutely necessary. Currently within the scope of numerous research projects, it is being studied whether CO_2 over a long term, does not again escape to the surface and enter the atmosphere. By using suitable measurement and monitoring methods, it must be explained if a safe sequestration of carbon dioxide can be guaranteed. These detection and control options will vary individually for each storage project.

4.5.2 *Enhanced Coal Bed Methane (ECBM)*

The search for storage possibilities nearby emitters – which would minimize the transport expenses for the separated CO_2 – has prompted considering coal deposits as an option. Thus, the storage of CO_2 in coal deposits has been discussed for several years. Hereby, the storage of CO_2 in coal seams can be differentiated into two categories: i.e., the storage in abandoned coal seams, and in deep, economically unfeasible coal seams (Radgen et al. 2006).

However, nowadays there are great uncertainties about securing methane utilization and about avoiding the uncontrolled release of methane and carbon dioxide through structural breaks in the geological formations. The uncontrolled release of methane would compromise the climate-protecting effect of the storage of CO_2, because the climatic effect of methane is many times greater than that of carbon dioxide.

Moreover, upon exploiting abandoned coal mines, problems may arise due to the partially wide mining networks and the resulting difficulties in sealing old shaft systems. Furthermore, the lower pressure in the coal deposit may lead to problems with the injection and to structural problems upon exceeding a maximum pressure.

Within the scope of the EU-sponsored project called RECOPOL ("Reduction of CO_2 Emissions by means of CO_2 Storage in Coal Seams in the Silesian Coal Basin of Poland"), the sorptive storage of CO_2 was investigated in coal seams upon simultaneous production of methane. Within the framework of this project conducted in a field test in the Upper Silesian coal mining area, around 700 t CO_2 were injected into a 1,200 m-deep coal seam and its spreading was monitored *via* a neighbouring production borehole. The results of the meanwhile completed project suggest that the option of a CO_2-storage in coal seams must be assessed clearly much more pessimistically.

4.5.3 *Methods for the Reversible Adsorption of CO_2*

Regarding the reversible adsorption of CO_2, there are several existing methods such as the absorption in suitable solvents, the chemical fixation or the physical adsorption to the surfaces of solids (D'Alessandro et al. 2010). Here, there is a gradual transition between a reversible adsorption and a reversible chemical fixation.

Mainly amine solutions are used for separating CO_2 from gaseous mixtures, whereby monoethanolamine and methyl diethanolamine are the most established reagents. Hereby it is being discussed whether the OH-groups are partially converted to carbamate groups which then are transformed to betaine structures.

The amines can also be supported on mesoporous silicate carriers such as SBA-15. Polyethylene amine and triethanol amine have been cited here (Liu et al. 2007; Son et al. 2008). The amino groups can also be chemically bound to the inorganic carrier, *e.g.*, by the reaction of aminosilanes with silicates (Hiyoshi et al. 2004). Membranes are also used to adsorb CO_2 and amines (Zou and Ho 2006). For CO_2-sequestration, such membranes can be structured from polyvinylalcohol and polyallylamine and cross-linked with formaldehyde.

The best known reversible chemical fixation is based on the formation of betaine from bicyclic amides such as DBN or DBU together with CO_2 (Endo et al. 2004). Mixtures of amidines, guanines, and alcohols have also been described such as amidines fixed to polymers. Moreover, for technical applications that have not yet been considered extensively, one possibility is the enzymatic fixation of CO_2 on pyrrol to form pyrrol-2-carbonic acid (Wiesner et al. 1998). This reaction, too, is reversible.

Furthermore, some complex organo-inorganic hybrides can reversibly adsorb CO_2. Especially attractive are so-called MOFs (metal organic frameworks), which are complexes consisting of copper, silver or nickel with bifunctional ligands which form highly porous three-dimensional networks (Li et al. 2011). But also mono- or multinuclear complexes with amine ligands may be good CO_2-adsorbents.

Inorganic systems are also known to reversibly adsorb carbon dioxide; for example, $CaO{-}CaCO_3$ in the temperature range of 700–900 °C or K_2CO_3-$KHCO_3$ in the temperature range of 150–200 °C. Even magnesium-aluminium mixed oxides are able to reversibly adsorb CO_2 (Fu et al. 2008). Other inorganic CO_2-absorbents are MnO_2, Fe_2O_3, calcium hydroxyapatite, hydrotalcite, zeolite (Na-ZSM-5) and activated charcoal.

4.5.4 Application in the Beverage and Food Industry

The purity of the applied carbon dioxide is the most important requirement for its applications in the beverage and food industry. Possible contaminants of carbon dioxide are COS, H_2S and benzene, for whose detection automatic sensor systems have been developed especially for the beverage industry (Duran et al. 2008). Strict purity criteria also apply for many of the extraction processes cited below. Thus, fermentation is the CO_2-source of choice, preferred over other CO_2-sources, such as from combustion processes. In principle, CO_2 can be purified with zeolites. Carbon dioxide is used to carbonate beverages and to produce de-oxygenated water. For applications on the interface between physical and chemical utilization, CO_2 is used to precipitate casein and to pretreat olives before they are pressed for their oil. In both cases, carbon dioxide is used as an acidifying agent.

4.5.5 Cleaning Agents and Extractants

Long-established is the extraction of hops with supercritical carbon dioxide (*sc*CO_2), whereby the extract is used to produce standardized hop aromas. Likewise, *sc*CO_2 is used to decaffeinate coffee. In some of the *sc*CO_2-extractions, the extract is utilized while in others, the purified substance, freed from an undesired residue, is used.

Through the extraction with *sc*CO_2, some substances are yielded in smaller quantities from plant raw materials such as terpenes from orange oil, fatty oils and carotene-like pigments from hawthorn, corn oil from corn, wine aromas from wine or fragrances and bioactive substances yielded from various plants (Awasthi and Trivedi 1997; Wai 2005; Theyssen 2005).

Mainly a purification process is involved in which the nonpolar contaminants are removed, *e.g.*, in the extraction of contaminated soils or of cardboard cartons. Lipophilic metal ions can likewise be extracted using supercritical carbon dioxide through the formation of complexes; this technique can be used in nuclear technology for separating metals, or in wastewater treatment for separating heavy metal pollutants (Erkey 2000; Ohashi and Ohashi 2008).

4.5.6 Use as an Impregnating Agent

The most frequently applied application of *sc*CO_2 is for impregnating wood, leather or textiles with fungicides, other biocides or hydrophobating agents. However, wood can also be loaded with organic dyes (for a thorough dyeing, *e.g.*, with Solvent Blue 35) or with monomers (for fabricating wood-polymer composites) (Kang et al. 2005; Lucas et al. 2007).

In addition, *sc*CO_2-impregnation can be used for catalytic processes. For example, it can be applied for impregnating aerogels and other carriers with complex metal ions, *e.g.*, silica-aerogels with ruthenium-sandwich complexes for the Fischer-Tropsch process (Wakayama and Fukoshima 2006). For coal liquefaction, coal, itself as carrier substance, can be impregnated with molybdenum hexacarbonyl. Magnetic aerogels can also be produced by impregnation with iron and nickel acetylacetonates. Platinum can also be incorporated as acetyl acetonate into silica-aerogels. Nano-porous glass can be loaded with organometallic precursors consisting of copper or erbium, whereby the metal ions are incorporated into the network *via* subsequent heating (Bagratashvili et al. 2004).

Also described is the modification of cement with monomers that are subsequently polymerized. Very interesting is the use of polymers as matrices that, for example, can be loaded with monomers to fabricate so-called interpenetrating polymer networks (semi-IPN) by subsequent polymerization. Styrene-divinylbenzene mixtures and N-cyclohexyl-maleimide-styrene mixtures can be incorporated in matrices such as PE in order to improve material properties (heat resistance) (Sun et al. 2004).

The lateral stability of Nafion® membranes for fuel cells can likewise be improved by impregnation with styrene-DVB and subsequent cross-linking *via* radical polymerization (Sauk et al. 2004). Moreover, precursors for inorganic networks, such as silicic acid esters or ortho-titanium acid esters, can be incorporated into polymers to yield organic-inorganic mixed-IPNs.

Lubricants can be loaded onto polymers, such as polyoxymethylene (POM), *via* *sc*CO_2-impregnation. Interesting is the impregnation of a polymer matrix, *e.g.*, polyurethane foam, with monomers, such as pyrrole, to yield conducting polymers and subsequent *in situ* polymerization to form conducting composites (Tang et al. 2003). Frequently described is also the impregnation of polymers with pharmacologically active ingredients to produce controlled-release drugs (Braga et al. 2008; Manna et al. 2007).

4.5.7 Inert Gas

The largest application of CO_2 is for the storage of fruit and vegetables under controlled atmosphere (CA) conditions (Nicolai et al. 2005). This means storing produce in a low-oxygen but CO_2-enriched atmosphere to slow down respiration processes and, thus, increase the shelf life of such food. Post-ripening (by considering the ripening hormone ethylene), retention of aroma substances and important components (*e.g.*, vitamins) play a large role in investigations about the storage of fruit and vegetables under CA-conditions. Even CO, being a respiration toxin, can be added to the CA, which greatly affects respiration and ethylene production.

One widespread CO_2-application is in fire extinguishers where CO_2 acts somewhat like an inert gas. By contrast, carbon dioxide has hardly proven itself as a suitable inert gas in chemical processes, because it chemically reacts with many organometallic compounds and functional groups such as isocyanates. Furthermore, carbon dioxide is also used in plasma- or arc welding as an additive to argon. Occasionally dry ice is used as a cooling agent.

4.5.8 Potential as a Solvent and Replacement of Volatile Organic Compounds

In the coating industry, there are many ways to reduce the emission of volatile organic compounds (VOC). However, water as a solvent considerably alters the paint resins due to hydrophiliating groups, which ultimately might adversely affect the coating properties. High solids, i.e., solutions in which there are possibly little organic solvents, preferably in reactive thinners, require low molecular masses of the coating resins, a fact which can adversely affect the elasticity of the coatings.

By contrast, powder coatings require a different type of application technique. By spray-applied coating, the partial substitution of solvents by carbon dioxide can result in low-VOC coatings. Here the coating resin is mostly unchanged, thereby retaining the good properties of the organic solvent-based coating systems. In addition, shellac, a bio-resin secreted by the female lac bud, is particularly suitable for applications in the foodstuff and pharmaceutical industries. It can be applied as a spray glaze from ethanol-CO_2 mixtures.

For nitrocellulose, Union Carbide has developed the UNICARB process in which large amounts of the solvent are replaced by CO_2. But carbon dioxide is interesting as a solvent also for other coating resins. Advantageous here is the fact that *sc*CO_2 greatly decreases the viscosity of the resin solutions so that the droplets have a narrower size distribution, are smaller and are parabolic in shape.

Not only is *sc*CO_2 interesting but also liquid CO_2 is promising for spray varnishes. Liquid carbon dioxide and ethanol are non-solvents for many coating resins such as polystyrene-MMA-glycidylmethacrylate-copolymers. Yet the solubility in mixtures is excellent. Powder particles are formed upon spraying; here, one speaks of a so-called RESS-N process (rapid expansion from supercritical solvent with non-solvent) (Hay and Khan 2002).

4.6 Evaluation of Especially Innovative Solution Approaches

There are many current innovative ideas and solution approaches regarding the material exploitation of CO_2 and beyond, such as how CO_2 might be used as a C1-source. Here, one needs to differentiate between the direct incorporation of CO_2 in products as well the technologies needed to do this.

4.6.1 *Material CO_2-Utilization and Innovative Products*

Especially in the area of products from CO_2, there are manifold solution approaches. The products presented in the following seem to be most promising from today's point of view (Fig. 4.4). The estimations of the amounts and period of CO_2-fixation are purely qualitative and do not portend completeness.

4.6.1.1 Polymers from Technically Fixated CO_2 (Duromers, Polycarbonates, Polycondensates)

The incorporation of CO_2 in polymers seems to be most promising (Müller 2008). In this context, especially those polymers are attractive that already boast a market

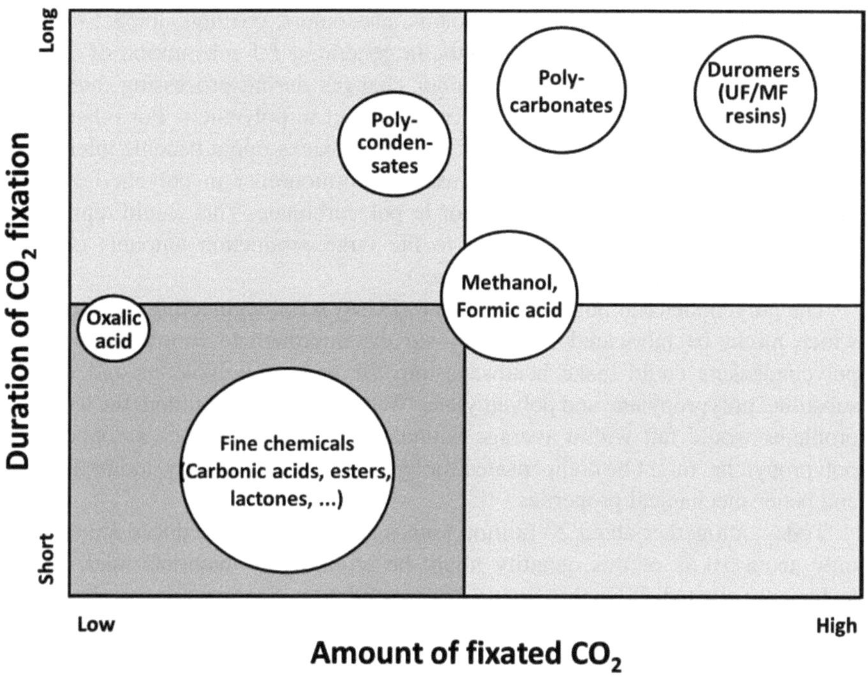

Fig. 4.4 Qualitative estimation of the potential of selected products for CO_2-fixation

in the range of hundreds of thousands to millions of tonnes and in which the transfer of the C1-source to fixated CO_2 suffices for constructing work materials extensively from technically fixated CO_2.

Duromers, mainly urea-formaldehyde resins (UF-resins) and melamine-formaldehyde resins (MF-resins), have the highest potential of the already established polymers for CO_2-fixation. The annual production amounts to several million tonnes, and the carbon can be made available in existing processes by fixated CO_2. Application areas, for example, might be for wood work materials and electric insulating elements (electric sockets and switches). Besides the storage of fixated CO_2, the increased use in the field of heat insulation could lead to a considerable reduction in carbon dioxide emissions as the result of energy savings.

Unlike most pathways to directly incorporate CO_2 into polymers, the alternating copolymerization of carbon dioxide with epoxides does not prescribe as a prerequisite *prior* reduction of the CO_2. Possible products are polyalkylene carbonates or polyethercarbonates. Polyethercarbonate polyols, which can contain fixated CO_2 *via* the direct incorporation (by non-alternating copolymerization of CO_2 with epoxides (Langanke et al. 2014)) or transesterification of diols with dimethyl carbonate, as well at the resulting polyurethanes, offer multiple application opportunities. They allow all the applications typical for polyurethane such as thermoplastic polyurethanes (TPU), foams, elastomers, coatings, impact-resistant modifiers, adhesives or filling materials. In general, a 1:1 substitution of existing products is feasible requiring only minor changes during processing due to the increased glass-transition temperature with respect to polyethers. For other applications, alternating copolymers from CO_2 and epoxides might become interesting.

p-Hydroxybenzoic acid might be used as co-monomer in polyethylene terephthalate, polybutylene terephthalate or in polycarbonate. This would represent a huge potential for CO_2-fixation due to the large production amounts of these plastics (in the range of millions of tonnes).

The polycondensate polyoxymethylene (POM) is an engineering thermoplastic which might be fabricated from CO_2 *via* the intermediate formaldehyde. This polycondensate could make headways into the area of polyolefins and thereby substitute polypropylene and polyethylene. With a partial substitution, the technical problems would fall within average boundaries. Extra costs, *e.g.*, as opposed to polypropylene, might be compensated for by POM's higher quality, longer lifespan and better mechanical properties.

Today, altogether about 20 million tonnes of plastics are produced annually. If only about 10 % of this quantity might be replaced by materials made out of technically fixated CO_2, this would correspond to a fixation rate of about four million tonnes CO_2 per year at a yearly incorporation rate of 20 % by weight.[2]

Due to the beneficial properties of polymers and the increasing capability of manufacturing polymers from CO_2, further market growth and diverse applications are forecasted. However, a comprehensive energy- and CO_2-balance would have to be prepared for every product. Note that for net CO_2-uptake, the production of polymers based on carbon dioxide will require the input of energy from non-fossil sources.

4.6.1.2 Fine Chemicals

Using CO_2 as a basis for synthesizing fine chemicals, various products are feasible, *e.g.*, carbonic acids, ester, lactones and heterocyclic compounds. With the exception of applications in the polymer field (antioxidants, light protection agents, softeners, deforming agents), the amount and period of carbon dioxide fixation is limited. Monomeric or oligomeric carbonate- or oxalate esters could be used for polymer softeners in order to apply technically fixated CO_2. Nonetheless, a high value creation is feasible in the successful production of fine chemicals.

[2] This would correspond to about the annual CO_2-emission of a modern coal-fired power plant.

4.6.1.3 Production of Methanol by Direct Hydrogenation of CO_2

The production of methanol is potentially the most important access to non-fossil C1-chemistry. With regard to the CO_2-balance, however, the prerequisite is that hydrogen be made available from non-fossil resources. The incorporation of secondary products of methanol in polymer chemistry promises a long-term and large amount of carbon dioxide fixation. The most important downstream product from methanol for the field of polymer chemistry (POM, duromers) is formaldehyde that is produced by an exothermic oxidation of methanol.

Furthermore, methanol can be reacted to synthesis gas that allows access to many products of organic chemistry. Besides the dry-reforming of biogas, the synthesis gas production is a good alternative to forego the use of fossil C-sources.

4.6.1.4 Oxalic Acid

Oxalic acid is found in many plants (rhubarb, cocoa, spinach, red beet, sorrel, wood sorrel). Technically, oxalic acid is used as a rust remover, bleaching agent and as an adjuvant in eloxal processes (electrochemical passivation). The diethyl ester or other diesters act as intermediates for producing pharmaceutically active ingredients and plant pesticides. However, the period and amount (ca. 140 t/a) of carbon dioxide fixation are very low for these applications.

Starting from CO_2, oxalic acid needs relatively few reduction equivalents in order to be produced. Nowadays, oxalic acid is manufactured by heating sodium formate, whereby sodium oxalate is formed along with hydrogen. The resulting hydrogen needs to be re-used in order not to waste reduction equivalents.

In principle, oxalic acid could offer a good introduction into the organic CO_2-chemistry and would complement other pathways into the non-fossil C2-chemistry. Nevertheless, for precisely evaluating the sense of this approach, the energy balances of the entire process chains for fabricating oxalic acid would have to be studied.

4.6.2 *Innovative Technologies for Material CO_2-Utilization*

Innovative technologies are needed for manufacturing products in which CO_2 might play a role as C1-source. Currently, four interesting technologies are being discussed: (1) incorporation of CO_2 in polymers; (2) hydrogenation of CO_2; (3) electrocatalytic activation of CO_2; and (4) photocatalytic activation of CO_2. Figure 4.5 depicts a qualitative assessment of these four technological routes regarding the time frame for realizing these technologies as well as their respective research demand.

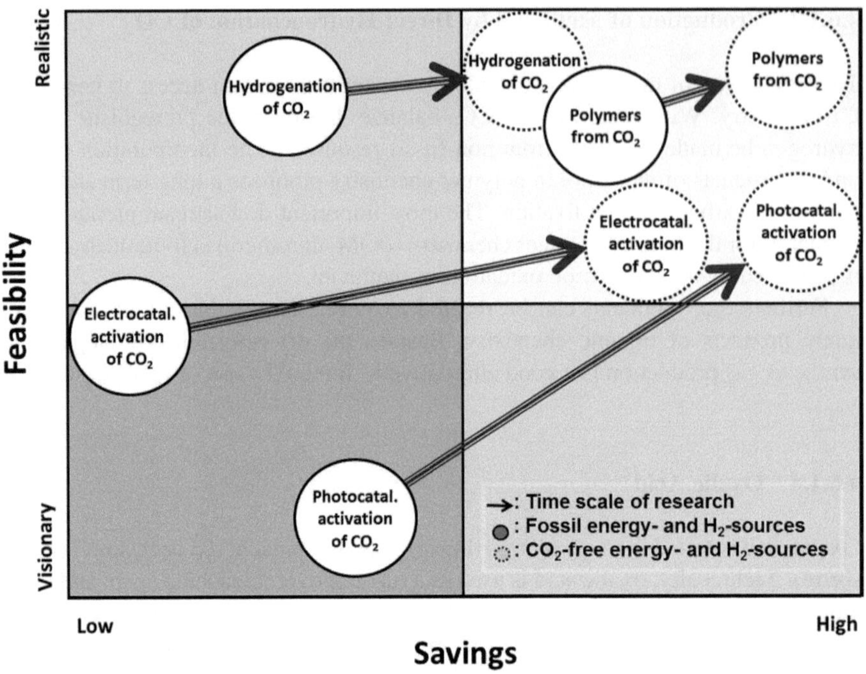

Fig. 4.5 Qualitative assessment of the potential of selected technologies for material CO_2-utilization (Source: Peters et al. 2011)

4.6.2.1 Polymers from CO_2

Most routes for incorporating CO_2 into polymers require the reduction of CO_2. One exception is the copolymerization of CO_2 with epoxides that leads to polyethercarbonates or to polyalkylene carbonates. Yet it has to be considered that producing epoxides likewise requires energy and results in CO_2-emissions. Ethylene oxide is produced by oxidation of ethylene. Ethylene can, in turn, be obtained from bioethanol through elimination of water. The oxidation to epoxide can occur in an exothermic reaction directly with air. Other olefins from renewable resources, *e.g.*, limonene, might likewise be interesting, although more elaborate processes are required to convert the olefin into the corresponding epoxide. Nonetheless, epoxides with high molecular mass naturally fixate less CO_2 during copolymerization.

4.6.2.2 CO_2-Hydrogenation

The hydrogenation of CO_2 represents another innovative technology for materially utilizing (Leitner 1995). The prerequisite for a significant reduction of CO_2-emissions with this type of CO_2-utilization is the availability of hydrogen from non-fossil sources. Were hydrogen made available from non-fossil sources, there would be various approaches for CO_2-fixation, *e.g.*, the production of formic acid (Hutschka et al. 1995, 1997; Wesselbaum et al. 2012) and CO. The methanol production would likewise be an interesting and economically relevant option to fixate carbon dioxide. Upon subsequently applying methanol, instead of fossil raw materials, as a starting material for producing formaldehyde, the market for CO_2-containing polymers could be raised sustainably.

Another possibility is to fabricate synthetic methane from hydrogen and CO_2 according to the catalytically directed Sabatier reaction (Sterner 2009; Bajohr et al. 2011). This reaction is a combination of several running individual reactions (water gas-shift reaction, conversion of CO and H_2 to CH_4). Typically applied catalysts are nickel or ruthenium (Bajohr et al. 2011). Moreover, hydrogen production *via* electrolysis with subsequent methanization is a feasible technical option, in particular, regarding the context of future surpluses of electricity from wind farms and photovoltaic generation of electricity where feeding into the grid is subjected to restrictions. The advantages of this option do not especially lie in the CO_2-balance but instead in energy-economic considerations. Thus, an indirect storage of large amounts of power (frequently called "power to gas") can take place over a longer period *via* the direct feeding of hydrogen or synthetic methane into the natural gas network; currently, this is impossible with conventional electricity storage devices (*e.g.*, batteries).

4.6.2.3 Electrochemical Activation of CO_2

Methods for electrochemically reducing carbon dioxide have long been investigated. Various products have been described such as CO, formic acid, methane, methanol, ethane, ethylene, ethanol, acetone, hydrocarbons as well as hydrogen as an undesired by-product.

In order to guarantee a sufficient electron-transfer, up to now rare metal cathodes on which high overvoltages were placed have been used in the laboratory for reducing CO_2 to C1-molecules (Ausfelder and Bazzanella 2008). The production of oxalic acid was reported already in 1981 (Fischer et al. 1981).

However, today's electrocatalysts are either inefficient for producing higher energy products or they need an additional sacrificial molecule as an electron donor (*e.g.*, alcohols, amines, or sulfite) (Ausfelder and Bazzanella 2008). Consequently, a large-scale industrial realization of electrochemical CO_2-reduction is not expected in the near future.

4.6.2.4 Photocatalytic Activation of CO_2

Attractive is the direct photoreduction of CO_2 to CO, methane, other hydrocarbons or methanol. There are various types of feasible photocatalysts for this reaction, *e.g.*, titanium dioxide-based systems (Kitano et al. 2007), indium tantalate and platinum metal complexes.

Titanium oxide (TiO_2)-based systems mostly use the photocatalytically more active mineral form anatase which is primarily formed during precipitation or sol-gel processes (Koci et al. 2009). The main products are hydrocarbons, mainly methane, along with methanol. Since the band gap of anatase is 3.2 eV, TiO_2 is sensitive to near UV-light. When also fractions of visible light should be used, one can use copper-, iron- or silver-doped titanium oxide systems. The literature cites a photoreactor whose fiber optic cables were coated with Cu/Fe-doped TiO_2 (Nguyen and Jeffrey 2008). Thereby a total energy efficiency of 0.04 % was determined for the products methane and ethylene.

Another photoactive semiconductor besides TiO_2 is indium tantalate ($InTaO_4$) (Chen et al. 2008). The product spectrum is similar to that of titanium dioxide, whereby methanol predominates. Moreover, the methanol selectivity can be raised by addition of enzymes such as methanol dehydrogenase (MDH) (Veeramalu and Mainardi 2008).

A fully other photocatalyst class are complexes of dipyridine ligands with platinum metals, but preferentially with ruthenium (Ru(II)) or rhenium (Re(I)) (Takeda et al. 2008). Carbon monoxide is primarily formed in the photoreduction of CO_2, in which hydrogen often occurs as by-product. In particular, high-performance catalysts of this type are heterotrimetallic complexes with tripodal, bipyridine-carrying units that simultaneously contain Re(I) und Ru(II) as metal ions (Bian et al. 2009).

Even though initial approaches have been described in the literature regarding photocatalytic reduction of carbon dioxide, more research is greatly needed (Osterloh 2008). Therefore, a short-term or mid-term technical realization of this seems hardly possible. Instead, successes that would pave the way for an industrially realizable application are expected in the long term.

Still prohibitive is the low energy efficiency of the technique. The photoreduction of CO_2 would have to be comparable to the reduction of CO_2 with hydrogen to methanol or hydrocarbons (Fischer-Tropsch). This hydrogen can be made available by electrolysis of water with power from renewable sources, *e.g.*, by means of water photolysis.

4.7 Conclusions

Quantitatively, the current contribution of the industrial utilization of CO_2 to climate protection is insignificant. While worldwide around 31 billion tonnes CO_2 are emitted annually nowadays, the industrial exploitation of CO_2 is only about 130 million tonnes/a. As yet, the use of CO_2 for organochemical and inorganic applications is also based mainly on industrial sources in which carbon dioxide is yielded as a coupled product. From an industrial political perspective, the utilization of CO_2 is gaining importance, because CO_2 can be used as a cost-effective raw material that, upon greater demand, can stem from CCS-sources.

There are many possibilities for CO_2-utilization that need to be analyzed in detail regarding their climate relevance as well as their value-creation potential. By considering the worldwide rising CO_2-emissions, one can assume that the exploitation of CO_2 does not represent an alternative, but rather a complementary approach to CO_2-sequestration. Nonetheless, one needs to keep in mind the value-creation potential of many promising types of CO_2-utilization; this potential should not be neglected.

The climate-protection relevance of an industrially and politically motivated CO_2-utilization depends on the quantity and period of CO_2-fixation. The fixation potential varies greatly depending on the particular application of the products and results from the combination of the respective amount of CO_2 bound and the period of the CO_2-fixation. At the same time, it has to be taken into account whether there are other resource- or energy-expenditures resulting from the activation and use of CO_2 (Elmas et al. 2013) that would adversely affect the overall energy balance. The Life Cycle Assessment (LCA) method is suited for the analysis of the complete energy- and CO_2-balance. This is an established approach for evaluating the environmental effects of processes or of products.

Even though a large quantity of CO_2 can be fixated during the production of an energy carrier (*e.g.*, methanol, synthetic methane), the corresponding fixation period is short. Although the production of fine chemicals allows the storage of small quantities of CO_2, under certain circumstances such production could result in a long fixation period for the bound CO_2. For both cases, the relevance to climate protection is rather low. By contrast, polymers, during their production, can potentially store much CO_2 and also fixate this for a long time. This infers a relatively high potential for climate protection.

Already in a qualitative analysis of the energy balance, it becomes clear that the potential for reducing CO_2-emissions depends very greatly on the production of the energy-rich reaction partner (*e.g.*, hydrogen, epoxides) and on the process energy (*e.g.*, heat, electricity, light). Thus, these aspects are important for a holistic view with respect to climate relevance. Preparing energy- and CO_2-balances for the individual options of CO_2-utilization is the prerequisite for the detailed evaluation of CO_2-utilization methods as well as of the CO_2-footprint of products. Moreover, the expected value-creation potential should be considered as a positive aspect in the evaluation of utilization strategies. The economic potential for chemical applications can vary. If one compares the amount of fixated carbon dioxide, the production of urea can potentially store more CO_2 than the production of salicylic acid. Salicylic acid in polycarbonates shows a longer CO_2-fixation period than that of salicylic acid in aspirin. When the evaluation is based on the criterion of value creation, salicylic acid scores highest.

Several innovative technologies are on the cusp of realization. The production of polycarbonates and carbonates from CO_2 enables an immediate access to substantial markets in the chemistry and plastics sector. Especially attractive hereby is the aspect that CO_2 is practically unlimitedly available as a building block for chemical synthesis. The use of catalysts is necessary for the technical realization of CO_2-based processes. In the 1990s and most recently, considerable advances have been made in the field of CO_2-related catalyst research in academia as well as in industry.

The hydrogenation of CO_2 to make important products such as methanol or formic acid and its derivatives is being intensively studied, and catalyst systems are principally known for the generally exothermic reactions. Nonetheless, considerable research efforts are needed on the interface between chemistry and engineering. In principle, the reverse water gas shift (RWGS) reaction is also a feasible option for producing carbon monoxide and, thus, another foothold for bringing CO_2, as a raw material, into the chemical value chain.

There are several active catalysts for the coupling of carbon dioxide with unsaturated substrates such as butadiene or acetylenes. With the exception of certain telomerization reactions (Kurarayi techniques), there is currently no market for the resulting products. Thus, an economic utilization would have to comprise production as well as product development.

The direct hydrocarbon carboxylation by the formal insertion of carbon dioxide into the C—H-bond of alkanes, aromatics or olefins counts among the so-called "*Dream Reactions*" of modern catalysis research and would open up an elegant pathway for producing fine chemicals. Among others, this route renders the respective conversions of methane and CO_2 to acetic acid, of benzene and CO_2 to benzoic acid, and of ethylene and CO_2 to acrylic acid possible. Although in many cases the processes are principally thermodynamically feasible and, *e.g.*, fundamental research into the organometallic chemistry of metal complexes gives important clues about the possibility of catalytic cycles, currently there are no efficient solutions to this problem.

The electro- and, in particular, the photocatalytic reduction of CO_2 would be basically the most elegant form of CO_2-utilization, since it mimics natural photosynthesis. Heterogeneous catalysts (mostly based on TiO_2 as photosensitizer) as well as homogeneous catalysts (mostly based on ruthenium und rhenium-bipyridyl-complexes) have been and are being investigated intensively. Yet conventional systems still need to be improved considerably before they can attain a technically usable efficiency.

In the context of greenhouse gas reduction, the potential of CO_2-utilization needs to be investigated in detail with regard to CO_2-fixation period and quantity. By considering all process steps for production as well as preliminary chains and possible substitution of CO_2 from other sources, an overall energy- and CO_2-balance has to be derived for each particular product. This can elucidate the possible greenhouse gas reduction along the entire cycle. The evaluation of value-creation potentials requires a detailed product-based market analysis that also accounts for possible substitution effects.

References

Almog J, Klein A, Sokol A, Sasson Y, Sonenfeld D, Tamiri T (2006) Urea nitrate and nitrourea: powerful and regioselective aromatic nitration agents. Tetrahedron Lett 47:8651–8652

Aresta M, Dibenedetto A (2007) Utilization of CO_2 as chemical feedstock: opportunities and challenges. Dalton Trans 28:2975–2992

Ausfelder F, Bazzanella A (2008) Verwertung und Speicherung von CO_2. DECHEMA, Frankfurt
Awasthi A, Trivedi RK (1997) A review on supercritical carbon dioxide extraction of natural products. Chem Eng World 32:65–71
Bagratashvili VN, Bogomolova LD, Gerasimova VI (2004) Doping of nanoporous glasses by supercritical fluid impregnation of β-diketonate Cu. J Non-Crystalline Solids 345 + 346:256–259
Bajohr S, Götz M, Graf F, Ortloff F (2011) Speicherung von regenerativ erzeugter elektrischer Energie in der Erdgasinfrastruktur. gwf-Gas, April, pp 200–210
Bazzanella A, Krämer D, Peters M (2010) CO_2 als Rohstoff. Nachrichten aus der Chemie 58 (12):1226–1230
Bian ZJ, Sumi K, Furue M, Sato S (2009) Synthesis and properties of a novel tripodal bipyridyl ligand tb-carbinol and its Ru(II)RE(I) trimetallic complexes: investigation of multimetallic artificial systems for photocatalytic CO_2 reduction. Dalton Trans 6:983–993
Boddien A, Loges B, Gärtner F, Torborg C, Fumino K, Junge H, Ludwig R, Beller M (2010) Iron-catalyzed hydrogen production from formic acid. J Am Chem Soc 132(26):8924–8934
Boddien A, Mellmann D, Gärtner F, Jackstell R, Junge H, Dyson PJ, Laurenczy G, Ludwig R, Beller M (2011) Efficient dehydrogenation of formic acid using an iron catalyst. Science 333 (6050):1733–1736
Braga ME, Vaz Pato MT, Costa Silva HS (2008) Supercritical solvent impregnation of opthalmic drugs on chitosan derivatives. J Supercrit Fluids 44:245–257
Brownscombe TE, Vaporciyan GG, Pfrehm SF (2001) Carboxylation of hydrocarbons to terephthalic acid or naphtalenedicarboxylic acid. Patent WO 2001016072
Bucherer HT, Steiner W (1934) Synthesis of hydantoins. Reactions of a-hydroxy and a-amino nitriles. Journal für Praktische Chemie 140:291–316
Centi G, Iaquaniello G, Perathoner S (2011) Can we afford to waste carbon dioxide? Carbon dioxide as a valuable source of carbon for the production of light olefins. ChemSusChem 4:1265–1273
Chen HC, Chou HC, Wu JC, Lin HY (2008) Sol-gel prepared InTaO4 and its photocatalytic characteristics. J Mater Res 23:1364–1370
Choi SK, Ko KS, Chun HD, Kim JG (2002) Utilization of carbon dioxide for neutralization of alkaline waste water. GHG6. Kyoto, Japan
D'Alessandro DM, Smit B, Long JR (2010) Carbon dioxide capture: prospects for new materials. Angew Chem Int Ed 49:6058–6082
Darensbourg DJ, Andreatta JR, Moncada AI (2010) Polymers from carbon dioxide: polycarbonates, polythiocarbonates, and polyurethanes. In: Aresta M (ed) Carbon dioxide as chemical feedstock. Wiley-VCH, Weinheim, pp 213–248
Davison J, Freund P, Smith A (2001) Putting carbon back into the ground. www.ieagreen.org.uk/publications.html
Dienes Y, Leitner W, Müller MGJ, Offermans WK, Reier T, Reinholdt A, Weirich TE, Müller TE (2012) Hybrid sol-gel double metal cyanide catalysts for the copolymerisation of styrene oxide and CO_2. Green Chem 14:1168–1177
Dittmeyer R, Keim W, Kreysa G, Oberholz A (eds) (2005) Chemische Technik, Prozesse und Produkte, Bd. 4 Energieträger. Wiley-VCH, Weinheim, pp 1–1198
Doll KM, Erhan EZ (2005) Improved synthesis of carbonated soybean oil in supercritical carbon dioxide. 229th ACS national meeting, San Diego, USA
Dorner RW, Hardy DR, Williams FW, Willauer HD (2010) Heterogeneous catalytic CO_2 conversion to value-added hydrocarbons. Energy Environ Sci 3:884–890
Doro F, Winnertz P, Leitner W, Prokofieva A, Müller TE (2011) Adapting a Wacker-type catalyst system to the palladium-catalyzed oxidative carbonylation of aliphatic polyols. Green Chem 13(2):292–295
Duex A, Ooms P, Rechner J, Boehm M, Hallenberger K, Ronge G, Eynde VJ (2009) Method for producing diarylcarbonates and or arylalkylarylcarbonates from dialkylcarbonates. Patent EP 2036880

Duran C, Vilanova X, Brezmes J, Llobete E, Correig X (2008) Thermal desorption pre-concentrator based system to assess carbon dioxide contamination by benzene. Sens Actuators B B131:85–92

Elmas S, Köhler B, Leitner W, Harrer M, Gürtler C, Müller TE, Sundermeyer J (2012) Verfahren zur Herstellung von linearen und/oder cyclischen Carbonatestern. Patent EP 12181442

Elmas S, Subhani MA, Vogt H, Leitner W, Müller TE (2013) Facile insertion of CO_2 into metal-phenoxide bonds. Green Chem 15(5):1356–1360

Endo T, Nagai D, Monma T, Yamaguchi H, Ochiai B (2004) A novel construction of a reversible fixation – release system of carbon dioxide by amidines and their polymers. Macromolecules 37:2007–2009

Erkey C (2000) Supercritical carbon dioxide extraction of metals from aqueous solutions: a review. J Supercrit Fluids 17:259–287

Fakeev AA (2004) Study and development of a technology for manufacture of special-purity calcium carbonate. Russ J Appl Chem 77:869–874

Fischer J, Lehmann T, Heitz E (1981) The production of oxalic acid from carbon dioxide and water. J Appl Electrochem 11:743–750

Fu X, Wu D, Zhao N (2008) Reversible CO_2 adsorption by MgAl-complex oxide materials. 235th ACS national meeting, New Orleans

Gao Y, Qin Y, Zhao X, Wang F, Wang X (2012) Selective synthesis of oligo(carbonate-ether) diols from copolymerization of CO_2 and propylene oxide under zinc-cobalt double metal cyanide complex. J Polym Res 19:9878

Garcia Deleon R, Kobayashi A, Yamauchi T, Ooshi J, Baba T, Sasaki M, Hiarata F (2002) Catalytic methoxycarbonylation of 1,6-hexanediamine with dimethyl carbonate to dimethyl hexane-1,6-dicarbamate using $Bi(NO_3)_3$. Appl Catalysis A General 225: 1–2, 43–49

Graham DC, Mitchell C, Bruce M, Metha G, Bowie J, Buntine M (2007) Production of acrylic acid through nickel-mediated coupling of ethylene and carbon dioxide: a DFT study. Organometallics 26:6784–6792

Gürtler C, Hofmann J, Müller TE, Wolf A, Grasser S, Köhler B (2011a) Method for the production of polyethercarbonate polyols. Patent WO 2011117332

Gürtler C, Müller TE, Ooms P, Rechner J, Risse F, Prokofieva A, Doro F, Köhler B, Leitner W, Wolf A (2011b) Bimetallic Schiff base-phenanthroline complexes as catalysts for oxidative carboxylation of phenols in production of diaryl carbonates. Patent WO 2011073087

Gürtler C, Müller TE, Ooms P, Risse F, Rechner J, Doro F, Prokofieva A, Winnertz P, Leitner W (2012) Preparation of cyclic carbonates by oxidatively carbonylating diols and polyols. Patent WO 2012045742

Gürtler C, Müller TE, Kermagoret A, Dienes Y, Barruet J, Wolf A, Grasser S (2013) Method for production of polyether-ester-carbonate polyols in the presence of double metal cyanide catalysts. Patent WO 2013087582

Hay JN, Khan A (2002) Review of environmentally friendly coatings using carbon dioxide as the carrier medium. J Mater Sci 37:4743–4752

Hischier R, Hellweg S, Capello C, Primas A (2005) Establishing life cycle inventories of chemicals based on differing data availability. Int J Life Cycle Assess 10(1):59–67

Hiyoshi N, Yogo K, Yashima T (2004) Reversal adsorption of carbon dioxide on amine-modified SBA-15 from flue gas containing water vapor. Stud Surface Sci Catal 153:417–422

Hofacker S (2006) Metal acetylacetonates as transesterification catalysts. Patent EP 1632512

Hofmann J, Gürtler C, Nefzger H, Hahn N, Lorenz K, Müller TE (2012) Method for producing polyether carbonate polyols with primary hydroxyl end groups for use in polyurethane soft foams. Patent WO 2012080192

Hölscher M, Gürtler C, Keim W, Müller TE, Peters M, Leitner W (2012) Carbon dioxide as a carbon resource – recent trends and perspectives. Z Naturforschung B J Chem Sci 67(10):961–975

Hull JF, Himeda Y, Wang W-H, Hashiguchi B, Periana R, Szalda DJ, Muckerman JT, Fujita E (2012) Reversible hydrogen storage using CO_2 and a proton-switchable iridium catalyst in aqueous media under mild temperatures and pressures. Nat Chem 4(5):383–388

Hutschka F, Dedieu A, Leitner W (1995) Metathesis as a critical step for the transition metal catalyzed formation of formic acid from CO_2 and H_2? An ab initio investigation. Angew Chem Int Ed Eng 34(16):1742–1745

Hutschka F, Dedieu A, Eichberger M, Fornika R, Leitner W (1997) Mechanistic aspects of the rhodium-catalyzed hydrogenation of CO_2 to formic acid – a theoretical and kinetic study. J Am Chem Soc 119(19):4432–4443

International Fertilizer Industry Association (2009) Fertilizer production. www.fertilizer.org/ifa/content/view/full/8873

Kang SM, Levien KL, Morell JJ (2005) Supercritical fluid impregnation of wood with biocides using temperature reduction to induce deposition. Wood Sci Technol 39:328–338

Kappe CO (2000) Recent advances in the Biginelli dihydropyrimidene synthesis. New tricks from an old dog. Acc Chem Res 33:879–888

Katmatani Y, Fujita N (1986) Reaction of an oxadiazinetrione derivative and its application to polyurethane coatings. J Coatings Technol 58:63–66

Kember MR, Buchard A, Williams CK (2011) Catalysts for CO_2/epoxide copolymerisation. Chem Commun 47:141–163

Kikelj D (2004) Product class 13: Quinazolines. Sci Synthesis 16:573–749

Kitano M, Matsuoka M, Ueshima M, Anpo M (2007) Recent developments in titanium oxide-based photocatalysts. Appl Catal Gen 325:1–14

Koci K, Obalova L, Matejva L, Placha D (2009) Effect of TIO_2 particle size on the photocatalytic reduction of CO_2. Appl Catal Environ 89:494–502

Koop U, Krummradt H, Schwarz M, Stoldt J, Eckstein J, Zehner S (2001) Process for reaction of carbonyl compounds with organometallic reagents especially Grignard reagents. Patent DE 10001317

Kuckshinrichs W, Linssen J, Markewitz P, Zapp P, Peters M, Köhler B, Müller TE, Leitner W (2009) Weltweite Innovationen bei der Entwicklung von CCS-Technologien und Möglichkeiten der Nutzung und des Recyclings von CO_2 Research Report

Lambertz J, Ewers J (2009) Optionen und Potenziale der CO_2-Umwandlung und -Nutzung. Energiewirtschaftliche Tagesfragen 59(1/2):112–117

Langanke J, Wolf A, Hofmann J, Böhm K, Subhani A, Müller TE, Leitner W, Gürtler C (2014) Carbon dioxide (CO2) as sustainable feedstock for polyurethane production. Green Chem 16:1865–1870

Lange S, Leitner W (2002) Mechanistic aspects of hydrogen addition during enantioselective rhodium-catalyzed reduction of C:C double bonds with formic acid/triethylamine or molecular hydrogen. J Chem Soc Dalton Trans 5:752–758

Laurenczy G (2011) Hydrogen storage and delivery: the carbon dioxide – formic acid couple. CHIMIA 65(9):663

Leigraf C, Besser J, Kleemann S (2002) CO_2 for pH-stabilisation in specialty paper production through using Adalka process. Wochenblatt für Papierfabrikation 130:772–781

Leitner W (1995) Carbon dioxide as a raw material: the synthesis of formic acid and its derivatives from CO_2. Angew Chem Int Ed Eng 34(20):2207–2221

Leitner W, Brown JM, Brunner H (1993) Asymmetric catalysis. Mechanistic aspects of the rhodium-catalyzed enantioselective transfer hydrogenation of a, b -unsaturated carboxylic acids using formic acid/triethylamine (5:2) as the hydrogen source. J Am Chem Soc 115 (1):152–159

Li J-R, Ma Y, Mccarthy MC, Sculley J, Yu J, Balbuena PB, Jeong H-K, Zhou H-C (2011) Carbon dioxide capture-related gas adsorption and separation in metal-organic frameworks. Coord Chem Rev 255:1791–1823

Liu X, Zhou L, Fu X, Su W, Zhou Y (2007) Adsorption and regeneration study of the mesoporous adsorbent SBA-15 adapted to capture/separation of CO_2 and CH_4. Chem Eng Sci 62:1101–1110

Liu A-H, Li Y-N, He L-N (2012) Organic synthesis using carbon dioxide as phosgene-free carbonyl reagent. Pure Appl Chem 84(3):581–602

Lucas S, Gonzales E, Calvo MP, Palencia C, Alonso E, Cocero MJ (2007) Supercritical CO_2 impregnation of radiata pine with organic fungicides: effect of operating conditions and two-parameters modeling. J Supercrit Fluids 40:462–469

Manna L, Banchero M, Solva D (2007) Impregnation of PVP microparticles with ketoprofen in the presence of supercritical CO_2. J Supercrit Fluids 44:245–257

Markewitz P, Kuckshinrichs W, Leitner W, Linssen J, Zapp P, Bongartz R, Schreiber A, Müller TE (2012) Worldwide innovations in the development of carbon capture technologies and the utilization of CO_2. Energy Environ Sci 5:7281–7305

Mizuno T (2002) Pharmaceutical materials from CO_2. New possibility. Eco Ind 7:18–27

Müller TE (2008) CO_2 fixation in polymer. Preprints Symp Am Chem Soc Division Fuel Chem 53:1

Müller TE, Gürtler C, Kermagoret A, Dienes Y, Busygin I, Köhler B, Leitner W (2012a) Method for producing aliphatic polycarbonate polyols by living polymerization of cyclic carbonates. Patent WO 2012059550

Müller TE, Gürtler C, Ooms P, Risse F, Rechner J, Doro F, Prokofieva A (2012b) Procedure for the production of diarylcarbonates from dialkyl carbonates. Patent DE 102010042937

Müller TE, Gürtler C, Subhani MA (2013a) Method for producing polyether carbonate polyols by catalytic addition of carbon dioxide and alkylene oxides to H-functional starter. Patent WO 2013087583

Müller TE, Gürtler C, Dienes Y, Barruet J, Köhler B, Leitner W (2013b) Method for activating double metal cyanide catalysts with heterocumulenes for production of polyether carbonate polyols. Patent WO 2013010987

Müller TE, Gürtler C, Wohak M, Hofmann J, Subhani MA, Cosemans M, Leitner W (2013c) Method for production of polyether polycarbonate polyols. Patent WO 2013011015

Müller TE, Gürtler C, Wohak M, Hofmann J, Subhani MA, Cosemans M, Leitner W (2013d) Method for activating double metal cyanide catalysts for use in production of polyether polycarbonate polyols. Patent WO 2013011014

Müller TE, Gürtler C, Vogt H, Keldenich A, Leitner W (2013e) Method for producing polyols by hydrogenating CO/olefin copolymers. Patent WO 2013034536

Müller, TE, Gürtler C, Vogt H, Köhler B, Leitner W (2013f) Process for production of CO/olefin copolymers for conversion to polyols and polyamines for use in polymer. Patent DE 102011053470

Müller TE, Gürtler C, Vogt H, Keldenich A, Leitner W (2013g) Method for producing polyols by hydrogenating CO/olefin copolymers. Patent DE 102011053466

Nguyen TV, Jeffrey CS (2008) Photoreduction of CO_2 in an optical-fiber photoreactor. Effects of metals addition and catalyst carrier. Appl Catal A General 335:112–120

Nicolai BM, Lammertyn J, Schotsmanns W (2005) Gas exchange properties of fruit and vegetables. Food Sci Technol 144:645–677

O'Connor WK, Dahlin DC, Nilsen DN, Rush GE, Walters RP, Turner PC (2000) CO_2 storage in solid form: a study of direct mineral carbonation. In: 5th International conference on greenhouse gas control technologies, pp 322–327, Cairns, QLD

Ohashi A, Ohashi K (2008) Supercritical carbon dioxide extraction of metal ions from aqueous solutions: a review of recent studies. Solvent Extraction Res Dev 15:11–20

Osterloh FE (2008) Inorganic materials as catalysts for photochemical splitting of water. Chem Mater 20:35–54

Peters M, Köhler B, Kuckshinrichs W, Leitner W, Markewitz P, Müller TE (2011) Chemical technologies for exploiting and recycling carbon dioxide into the value chain. ChemSusChem, Special Issue: Carbon Dioxide Recycling 4:9, 1216–1240

Radgen P, Crenmer C, Warkentin S, Gerling P, May F, Knopf S (2006) Verfahren zur CO_2-Abscheidung und -Speicherung. Climate Change 07/06, Berlin/Dessau

Ritter SK (2007) What can we do with carbon dioxide. Chem Eng News 85:11–17

Sakakura T, Choi J, Yasuda H (2007) Transformation of carbon dioxide. Chem Rev 107:2365–2387

Sanuki S, Matsuchita K, Nishiwaki M, Majima H (2000) Preparation of Nd(III) carbonate by precipitation stripping of Nd(III)-loaded VA10. Metallurgical Mater Trans B 31B:5–13
Sauk J, Byun J, Kim H (2004) Grafting of styrene on to nafion membranes using supercritical CO_2 impregnation for direct methanol fuel cells. J Power Sources 132:59–63
Sawada T (1990) Preparation and properties of high purity phthalocyanines. Phys Sci Eng 26:47–59
Son WJ, Choi JS, Ahn WS (2008) Adsorptive removal of carbon dioxide using polyethyleneimine-loaded mesoporous silica materials. Microporous Mesoporous Mater 113:31–40
Song C (2006) Global challenges for control, conversion and utilization of CO_2 for sustainable development involving energy, catalysis, adsorption and chemical processing. Catalysis Today 115:2–32
Sterner M (2009) Bioenergy and renewable power methane in integrated 100 % renewable energy systems – limiting global warming by transforming energy systems, vol 14. Kassel University Press, Kassel, pp 1–234
Sun D, Wang B, He J, Zhang R, Liu Z, Han B, Huang Y (2004) Grafting of polypropylene with N-cyclohexylmaleimide and styrene simultaneously using supercritical CO_2. Polymer 45:3805–3810
Takeda H, Koke K, Inoue H, Ishitani O (2008) Development of an efficient photocatalytic system for CO_2 reduction using Rhenium(I) complexes based on mechanistic studies. J Am Chem Soc 130:2023–2031
Tamura K, Tsuge H (2006) Characterisations of multistage column crystallizer for gas-liquid reactive crystallization of calcium carbonate. Chem Eng Sci 61:5818–5826
Tang M, Wen TY, Du TB, Chen YP (2003) Synthesis of electrically conductive polypyrrole-polystyrene composites using supercritical carbon dioxide: II. Effects of the doping conditions. Eur Polym J 39:151–156
Theyssen N (2005) Chapter 6.2. In: Cornils B, Herrmann WA, Vogt D, Horvath I, Olivier-Bourbigon H, Leitner W, Mecking S (eds) Multiphase homogeneous catalysis. Wiley-VCH, Weinheim, pp 630–643
Tundo P, Memoli S, Merault D, Hill K (2004) Synthesis of methylethers by reaction of alcohols with dimethylcarbonate. Green Chem 6:609–612
Veeramalu R, Mainardi DS (2008) Photochemical reduction of carbon dioxide to methanol with the assistance of methanol dehydrogenase enzyme. 64th Southwest regional meeting of the American Chemical Society, Little Rock, USA
Wai CM (2005) Extraction of natural products from herbs using high-pressure cold water and supercritical carbon dioxide. Abstracts of papers, 230th ACS national meeting, Washington, AGFD-066
Wakayama H, Fukoshima Y (2006) Supercritical CO_2 for making nanoscale materials. Indus Eng Chem Res 45:3328–3331
Wang B, Lan DQ (2010) Biofixation of carbon dioxide (CO_2) by microorganisms. In: Maroto-Valer MM (ed) Developments and innovation in carbon dioxide (CO_2) capture and storage technology. Woodhead Publishing Ltd, Oxford, pp 411–432
Wesselbaum S, Hintermair U, Leitner W (2012) Continuous-flow hydrogenation of carbon dioxide to pure formic acid using an integrated *sc*CO_2 process with immobilized catalyst and base. Angew Chem Int Ed 51:8585–8588
Wielicka J, Ptasiewicz-Malinowskaja A, Jedrych T, Minda D (2003) Disinfacts containing active oxygen: method for the preparation of urea hydroperoxide and melamine hydroperoxide. Polisch J Chem Technol 5:19–21
Wiesner M, Fujii N, Yoshida T, Nagsawa T (1998) Carbon dioxide fixation by pyrrole-2-carboxylate decarboxylate from baccillus megaterium PYR2910. Eur J Biochem 257:495–499
Wirz B, Kueng W (1983) Reaction of carbon dioxide 14-C with Grignard reagents to form either carboxylic acids or ketones. J Label Compd Radiopharm 20:635–653
Xu X, Zhu T (2005) Coupled process of reaction and solvent extraction: I. The reaction between CO2 and SrCl2 coupled with solvent extraction of HCl. Hydrometallurgy 76:11–17

Yajima K, Kawashima M, Hashimoto N, Miyake Y (1996) Kinetic study on the reaction of heptylmagnesium bromide with carbon dioxide using non-carrier-added C-11 labeled carbon dioxide. J Phys Chem 100:14936–14940

Yalfani MS, Lolli G, Wolf A, Mleczko L, Müller TE, Leitner W (2013) Methyl formate as a carbonylating agent for the catalytic conversion of phenol to methyl phenyl carbonate. Green Chem 15(5):1146–1149

Yon Z, Mata V, Rodriguez AE (2002) Adsorption of carbon dioxide at high temperature – a review. Sep Purif Technol 26:195–205

Yu KMK, Curcic I, Gabariel J, Tsang SCE (2008) Recent advances in CO2 capture and utilization. ChemSusChem 1:893–899

Zevenhoven R, Fagerlund J (2010) Mineralisation of carbon dioxide (CO_2). In: Maroto-Valer MM (ed) Developments and innovation in carbon dioxide (CO_2) capture and storage technology. Woodhead Publishing Ltd, Oxford, pp 433–462

Zevenhoven R, Eloneva S, Teir S (2006) Chemical fixation of CO_2 in carbonates: routes to valuable products and long-term storage. Catalysis Today 115:73–79

Zou J, Ho WS (2006) CO_2-selective polymeric membranes containing amines in crosslinked poly (vinyl alcohol). J Membr Sci 286:310–321

Chapter 5
Environmental Aspects of CCS

Andrea Schreiber, Petra Zapp, and Josefine Marx

Abstract The use of CO_2 capture technologies causes efficiency losses which leads to an additional demand of fuel and related other emissions. Also necessary operating materials and a change in waste composition are consequences of this utilisation. Life Cycle Assessment (LCA) has proved to be a helpful tool to investigate the different environmental consequences associated with the introduction of CCS. For all capture routes environmental effects of conventional capture technologies are analyzed. Additionally, the impacts of a second generation capture technology, ceramic membranes, are investigated. The share of life cycle segments, such as power plant operation, fuel supply or CO_2 transport and sequestration, can be identified for the different impact categories. Generally, the intended decrease of CO_2 emissions goes along with an increase in most other impact categories regardless of technology or fuel used.

Keywords Life cycle assessment • Environmental effects • Capture technologies • Membrane-based air separation

5.1 Introduction

Scientific discussions are focusing more and more frequently on additional environmental impacts triggered by the targeted reduction of CO_2. A greater demand for fuel due to efficiency losses, toxic effects of chemical detergents, and new compositions of waste streams, to name but a few topics, make clear that a comprehensive consideration of the whole system is required. Only in this way can it be ensured that an environmental impact is not improved at the expense of other environmental effects and that there is not simply a shift towards other phases in the life cycle. One approach to a holistic evaluation of environmental effects is the life cycle assessment (LCA), which has been employed for many years to identify the environmental impacts of products or technologies and to compare the impacts of different systems. In recent years, this methodology has been widely used to map the

A. Schreiber (✉) • P. Zapp • J. Marx
Institute of Energy and Climate Research – Systems Analysis and Technology Evaluation (IEK-STE), Forschungszentrum Jülich GmbH, D-52425 Jülich, Germany
e-mail: a.schreiber@fz-juelich.de; p.zapp@fz-juelich.de; j.marx@fz-juelich.de

W. Kuckshinrichs, J.-F. Hake (eds.), *Carbon Capture, Storage and Use*,
DOI 10.1007/978-3-319-11943-4_5

environmental effects of carbon capture. At the same time, LCA makes it possible to define targets for the development of new technologies by using benchmarks from existing systems.

5.2 Life Cycle Assessment as an Ecological Evaluation Method

The central function of an LCA is the compilation and assessment of relevant input and output flows of product systems and their potential environmental impacts throughout the life cycle of the product (DIN EN ISO 14040 2006). The environmental impacts of a product or a technology are identified and evaluated from the 'cradle to the grave': from the extraction of raw materials, production, and use, up to final disposal of the product.

The International Organization for Standardization (ISO) has standardized the procedure for LCAs in the international standards ISO 14040 and 14044 (DIN EN ISO 14040 2006; DIN EN ISO 14044 2006).

An LCA is accordingly divided into four phases (Fig. 5.1): (1) Goal & Scope Definition, (2) Inventory Analysis, (3) Impact Assessment, and (4) Interpretation.

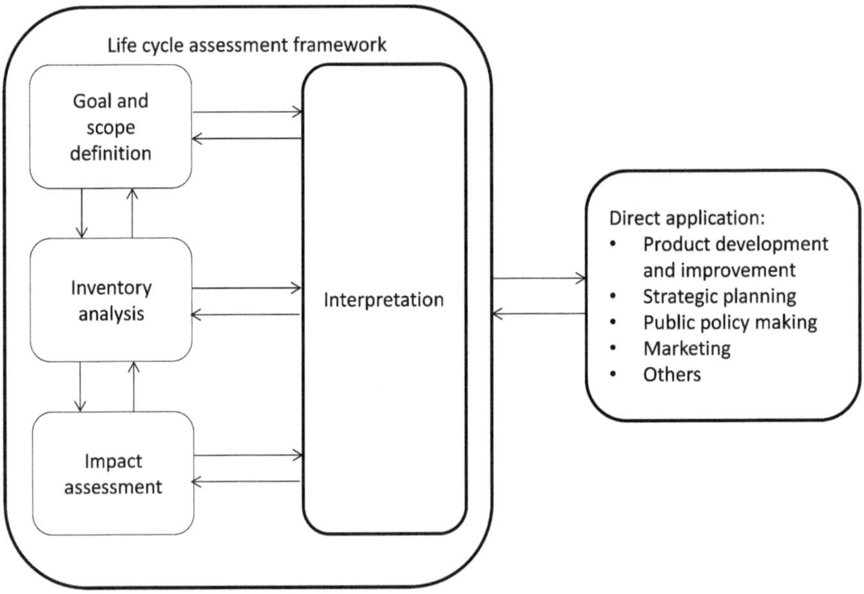

Fig. 5.1 Phases of life cycle assessment according to ISO (Source: DIN EN ISO 14040 2006)

The first phase defines the goals of the analysis. To this end, the object to be investigated is precisely defined; the spatial and temporal boundaries of the system laid down, the processes to be considered are described, and the environmental effects to be included in the interpretation are defined.

In the following life cycle inventory (LCI), all of this defined system's material and energy flows and emissions are recorded in detail. These aspects are then related to the function of the product or technology. This function must therefore be unambiguously defined and measurable. It is only permissible to compare systems on the basis of the same function. The subsequent life cycle impact assessment (LCIA) serves to assign these flows to potential environmental impacts and to summarize them in impact categories.

In Table 5.1 typical effects in an LCA are mapped and briefly explained, giving examples of relevant emissions and the related indicators describing the effect.

The results are evaluated in the final phase (interpretation) with respect to the issue under investigation. Ultimately, this may lead to recommendations for action on the part of various actors. In this way, the environmental effects associated with a product can be systematically identified and used to compare or optimize products.

5.3 Environmental Effects of Conventional Capture Technologies

In recent years, a number of LCA studies have been conducted on different methods of carbon capture as well as on transportation and storage. Due to the large number of systems considered and the divergent goals of the studies, it is difficult to derive a consistent picture of the environmental effects of CCS. Nevertheless, general aspects and trends can be identified, as well as parameters that considerably influence the results (IEA/GHG 2010).

The analysis is based on 17 studies performed since the year 2000 (Table 5.2). Nine studies focus on Europe, three consider the situation in the USA or Japan, and four take a global approach. Although all the studies relate their evaluation to the functional unit of 1 kWh of electricity generated and some studies even consider the same CCS technologies, nevertheless due to the different technical data and approaches it is necessary to make a detailed consideration of the studies.

5.3.1 Technology-Related Differences

5.3.1.1 Capture Technologies

Studies are already available for all three capture routes and almost always include post-combustion processes. Nine studies additionally focus on pre-combustion processes, whereby the oxyfuel variant is considered least frequently (five studies).

Table 5.1 Brief description of impact categories

Impact category	Abbreviation	Brief description	Examples of relevant emissions	Impact indicators
Global warming potential	GWP	Impacts of anthropogenic emissions on reflection into the atmosphere contributing to a temperature rise, greenhouse effect	CO_2, N_2O, CH_4, SF_6, $CHCl_3$, CF_4, CFCs, HCFCs, CH_3Br	kg CO_2 equivalent
Acidification potential	AP	Emissions of acidifying substances, acid rain	SOx, NOx, HCl, HF, NH_3, HNO_3, H_2SO_4, H_3O_4P	kg SO_2 equivalent
Eutrophication potential	EP	Excessive nutrient supply, over-fertilization	PO_4^{3-}, N2, NO_2, HNO_3, NH_3, H_3PO_4, COD	kg PO_4^{3-} equivalent
Photochemical ozone creation potential	POCP	Formation of reactive chemical compounds by the effect of sunlight on primary pollutants, summer smog	PAH, NOx, NMVOC, CH_4	kg C_2H_4 equivalent
Ozone depletion potential	ODP	Amount of depleted atmospheric ozone, ozone hole	CFCs, HCFC, CH_3Br	kg CFC equivalent
Human toxicity potential	HTP	Impact of toxic substances on human health	PM10, SO_2, NOx, CH_4, CH_2O, C_6H_6 PAH, As, Cd, dioxin	kg 1,4-DCB equivalent
Freshwater aquatic ecotoxicity potential	FAETP	Effect of toxic substances on freshwater	Heavy metals	kg 1,4-DCB equivalent
Terrestrial ecotoxicity potential	TETP	Effect of toxic substances on soils	Heavy metals	kg 1,4-DCB equivalent
Abiotic resource depletion potential	ADP	Abiotic resource depletion to obtain raw materials		kg Sb equivalent
Cumulative energy demand	CED	Quantity of primary energy used		MJ

Source: Guinee et al. (2002)

All the studies include MEA scrubbing as a post-combustion process, whereas only two studies also consider other processes (D'Addario et al. 2003; Khoo and Tan 2006) or other scrubbing substances (Muramatsu and Iijima 2003). Although several demonstration facilities with MEA scrubbing are already in existence, to date all the studies use exclusively generic data. Results of actual measurements are not taken into consideration. Cryogenic air separation is always selected to supply oxygen for the oxyfuel route. Various scrubbing substances are considered for the

Table 5.2 CCS systems considered in the individual studies

			Capture			Fuel		
Study	Region	Time	Post-comb.	Pre-comb.	Oxy-fuel	Hard coal	Lignite	Nat. gas
D'Addario et al. (2003)	Italy	Today	×	×				×
Doctor et al. (2001)	USA	Today		×		×		
IEA/GHG (2006)	Global	Today-2050	×	×		×		×
Khoo and Tan (2006)	USA	Today	×			×		
Koornneef et al. (2008)	The Netherl.	2000/2020	×			×		
Korre et al. (2009)	Global	N.d.	×			×		
Nie et al. (2011)	Global	N.d.	×		×	×		
Lombardi (2003)	Hypothetical	N.d.	×	×		×		×
Modahl et al. (2009)	Norway	N.d.	×					×
Muramatsu and Iijima (2003)	Japan	Today	×			×		
Bauer et al. (2009)/ NEEDS Study	Europe	Today, 2025, 2050	×	×	×	×	×	×
Odeh and Cockerill (2008)	UK	2005	×	×		×		×
Pehnt and Henkel (2009)	Germany	2020	×	×	×		×	
Schreiber et al. (2009)	Germany	2020	×			×	×	
Singh et al. (2011)	Global	N.d.	×	×	×	×		×
Spath and Mann (2004)	USA	Today	×			×		×
Viebahn et al. (2007)	Germany	2020	×	×	×	×	×	×

N.d. no data

pre-combustion route. New process technologies such as carbonate looping or membranes have not yet been examined with respect to their ecological effects.

Hard coal is most frequently considered as a fuel (14 studies). Particularly in German studies, consideration is given to lignite. More recent studies and those with a wider European or global focus analyse gas-fired electricity generation.

Table 5.2 gives an overview of the capture routes and fuels considered for the various studies.

Since all the CCS technologies involve more or less advanced future technologies, the development of the efficiencies of the electricity generating techniques and the expected efficiency losses due to carbon capture have to be estimated. Careful comparison reveals that there is no uniform picture of the development neither of efficiencies nor of the expected efficiency losses. Moreover, the individual studies consider different time horizons. It is often unclear what technical assumptions, such as the state of the art in power plant engineering or emission reduction measures, the studies are based on. The expected rise in efficiencies for technology models referring to the distant future is therefore not always included. Nor do these technologies necessarily have the lowest efficiency losses.

Figure 5.2 plots the net efficiencies and the associated efficiency losses for the respective fuels in the studies, broken down according to three process routes.

For hard-coal-fired power plants, the efficiencies are reduced by post-combustion capture to between 29.6 % (retrofitting variant 2010 (Schreiber et al. 2009)) and 49 % (values for 2050 in the NEEDS Study (Bauer et al. 2009)). In the case of lignite, the difference between the lowest (26.3 %, retrofitting variant 2010 (Schreiber et al. 2009)) and the highest efficiency (49 %, values for 2050 (Bauer et al. 2009)) is even greater. In the oxyfuel process, a high proportion of the energy is required for oxygen production. The NEEDS Study with the longest time horizon (2050, (Bauer et al. 2009)) displays the highest net efficiencies. In general, the expected efficiency losses are lowest for pre-combustion and highest for post-combustion.

The carbon capture rate is another important technical parameter. In their report, Odeh & Cockerill (Odeh and Cockerill 2008) show that a reduction of 5 % in the rate leads to a rise in the global warming potential (GWP) of 11.3–25.6 %. Koornneef et al. (2008) estimate the change in the GWP as ±20 % if the carbon capture rate is varied by ±5 %.

The various capture technologies produce CO_2 with different degrees of purity and different aggregate states (see chapter on CO_2 transportation). These different properties of CO_2 influence the amount of energy required and the resulting emissions. The different CO_2 quality levels inevitably lead to different process requirements with respect to carbon capture and should be considered in the analyses. These aspects have not been included to date.

5.3.1.2 CO_2 Transportation and Storage

The effect of CO_2 transportation and storage on the overall environmental impacts is regarded as relatively low even if the proportion varies considerably (1–10 %) depending on the carbon capture rate and the fuel used. Large volumes of CO_2, for example in the case of old lignite-fired power plants or high carbon capture rates, lead to increased expenditure for transportation and storage.

Furthermore, the type of storage site (gas field, aquifer), the depth of the injections, the transportation distance, and the number of recompression steps influence the results. For example, Wildbolz (Wildbolz 2007) found that energy

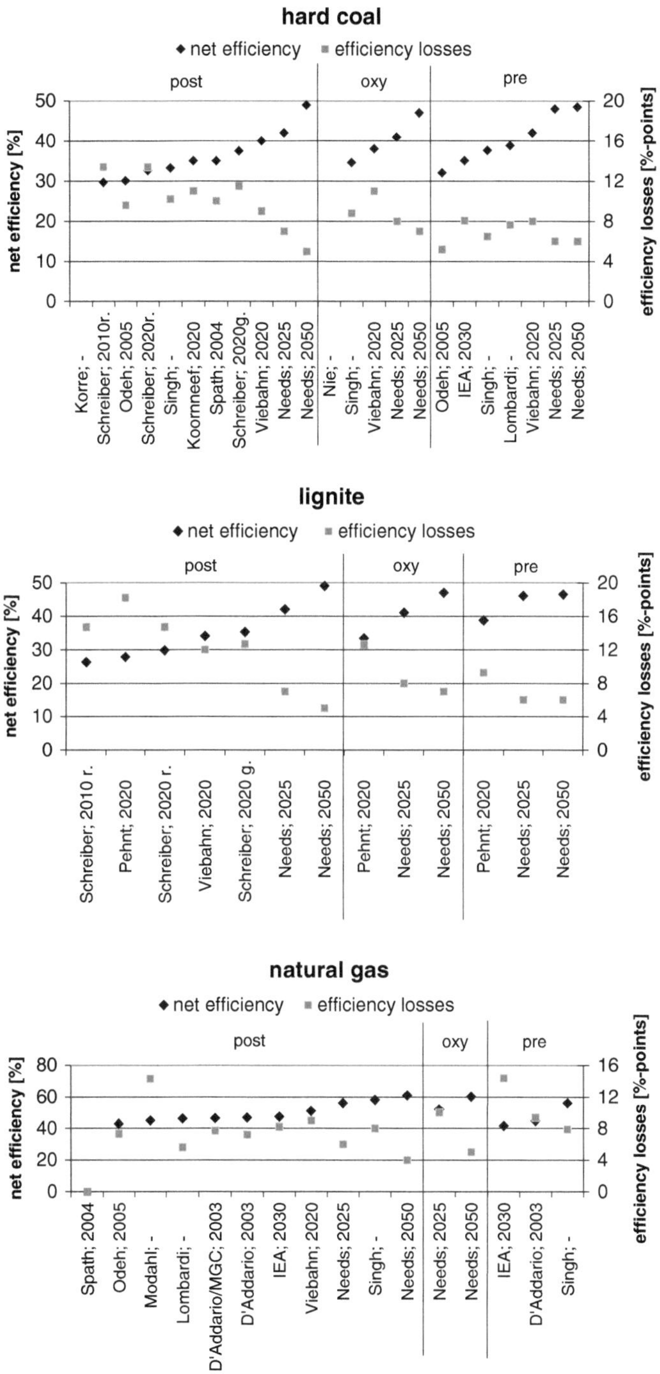

Fig. 5.2 Net efficiencies and efficiency losses due to capture (Values for Korre et al. 2009; Nie et al. 2011; Spath and Mann 2004 not available; *r* retrofit, *g* greenfield, NEEDS Study is cited as Bauer et al. 2009) (Source: Adapted from Schreiber et al. 2012

requirements for injection and storage were three times as high in an exhausted gas field at a depth of 2,500 m than in a saline aquifer at a depth of 800 m.

The length of the CO_2 pipeline has little impact on environmental effects (Odeh and Cockerill 2008; Schreiber et al. 2009; Wildbolz 2007). Spath and Mann (2004) calculate the increase in the share of total GWP arising from CO_2 transportation as ranging from 0.1 % for a pipeline 300 km in length to 1 % for 1,800 km.

CO_2 leak rates during long-term storage are only considered marginally in three of the studies (Khoo and Tan 2006; Nie et al. 2011; Viebahn et al. 2007) (a detailed consideration of safety and risks associated with storage can be found in the chapter on CO_2 storage).

5.3.1.3 Origin and Composition of Fuels

The type, origin, and composition of fuels have a very great influence on the findings of LCA studies (Odeh and Cockerill 2008; Schreiber et al. 2009). Emissions result directly from the fuel composition and from the efficiency of the combustion reaction, and may thus vary for the same technology. Without knowing all the background data it is, however, impossible to decide how great this fuel-related fraction of the findings is, and what fraction can be attributed to different technologies. The origin of the fuel determines the composition, but it has an additional environmental effect due to the resulting transportation distances. Fuel transportation causes additional emissions (see also the chapter on CO_2 transportation).

5.3.2 Differences Arising from the LCA Methodology

Although an LCA standard has been defined (DIN EN ISO 14040 2006), the flexibility in implementing a life cycle assessment is still relatively great. Depending on the objectives of the study and the choice of system under consideration, different decisions are taken on the parameters to be employed. Some decisions have a significant influence on the overall result of an LCA.

5.3.2.1 Impact Categories

In impact assessment, environmental effects are assigned to emissions from the inventory. However, for some impact categories there are different characterization models and indicators for the description. It may therefore be the case that different studies cannot be directly compared although they investigate the same impact categories. Typical impact categories addressed in a number of studies on CCS are the global warming potential (GWP), the acidification potential (AP), the eutrophication potential (EP), the photochemical ozone creation potential (POCP), the human toxicity potential (HTP), the freshwater aquatic ecotoxicity potential, and the terrestrial ecotoxicity potential (FAETP, TETP), as well as the cumulative energy demand (CED) (see also Table 5.1). After assigning and evaluating the impact categories, the impact assessment should then be complemented by normalization in order to better classify the environmental effects analysed. To this end, each effect is related to the values of a reference system, such as those of a certain region or a citizen of a country. In many studies, this analysis is not undertaken.

5.3.2.2 Time Horizon

The time horizon has a major influence on the definition of future technological parameters both for CCS systems and also for competing technologies. In almost all cases, current and future power plants and CCS facilities up to the year 2020 are taken into consideration. Only in two studies (Bauer et al. 2009; IEA/GHG 2006) are values extrapolated to the year 2030 or 2050, which leads to improved efficiency assumptions in the technology description.

The choice of time horizon is also of significance for CO_2 storage, and thus for associated leakages. In life cycle assessments, long-term emissions are not normally analysed but are rather truncated by the cut-off criterion (DIN EN ISO 14040 2006). A 'fair' comparison between current emissions and long-term emissions remains unresolved in the LCA methodology (Hellweg et al. 2003). On the basis of different concepts of future climatic conditions, such as the CO_2 buffer capacity of oceans or the biosphere, it is inadvisable to extrapolate the present negative effects of CO_2 emissions into the future. The environmental impacts of possible leakages which may only appear after 100 or more years, cannot, therefore, be quantified. In order to ensure that they are not underestimated, Khoo and Tan (2006) and Viebahn et al. (2007) consider CO_2 leakages in a sensitivity analysis.

5.3.2.3 Spatial Representation

Some steps in the process chain are particularly locally or spatially specific. One example is, once again, CO_2 storage. With the exception of transportation distances, hardly any site-specific data on energy consumption during recompression or CO_2 injection or data on the number of boreholes are found as yet in studies of CO_2 storage.

Moreover, the process chains of upstream fuel processing and electricity generation are strongly regionalized and have a considerable influence on findings (see also Sect. 5.3.1.3). Site-specific features are frequently not explicitly identified or generalized data may be used making it as a rule difficult to assess the related impacts.

The different environmental effects have different impacts. For example, the greenhouse effect and ozone depletion have a global impact, whereas eutrophication and acidification occur regionally or locally. Depending on the geographical position and nature of the ecosystem affected, the effects can be assessed quite differently. Although methodological approaches have been available for some time to integrate special regional features in LCAs (Posch et al. 2008; Seppälä et al. 2006), these methods have not yet been applied in the field of LCA/CCS. Normalization represents the first step towards incorporating regional effects. Each of the environmental effects caused by a technology can then be related to the annual totality of this effect for the whole region. Studies that perform a normalization (Koornneef et al. 2008; Modahl et al. 2009; Schreiber et al. 2009) use the same approach (CML 2001 (Guinee et al. 2002)), but apply reference data from different countries.

5.3.2.4 Upstream and Downstream Process Chains

All studies show a significant proportion of upstream and downstream process chains in the environmental impacts of the entire process chain. This proportion is even higher for CCS power plants than for conventional power plants, which is due in particular to the efficiency loss which has to be compensated by increased fuel input per kWh generated. However, the proportion of upstream and downstream process chains varies for the individual impact categories.

Using the example of the global warming potential, Fig. 5.3 shows the proportion of emissions of a power plant for the sections of the process chains during operation, fuel processing, CO_2 transportation and storage, as well as other upstream and downstream process chains (e.g. provision of chemicals or waste treatment) for various CCS technologies. The differences between the individual studies and the various fuels thus become quite apparent. The proportion of upstream and downstream chains varies between approx. 10 % for a lignite-fired pre-combustion power plant in the NEEDS Study (Bauer et al. 2009) and 90 % for a hard-coal-fired oxyfuel power plant in (Viebahn et al. 2007).

With respect to the fuels, the supply chains for lignite have the lowest proportion since the effects arising from the provision of fuel are relatively small due to the short transportation distances. Electricity generation from natural gas displays the highest proportions of upstream and downstream process chains. It should be noted that this sequence does not necessarily also apply to the absolute values. This shows that optimization measures at the power plant appear a promising option for lignite-fired facilities whereas improving fuel processing and transportation appears appropriate for plants fired by hard coal and natural gas.

Within one fuel class, emissions directly at the power plant during operation are lowest for oxyfuel power plants, followed by post-combustion and pre-combustion plants with similar contributions.

The hard coal processing chain also has a considerable influence on other impact categories (AP, EP, POCP). In particular, the ocean transportation of coal is a major cause of acidification contributing 30 % to the entire AP. The reason for this is that ocean-going cargo vessels use heavy fuel oil containing sulphur.

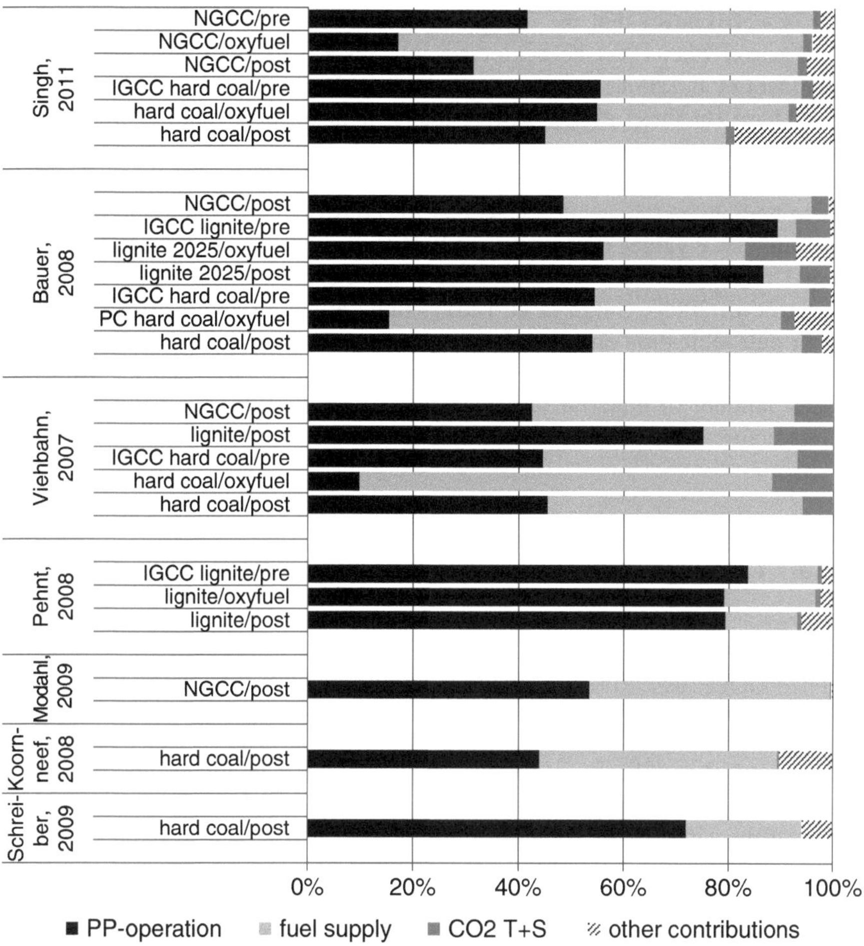

Fig. 5.3 Proportions of the respective process chains with the example of GWP (Source: Zapp et al. 2012)

The production of MEA scrubbing liquid for post-combustion processes is responsible for an extremely steep increase of the toxic values in air and water since the toxic ethylene oxide is emitted during MEA production. In contrast, the disposal of soil reclaimer waste has very little influence on the HTP (0.005 %) (Koornneef et al. 2008). Recent studies have concerned themselves with the impact of direct MEA emissions and possible degradation products during carbon capture (Knudsen et al. 2009), and have analysed these effects in detail. In future, it may therefore be possible to indicate a greater effect.

The process chains of the upstream and downstream process steps are frequently not analysed in the same detail as the main process chain. In some cases, recourse is taken to old data (e.g. MEA production) or only rough estimates are made. Upstream and downstream process chains may have a considerable influence on the overall result so that these data should be verified more thoroughly.

It is well known that the construction and dismantling phases of conventional power plants can be neglected with respect to environmental impacts. Koornneef et al. (2008) and Pehnt and Henkel (Pehnt and Henkel 2009) put the GWP share for these phases at less than 0.2 % of the overall GWP. In CCS power plants, the percentage of environmental impacts for construction and dismantling increases because the absolute power plant emissions are reduced due to carbon capture while emissions for construction remain approximately constant. GWP values range between 0.34 % for a hard-coal-fired IGCC power plant (Lombardi 2003) and 4.9 % for a lignite-fired oxyfuel power plant (Bauer et al. 2009). Singh et al. (2011) draw attention above all to the much higher eutrophication and toxicity values for CCS power plants. These values can be attributed to elevated emissions of heavy metals resulting from the greater amount of material (especially steel) required for the additional CCS infrastructure.

5.3.3 CCS Technologies and Their Environmental Impacts

In the following, the CCS technologies will be compared on the basis of capture technology and fuel type. The first diagram of the figure, (a), shows the environmental impacts of a power plant without CCS in absolute values, broken down according to fuel. The values shown are the arithmetic mean derived from existing studies and also indicate scatter by representing the minimum and maximum values. The subsequent graphics (b) then show the relative difference between a power plant with carbon capture and a reference power plant without CCS. The mean values are also shown here.

The representation of relative changes may lead to impact categories with large changes being overestimated although their contribution to the overall environmental impacts is still small. For this reason, the results of the individual environmental impacts are related to those of a certain region (normalization). Since the studies cover different regions, the global average for each environmental category is given as a reference. The reference year is 2000, for which the most recent results are available. In a comprehensive study, Sleeswijk et al. (2008) compiled reference values for a large number of impact categories from various regions (Table 5.3).

For reasons of consistency, the figures for global electricity generation from different energy sources are also given for the year 2000, assuming exclusively CCS technology (hard coal: 5,136 TWh, lignite: 749 TWh, natural gas: 2,677 TWh; (OECD/IEA 2002)). Consequently, in Figs. 5.4, 5.5 and 5.6, the emissions from the assumed electricity generation with CCS technology is given on the right-hand axis in relation to total global emissions in the year 2000 in order to illustrate the significance of the different impact categories.

Table 5.3 Normalization factors world 2000

Impact category	World 2000
Global warming potential (GWP)	4.18 E + 13 $kgCO_2$-eq
Acidification potential (AP)	2.39 E + 11 $kgCO_2$-eq
Eutrophication potential (EP)	1.58 E + 11 $kgPO_4^{3-}$-eq
Photochemical ozone creation potential (POCP)	2.90 E + 10 kgC_2H_4-eq
Human toxicity potential (HTP)	3.63 E + 12 kg1,4DCB-eq
Freshwater aquatic ecotoxicity potential (FAETP)	3.47 E + 12 kg1,4DCB-eq
Terrestrial ecotoxicity potential (TETP)	1.09 E + 12 kg1,4DCB-eq

Source: Sleeswijk et al. (2008)

5.3.3.1 Hard Coal and Lignite

The global warming potential of hard-coal-fired power plants without carbon capture varies between 765 gCO_2-eq/kWh and 1,092 gCO_2-eq/kWh, depending on the efficiency and the type of coal used (Fig. 5.4a, left). The mean value is 875.3 g CO_2-eq/kWh. The scatter of the values of the acidification potential is greater, depending on the assumed flue gas purification and coal composition. Koornneef et al. (2008), for example, assume a very high value (2.8 gSO_2-eq/kWh) for an 'old' hard coal power plant from the year 2000, whereas the lowest value is 0.4 gSO_2-eq/kWh (Korre et al. 2009). The EP, POCP, CED, and the toxic values (HTP, FAETP, TETP) do not vary greatly. The fluctuations depend directly on the efficiency selected.

As expected, the GWP of a lignite-fired power plant without a CCS facility (Fig. 5.4a, right) is somewhat higher than for the hard-coal-fired plant (887 $kgCO_2$-eq/kWh). The APs are comparatively low and vary between 0.66 gSO_2-eq/kWh and 1.6 gSO_2-eq/kWh because of the lower calorific value. The reason for this is that lignite is mined close to the power plants and therefore does not require long transportation paths, which would lead to emissions of NO_x and SO_2.

The normalization shows that hard-coal-based electricity generation without CCS contributes more than 10 % and thus represents a considerable proportion of the total global greenhouse gas emissions. Acidifying emissions from power plants contribute 2.6 %. The impacts on EP, POCP, HTP, FAETP, and TETP are even lower. Due to the considerably smaller volume of electricity generated by lignite-fired power plants worldwide (749 TWh) the share of this form of electricity production is globally much smaller than hard coal. The GWP amounts to a maximum of 1.6 % without CCS. All other categories are negligible (<1 %).

As expected, the results for CCS electricity generation based on hard coal and lignite show a significant reduction in GWP. However, this reduction is not as great as the carbon capture rates of generally more than 90 % would lead us to expect. The reason for the reduced effect is, firstly, the larger volume of CO_2 produced due

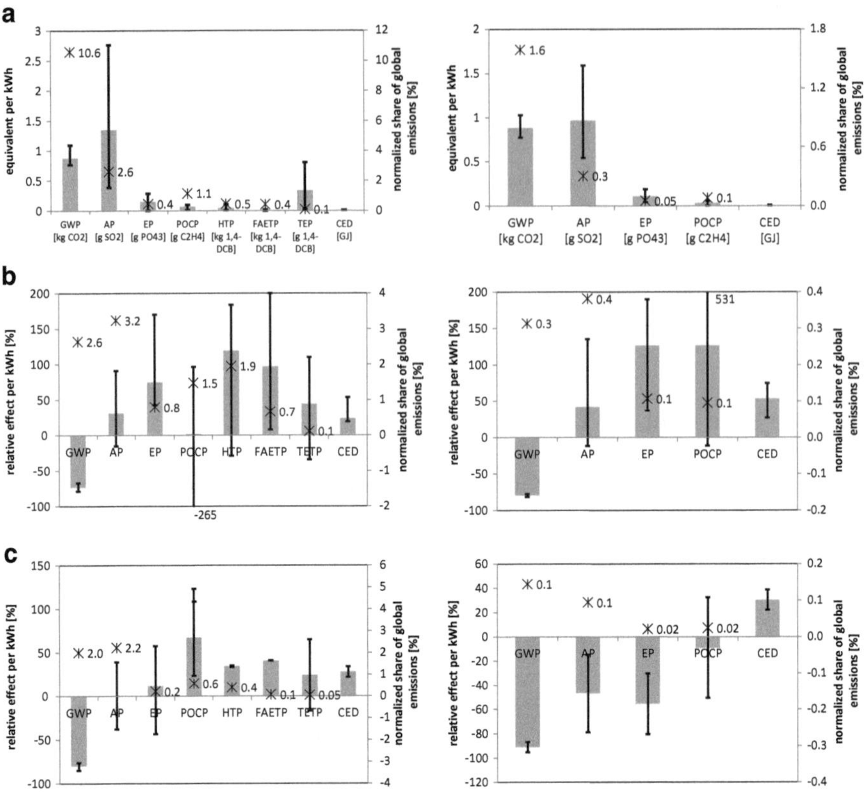

Fig. 5.4 Environmental impacts of hard-coal-fired (*left*) and lignite-fired (*right*) power plants for post-combustion/MEA scrubbing or oxyfuel technology and normalized environmental impacts relative to global emissions in 2000 (**a**) without carbon capture (**b**) relative environmental impacts for post-combustion/MEA (**c**) relative environmental impacts for oxyfuel (Source: Zapp et al. 2012)

to the poorer efficiency. However, the increased methane and CO_2 emissions during mining and transportation of coal have a much greater influence. All the other impact categories (AP, EP, POCP, CED, HTP, FAETP, and TETP) are increased. The acidification and eutrophication potentials increase even though improved flue gas purification for NO_x and especially for SO_2 can be assumed for MEA separation. Apart from minor quantities during MEA fabrication, the release of SO_2 and NO_x during the transportation of additional coal supplies is mainly responsible for this rise. This highlights the regional shift in environmental effects. Whereas local emissions are being reduced at power plant sites, they now increasingly occur in other parts of the world (e.g. South America, Australia or in the oceans) due to upstream process chains.

POCP emissions are also increased by fuel extraction and transportation. Ethylene oxide emissions during MEA fabrication additionally contribute to this effect.

The release of these emissions is, moreover, a major reason for the considerable increase of toxic effects in water and air. Another aspect is pollution with heavy metals and phosphate emissions during the dumping of hazardous waste and coal ash. For commercial-scale facilities, implementing proper waste management could be of assistance.

Nevertheless, the normalization shows that the steep rise in toxic effects does not lead to electricity generation causing any significant increase in the global values for FAETP and TETP (0.7 and 0.1 %, respectively). A perceptible rise can, however, be seen for the human toxicity values. In this case, the use of CCS technology would considerably increase the percentage of global emissions from 0.5 to 1.9 %. The percentages of the global AP for hard coal power plants with CCS increase from 2.6 % (without CCS) to 3.2 %, for EP and POCP to 0.8, and 1.5 %, respectively.

Except for the reduction in greenhouse gases, LCA analyses of hard-coal-fired oxyfuel power plants do not display consistent results. The values for the AP are between −38 % and 40 %, for the EP between −43 % and 58 %, and for the POCP between 23 and 123 %. Half of the studies expect a reduction for the AP and EP, and the other half a rise in emissions. The human toxicity and ecotoxicity effects in water are regarded as more uniform (approx. 34 % and 41 %, respectively), whereas opinions on terrestrial ecotoxicity diverge again (−17 % and −65 %). The reason for the different expectations could be differences in the efficiency or efficiency loss considered or an improvement in flue gas purification. Another reason could be the different assignment of SO_2 and NO_x during compression. Whereas some studies assign these components to the compressed CO_2 stream, others assume that they are emitted. The studies do not provide a uniform interpretation. This is why it has not yet been possible to draw any general conclusions for the environmental assessment of oxyfuel power plants.

In the case of lignite-fired oxyfuel power plants, all other impact categories in addition to the global warming potential are reduced, even if the extent of the effect is considered to be different in the two studies (AP −15 % to −80 %; EP −30 % to −80 %). Only the CED increases due to the efficiency losses. Once again, the reduction in the AP and EP in contrast to hard-coal-fired power plants can be attributed to the short transportation paths. If the use of lignite-fired power plants remains low, then the environmental impacts of the AP, EP and POCP will also be negligible in comparison to overall global emissions. Even if the POCP were to increase by more than 500 %, this would have no perceptible effect on the global POCP value.

The absolute numbers for the IGCC systems without CCS are generally lower than for conventional coal-fired power plants (Fig. 5.5). IGCCs with pre-combustion show the same trend as post-combustion, but on a lower level. Although various CO_2-scrubbing substances are taken into consideration, the increase of the AP, EP, POCP, and CED is usually smaller than 40 % and is very low compared with global EP and POCP values. Their share of the global AP increases from 1.5 to 1.8 %, and the GWP drops from 11.5 to 2.9 %.

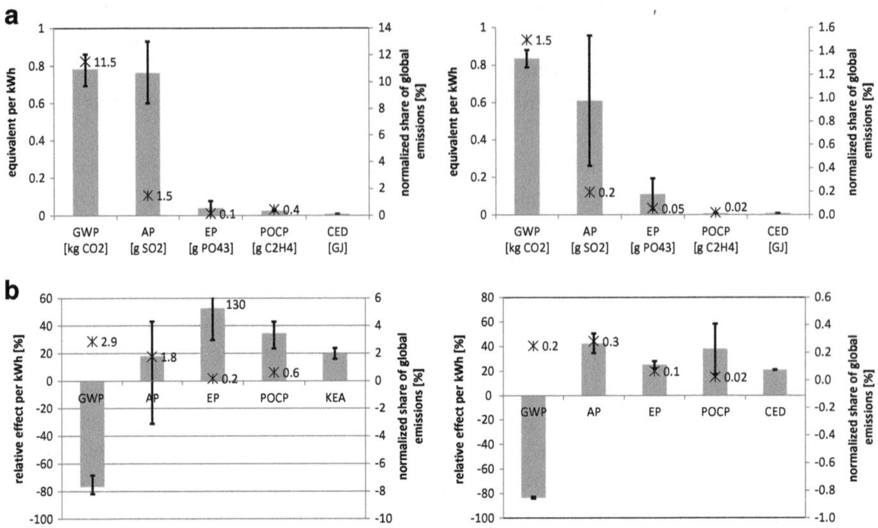

Fig. 5.5 Environmental impacts of IGCC power plants (hard coal on the *left*, lignite on the *right*) and normalized environmental impacts relative to global emissions in 2000 (**a**) IGCC without carbon capture (**b**) relative environmental impacts for IGCC with pre-combustion (Source: Zapp et al. 2012)

5.3.3.2 Natural Gas

To date, post-combustion facilities have been investigated mainly for natural-gas-fired plants. Singh et al. (2011) are the only researchers to have analysed a natural-gas-fired oxyfuel plant. The higher efficiencies of natural-gas-fired power plants without capture mean that the GWP is much lower than for coal-fired plants. Since, moreover, the proportion of natural-gas-fired electricity generation is only about half that of coal-fired plants, the GWP makes a much smaller contribution (2.7 %) to the global GWP (Fig. 5.6). With CCS, the share of GWP is reduced to 1.0 %. The analyses for the reduction of the GWP are also fairly consistent with respect to natural gas. However, the fluctuations are greater for the remaining impact categories. The increase ranges between 15 and 50 % for the AP, EP, POCP, and CED. All the normalized values are much lower than 1 %, even if there is a considerable increase in the absolute values for some environmental impacts.

In summary, it can be said that for all fuels and capture technologies, only the GWP is a robust environmental effect that can be used to compare life cycle assessments. More well-documented LCAs are necessary in order to obtain reliable statements on the other environmental impacts of CCS systems since existing studies are difficult to compare.

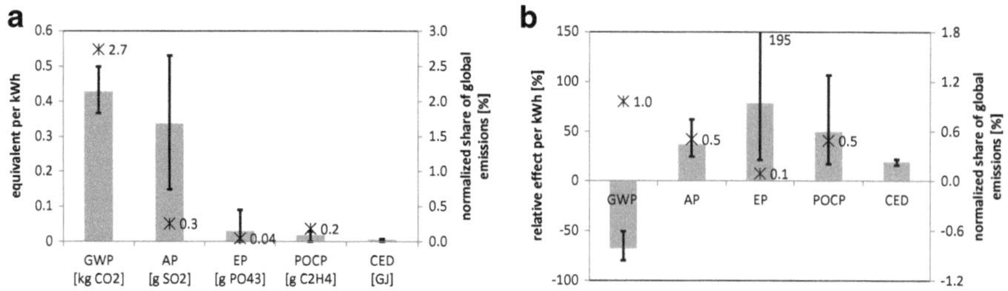

Fig. 5.6 Environmental impacts of an NGCC and normalized environmental impacts relative to global emissions in 2000 (**a**) NGCC without carbon capture (**b**) relative environmental impacts for NGCC with post-combustion (Source: Zapp et al. 2012)

5.4 Environmental Aspects of Future Capture Technologies of the 2nd Generation

Technologies are under investigation that involve lower efficiency losses or that operate without chemicals in order to reduce the environmental aspects as discussed above. A promising option provided by these second-generation technologies is membrane technology. High-temperature membranes made of $Ba_{0.5}Sr_{0.5}Co_{0.8}Fe_{0.2}O_{3-\delta}$ (BSCF) perovskite material are being developed for the oxyfuel process (Czyperek et al. 2010). They can thus be directly compared with cryogenic air separation units (see chapter on carbon capture). The ecological effects of membrane application remain to be investigated.

As discussed above, uniform parameters must be selected in order to compare technologies. Since no uniform opinion has yet been established on environmental impacts for oxyfuel systems and the selection of parameters is decisive, in the following the environmental impacts of three hard-coal-fired power plants will be uniformly modelled and then compared. The investigations are based on a supercritical (SC) German hard coal power plant (state-of-the-art) without carbon capture. Its technical parameters serve as the basis for modelling the power plants. This will be compared with the ecological effects of a conventional power plant with cryogenic air separation unit (C ASU) and the effects of a possible future power plant with membrane-based air separation unit.

5.4.1 *Power Plant Concepts*

The same basic assumptions were applied for all power plant concepts. System components were defined as unit processes for the energy and material flows. Figure 5.7 shows the defined process chain structures and system boundaries. In the coarse structure, the process chain consists of 'coal conditioning', 'electricity generation', 'desulfurization', 'dedusting' and 'denitrification'.

Table 5.4 Parameters for the three power plant types

Parameter	RPP SC	C ASU	HTM ASU
Net capacity [MW]	555.3	440.9	479.5
Net efficiency LHV [%]	45.9	36.4	39.6
Efficiency losses [% points]	–	9.5	6.3
Membrane area [thousand m^2]	–	–	254
Carbon capture rate [%]	–	90.2	90.1

5.4.1.1 Reference Power Plant (RPP SC) Without CCS

A consortium of industry and science defined basic parameters (VGB-PowerTech 2004) in order to describe an SC power plant for 2004. This theoretical SC power plant achieved a net efficiency of almost 46 % (Table 5.4). Flue gas purification achieved emission values below the present limits set by German legislation (13. BImSchV 20.07.2004), (SOx: 150 mg/Nm^3, NOx: 100 mg/Nm^3). The analysis considers South African hard coal of the 'Kleinkopje' type transported to Germany by sea.

5.4.1.2 Oxyfuel Concept

In addition to the above-mentioned processes the oxyfuel process also involves 'the provision of oxygen by cryogenic or membrane-based air separation', 'H_2O condensation, CO_2 compression and liquefaction', 'CO_2 transportation by pipeline (400 km onshore)' and 'CO_2 storage in a saline aquifer (800 m)' (see also Fig. 5.7). The necessary oxygen is either supplied by a cryogenic air separation unit or by membrane-based O_2 separation.

5.4.1.3 Cryogenic Air Separation (C ASU)

Cryogenic air separation is a state-of-the-art technology for producing oxygen on a commercial scale for power plants. While the energy consumption of an ASU increases with higher demands on O_2 purity, higher O_2 purity reduces energy requirements for CO_2 compression and purification. Several studies (Castle 2002; Dillon et al. 2005) specify an O_2 purity of 95 % (with 3.8 % Ar and 1.2 % N_2 (Hausen and Linde 1985)) as the optimum value. This requires approx. 200 kWh/tO_2 of electrical energy. The net efficiency of a cryogenic oxyfuel power plant is 36.4 %, thus yielding an efficiency loss of 9.5 %-points.

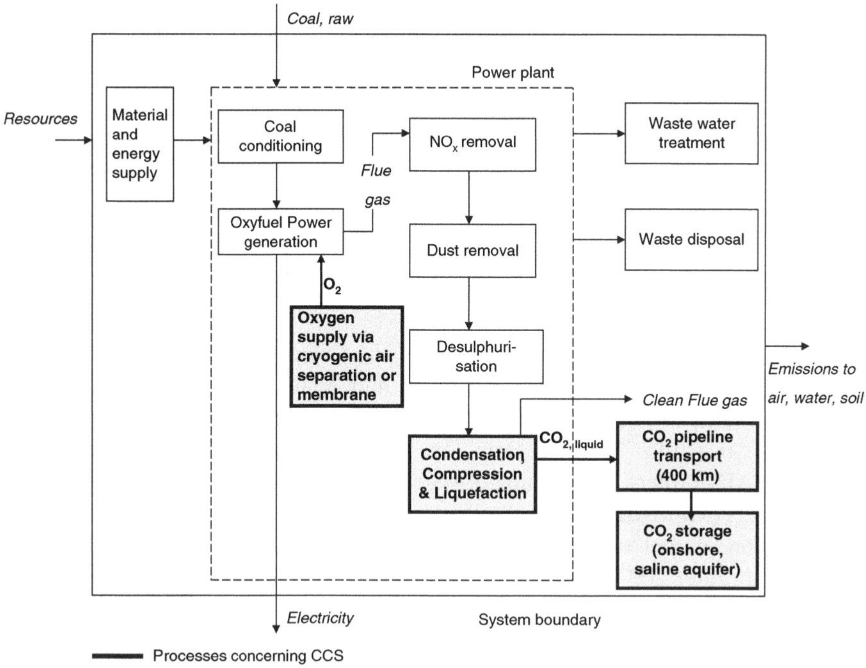

Fig. 5.7 LCA system boundaries for a conventional power plant without CCS and oxyfuel power plants with cryogenic or membrane-based ASU

5.4.1.4 Membrane-Based Air Separation (HTM ASU)

In recent years, great efforts have been made to develop and integrate novel gas separation membranes to improve the efficiencies of oxyfuel power plants (Czyperek et al. 2010). One possibility of separating oxygen is by using ceramic high-temperature membranes (HTMs). A popular membrane material is $Ba_{0.5}Sr_{0.5}Co_{0.8}Fe_{0.2}O_{3-\delta}$ (BSCF), because its permeation rate is very high due to its high ionic and electronic conductivity (Engels et al. 2010). However, this material experiences problems with chemical stability in the hot CO_2 and SO_2 environments in the flue gas employed in the circuit as a sweep gas on the permeate side in the four-end concept (Castillo 2011a). This is why a three-end concept is currently preferred in which the membrane is not in direct contact with the flue gas and the driving force is provided by an additional vacuum pump. Thermodynamic modelling of a highly integrated membrane module identifies a net efficiency of 39.6 % (Castillo 2011b). With an assumed membrane thickness of 0.6 mm and an average oxygen permeation rate of 1.75 ml/(min*cm^2), the membrane area required for this power plant is 254,000 m^2. In an initial analysis, a lifetime of 40 years is assumed for the membrane and the associated modules.

Additional upstream processes, such as the provision of fuel or chemicals, and downstream processes, such as waste treatment, were included for all three power plant types. Consideration was also given to power plant design and the cryogenic or membrane-based air separation unit. Table 5.4 lists the parameters selected for the three different power plant types.

5.4.2 *Results of the Life Cycle Inventory*

The inputs and outputs of the three power plant concepts are recorded and then added up throughout the entire life cycle and related to a kWh of electricity produced. The power plant parameters are derived from detailed thermodynamic modelling (Castillo 2011b). Data on the membrane are taken from the MEM-BRAIN project, which aimed to develop membranes for gas separation in power plants (MEM-BRAIN 2011). Upstream and downstream process chains such as provision of fuel or waste treatment are taken from the ecoinvent database (EcoinventCenter 2007). LCA information on CO_2 transportation and storage are derived from the study by Wildbolz (2007).

Table 5.5 shows the most important inputs and outputs for all power plants per kWh electricity. The CO_2 emissions drop considerably when CCS technology is used. Due to the different process management, the NO_x emissions are also reduced in the oxyfuel process accompanied by a reduction in the need for ammonia for denitrification. The efficiency losses mean that more coal is required, which also leads to a rise in certain emissions such as CO, particulates or ash. A decrease in SO_2 emissions is achieved by using increased amounts of limestone for desulfurization. This is, however, associated with an increased production of gypsum.

Table 5.5 Specific key inputs and outputs (including upstream and downstream processes)

	RPP SC	C ASU	HTM ASU
Input (g/kWh)			
Coal	401	508	467
Limestone	5.2	7.1	6.6
Ammonia	0.9	0.15	0.2
Output (g/kWh)			
CO_2	803	165	151
CO_2 separated	–	886.7	811.9
SO_2	0.74	0.47	0.43
NO_x	0.91	0.86	0.79
CO	0.16	0.21	0.19
Particulates	1.17	1.49	1.37
Ash (boiler, ESP)	44	55.7	51.2
Gypsum	8.9	12.3	11.3

5.4.3 *Results of the Impact Assessment*

For the environmental impacts, the three power plant types are distinguished in Fig. 5.8 according to effects at the power plant, supply of coal, other upstream or downstream chains, and power plant construction. For the oxyfuel power plants, the proportion of CO_2 transportation and storage is shown as well as the air separation unit, including membrane fabrication for the HTM ASU variant.

The results show that the contributions to the greenhouse effect and to acidification drop for both the oxyfuel power plants in comparison to the reference power plant without CCS. However, the eutrophication potential increases by 25 % for cryogenic air separation and by 15 % for membrane-based air separation unit.

The human toxicity potential also rises from 132 gDCB-eq/kWh for the reference plant to 166 gDCB-eq/kWh for the membrane-based oxyfuel power plant. The POCP remains almost unchanged (reference 0.103 gC_2H_4-eq/kWh, membrane 0.094 gC_2H_4-eq/kWh). Due to the lower efficiency losses, the membrane-based oxyfuel power plant is always superior to the cryogenic type.

For almost all environmental effects, emissions are considerably reduced at oxyfuel power plants even in comparison to the reference power plant. On the other hand, the effects due to fuel provision rise so dramatically for the EP that the total value of the oxyfuel power plants is higher. The reasons for this are once again dumping processes during mining, higher SO_2 and NO_x emissions during coal transportation, and the downstream disposal of boiler ash.

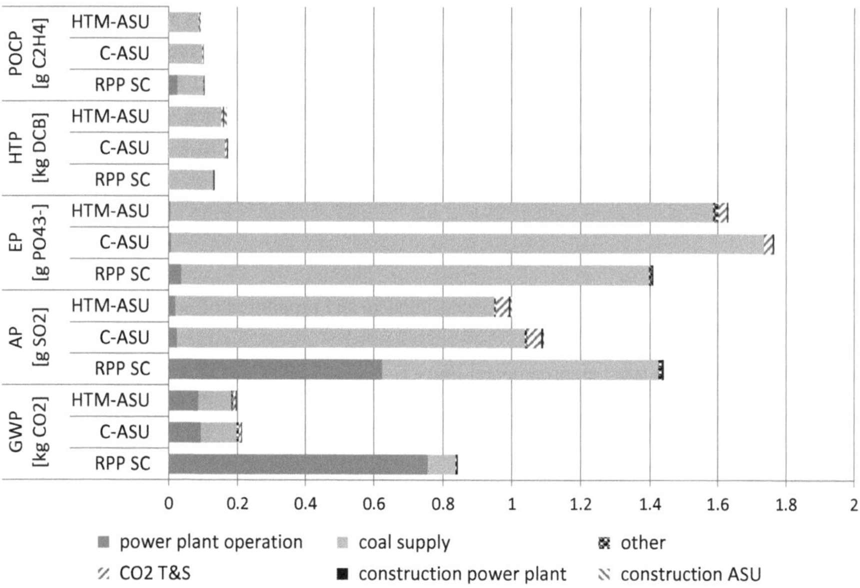

Fig. 5.8 Selected environmental impacts of electricity generation for three power plant types (including upstream and downstream processes) broken down into sections of the process chain

The proportions of CO_2 transportation and storage and also the construction of the power plants and air separation units are low for all impact categories. With respect to the HTP, these sections of the process chain contribute the greatest share (10 %) for the HTM ASU plant. The HTP is mainly determined by the fabrication of the membrane module, especially the housing with large quantities of steel. The sintering processes during membrane fabrication have a great influence on POCP results since a considerable amount of NMVOC and other organic emissions (binders for membrane fabrication) are released here.

The results for membrane-based air separation are therefore greatly dependent on the assumed lifetime of the module and the membranes. An analysis reveals a sharp rise in the HTP when the lifetime of the membrane module is reduced from an optimistic 40 years to a currently realistic assumption of 5 years (from 166 gDCB-eq/kWh for 40 years to 230 gDCB-eq/kWh for 5 years). According to this assumption, the performance of the membrane-based power plant would be inferior to that of the cryogenic air-separation plant. The same is true of the POCP.

In order to assess the significance of the different environmental impacts, each environmental effect is compared with the total environmental effects worldwide in 2010. It is assumed that all electricity production based on hard coal (5,790 TWh (OECD/IEA 2012)) is exclusively generated by one of the three power plant types (Table 5.6). The world was selected as the reference system since many environmental effects, as described above, result from upstream chains which are not located in Germany. Only in the case of GWP up to 50 % of the total emissions is generated at power plants in Germany (132 TWh), which is why the German GWP is additionally shown in Fig. 5.6.

Table 5.6 shows that hard-coal-fired electricity generation is responsible for a considerable proportion of the global and German greenhouse effect, and that this is significantly reduced by CCS. The acidification share is reduced for oxyfuel power plants, even if not so strikingly as for the GWP. The introduction of CCS increases the share of eutrophication. The HTP similarly increases slightly while the POCP continues to be insignificant.

Table 5.6 Total of the German environmental impacts and proportion of normalized effects for the year 2010

	Total global environmental impacts [kg equivalents]	RPP SC [%]	C ASU [%]	HTM ASU [%]
GWP	4.18 E13	11.6	2.9	2.7
GWP (Germany)	1.14 E12	9.7	2.5	2.3
AP	2.39 E11	3.5	2.6	2.4
EP	1.58 E11	5.2	6.4	6.0
HTP	3.63 E13	2.1	2.7	2.6
POCP	3.68 E10	1.6	1.6	1.5

Source: GaBi (2012)

5.4.4 Interpretation

Oxyfuel technology can make an important contribution to more environmentally friendly electricity generation. However, due to the efficiency losses the amount of hard coal required increases by 25 % for cryogenic air separation and by 15 % for membrane-based air separation. The introduction of oxyfuel technologies considerably reduces the proportion of GWP caused by electricity generation from more than 10 % to about 2.5 %. Membrane-based air separation processes display a higher potential since the efficiency losses are lower. However, there is still a need for more research. For example, long-term tests performed under operating conditions are required to demonstrate that the modelled performance data are realistic. Above all, the lifetime and the size of the housing of the membrane module are decisive.

5.5 Summary and Conclusions

The introduction of carbon capture technologies to reduce globally increasing greenhouse gas emissions often leads to an amplification of other environmental effects. A comprehensive consideration is therefore required. The rise in other environmental effects is usually triggered by the decline in net efficiency and the related additional requirements for fuels and chemicals (e.g. scrubbing substances) as well as increased volumes of waste. A detailed analysis of the reasons shows that optimizing the reduction of emissions at the power plant is in itself not sufficient to prevent this rise. In particular, the provision of fuel often involves a high proportion of different environmental impacts. If scrubbing substances are additionally used, the human and ecotoxicity potential rises mainly because of emissions during production. Heavy metal emissions during the dumping of hazardous waste and ash also contribute to increased toxicity. A comparison of the studies shows that the processes of the upstream and downstream chains are often not represented in the same detail as the electricity generation and subsequent carbon capture processes. These processes should therefore be investigated in more detail.

A consideration of the entire life cycle also shows that there may be local or regional environmental effects upstream. While acidification and eutrophication are reduced at the power plant site, they increase in regions where the fuel is extracted and along transportation paths.

The LCA helps to identify various environmental effects and to specify their causes. Furthermore, a comparison with the overall effects of a region helps to relate different effects to each other. The desired effect of reducing greenhouse gas emissions is obvious. However, more detailed consideration must be given to emissions promoting acidification and human toxicity, especially for post-combustion plants.

The paramount goal must therefore be to reduce efficiency losses. New technological developments such as membranes are promising. Nevertheless, further analyses including a detailed description of the system boundaries and the parameters are required in order to provide robust information on the respective environmental impacts of the different technologies.

References

Bauer C, Heck T, Dones R, Mayer-Spohn O, Blesl M (2009) Final report on technical data, costs, and life cycle inventories of advanced fossil power generation systems. NEEDS (New Energy Externalities Developments for Sustainability). Paul Scherrer Institut (PSI) and Institut für Energiewirtschaft und Rationelle Energieanwendung, Univ. Stuttgart (IER)

BImSchV 20.07.2004 Verordnung über Großfeuerungs- und Gasturbinenanlagen – 13. BImSchV

Castillo R (2011a) Thermodynamic analysis of a hard coal oxyfuel power plant with high temperature three-end membrane for air separation. Appl Energy 88:1480–1493

Castillo R (2011b) Thermodynamic evaluation of membrane based oxyfuel power plants with 700 °C technology. Energy Procedia 4:2026–2034

Castle WF (2002) Air separation and liquefaction: recent developments and prospects for the beginning of the new millennium. Int J Refrigeration 25:158–172

Czyperek M, Zapp P, Bouwmeester HJM, Modigell M, Ebert K, Voigt I, Meulenberg WA, Singheiser L, Stöver D (2010) Gas separation membranes for zero-emission fossil power plants: MEM-BRAIN. J Membr Sci 359:149–159

D'Addario E, Clerici G, Musicanti M, Pulvirenti G, Serenellini S, Valdiserri MG (2003) Environmental analysis of different options of CO_2 capture in power generation from natural gas. In: Gale J, Kaya Y (eds) Greenhouse gas control technologies. Elsevier Science Ltd, Amsterdam

Dillon DJ, White V, Allam RJ, Wall RA, Gibbings J (2005) Oxy combustion processes for CO_2 capture from power plants. IEA Greenhouse Gas R&D Programme and Mitsui Babcock

DIN EN ISO 14040 (2006) Umweltmanagement – Ökobilanz – Grundsätze und Rahmenbedingungen. Beuth Verlag GmbH, Berlin

DIN EN ISO 14044 (2006) Umweltmanagement – Ökobilanz – Anforderungen und Anleitung. Beuth Verlag GmbH, Berlin

Doctor RD, Molburg JC, Brockmeier NF, Manfredo L, Gorokhov V, Ramezan M, Stiegel GJ (2001) Life-cycle analysis of a shell gasification-based multi-product system with CO_2 recovery. Argonne National Laboratory, Argonne

EcoinventCenter (2007) Ecoinvent data v2.0, 2007 [Online]. Available: www.ecoinvent.ch

Engels S, Beggel F, Modigell M, Stadler H (2010) Simulation of a membrane unit for oxyfuel power plants under consideration of realistic BSCF membrane properties. J Membr Sci 359:93–101

GaBi (2012): GaBi Product Sustainability software Version 4.4, PE Product Engineering GmbH, Leinfelden-Echterdingen

Guinee J, Gorree M, Heijungs R, Huppes G, Kleijn R, Koning AD, Oers LV, Wegener Sleeswijk A, Suh S, Haes HAUD, Bruijn HD, Duin RV, Huijbregts MAJ (2002) Handbook on life cycle assessment: operational guide to the ISO standards. Kluwer Academic Publishers, Dordrecht

Hausen H, Linde H (1985) Tieftemperaturtechnik: Erzeugung sehr tiefer Temperaturen, Gasverflüssigung u. Zerlegung von Gasgemischen. Springer, Berlin

Hellweg S, Hofstetter TB, Hungerbühler K (2003) Discounting and the environment – should current impacts be weighted differently than impacts harming future generations? Int J Life Cycle Assess 8:8–18

IEA/GHG (2006) Environmental impacts of solvent scrubbing of CO_2. IEA Greenhouse R&D Programme

IEA/GHG (2010) Environmental evaluation of CCS using life cycle assessment (LCA). IEA Greenhouse Gas R&D Programme

Khoo HH, Tan RBH (2006) Life cycle investigation of CO_2 recovery and sequestration. Environ Sci Technol 40:4016–4024

Knudsen S, Karl M, Randall S (2009) Summary report: amine emissions to air during carbon capture; Phase I: CO_2 and amines screening study for effects to the environment. Norwegian Institute for Air Research (NILU OR 8/2009), Kjeller, Norway

Koornneef J, VAN Keulen T, Faaij A, Turkenburg W (2008) Life cycle assessment of a pulverized coal power plant with post-combustion capture, transport and storage of CO_2. Int J Greenh Gas Control 2:448–467

Korre A, Nie Z, Durucan S (2009) Life cycle modelling of fossil fuel power generation with post combustion CO_2 capture. Energy Procedia 1:3771–3778

Lombardi L (2003) Life cycle assessment comparison of technical solutions for CO_2 emissions reduction in power generation. Energy Convers Manag 44:93–108

MEM-BRAIN (2011) Available: www.helmholtz.de/pakt_fuer_forschung_und_innovation/impuls_und_vernetzungsfonds/helmholtz_allianzen/mem_brain

Modahl IS, Nyland CA, Raadal HL, Karstad O, Torp TA, Hagemann R (2009) LCA as an ecodesign tool for production of electricity, including carbon capture and storage – a study of a gas power plant case with post-combustion CO_2 capture at Tjeldbergodden. Joint actions on climate change, Aalborg, Denmark

Muramatsu E, Iijima M (2003) Life cycle assessment for CO_2 capture technology from exhaust gas of coal power plant. In: Gale J, Kaya Y (eds) Greenhouse gas control technologies – 6th international conference. Pergamon, Oxford

Nie Z, Korre A, Durucan S (2011) Life cycle modelling and comparative assessment of the environmental impacts of oxy-fuel and post-combustion CO_2 capture, transport and injection processes. Energy Procedia 4:2510–2517

Odeh NA, Cockerill TT (2008) Life cycle GHG assessment of fossil fuel power plants with carbon capture and storage. Energy Policy 36:367–380

OECD/IEA (2002) World energy outlook 2002, International Energy Agency, Paris

OECD/IEA (2012) Electricity information 2012, International Energy Agency, Paris

Pehnt M, Henkel J (2009) Life cycle assessment of carbon dioxide capture and storage from lignite power plants. Int J Greenh Gas Control 3:49–66

Posch M, Seppälä J, Hettelingh J-P, Johansson M, Margni M, Jolliet O (2008) The role of atmospheric dispersion models and ecosystem sensitivity in the determination of characterisation factors for acidifying and eutrophying emissions in LCIA. Int J Life Cycle Assess 13:477–486

Schreiber A, Zapp P, Kuckshinrichs W (2009) Environmental assessment of German electricity generation from coal-fired power plants with amine-based carbon capture. Int J Life Cycle Assess 14:547–559

Schreiber A, Zapp P, Marx J (2012) Meta-analysis of life cycle assessment studies on electricity generation with carbon capture and storage. J Ind Ecol 16:S155–S168

Seppälä J, Posch M, Johansson M, Hettelingh J-P (2006) Country-dependent characterisation factors for acidification and terrestrial eutrophication based on accumulated exceedance as an impact category indicator. Int J Life Cycle Assess 11:403–416

Singh B, Strømman AH, Hertwich EG (2011) Comparative life cycle environmental assessment of CCS technologies. Int J Greenh Gas Control 5:911–921

Sleeswijk AW, VAN Oers LFCM, Guinée JB, Struijs J, Huijbregts MAJ (2008) Normalisation in product life cycle assessment: an LCA of the global and European economic systems in the year 2000. Sci Total Environ 390:227–240

Spath P, Mann M (2004) Biomass power and conventional fossil systems with and without CO_2 sequestration comparing the energy balance, greenhouse gas emissions and economics. National Renewable Energy Laboratory, Golden

VGB-PowerTech (2004) VGB-PowerTech e.V. (ed) Konzeptstudie Referenzkraftwerk Nordrhein-Westfallen (RKW NRW), Essen

Viebahn P, Nitsch J, Fischedick M, Esken A, Pastowski A, Schuwer DEA (2007) RECCS Strukturell-ökonomisch-ökologischer Vergleich regenerativer Energietechnologien (RE) mit Carbon Capture and Storage (CCS), Wuppertal-Institut für Klima, Umwelt, Energy GmbH, Wuppertal

Wildbolz C (2007) Life cycle assessment of selected technologies for CO_2 transport and sequestration. Diplom Diplomarbeit, Eidgenössische Technische Hochschule Zürich (ETH), Institut für Umwelt-Engineering (IfU), Diplomarbeit

Zapp P, Schreiber A, Marx J, Haines M, Hake J-F, Gale J (2012) Overall environmental impacts of CCS technologies – a life cycle approach. Int J Greenh Gas Control 8:12–21

Chapter 6
Safe Operation of Geological CO_2 Storage Using the Example of the Pilot Site in Ketzin

Michael Kühn, Axel Liebscher, Sonja Martens, Fabian Möller, Thomas Kempka, and Martin Streibel

Abstract Reservoir rocks with the potential for storing CO_2 are mainly sandstones. In them, four trapping mechanisms facilitate permanent and safe storage: (i) structural trapping below an impermeable caprock, (ii) immobilization via capillary forces in the pore space, (iii) dissolution of CO_2 in the formation water, and (iv) mineral trapping via carbonization. Because leaks can occur monitoring of CO_2 storage sites is essential. However, the technological risks appear to be manageable. This is emphasized by the experience from the first continental European field laboratory in Ketzin, Germany. The results show that: (i) the geological storage of CO_2 is safe and reliable, and poses no danger to humans or the environment, (ii) a well-thought-out combination of different geochemical and geophysical monitoring methods can detect small amounts of CO_2 and image its spatial distribution, (iii) the interactions between fluid and rock induced by CO_2 injection at the pilot site in Ketzin have no significant impacts and do not influence the integrity of the reservoir or the caprock, and (iv) numerical simulations can depict the temporal and spatial behaviour of injected CO_2. In addition, results from studies at Ketzin provide basic and transferable knowledge which is of value for a new integrated concept of CO_2 mitigation and utilization in combination with the power-to-gas concept based on a closed carbon cycle approach.

Keywords CO_2 storage • Leakage • Trapping mechanisms • Ketzin pilot site • Field experiment • Monitoring • Modelling • Power-to-gas-to-power concept

6.1 Introduction and Motivation

Carbon dioxide (CO_2) storage research at the GFZ German Research Centre for Geosciences focuses on whether the long-term and safe storage of CO_2 is possible in geological formations and whether this could help to mitigate greenhouse gas emissions into the atmosphere. Technologies for monitoring and predicting CO_2

M. Kühn (✉) • A. Liebscher • S. Martens • F. Möller • T. Kempka • M. Streibel
GFZ German Research Centre for Geosciences, Telegrafenberg, D-14473 Potsdam, Germany
e-mail: michael.kuehn@gfz-potsdam.de; axel.liebscher@gfz-potsdam.de; sonja.martens@gfz-potsdam.de; fabian.moeller@gfz-potsdam.de; thomas.kempka@gfz-potsdam.de; martin.streibel@gfz-potsdam.de

W. Kuckshinrichs, J.-F. Hake (eds.), *Carbon Capture, Storage and Use*,
DOI 10.1007/978-3-319-11943-4_6

storage in porous rocks in the deeper subsurface are being tested and new technologies developed. Near the town of Ketzin/Havel in Brandenburg, the first continental European field laboratory for on-shore CO_2 storage was set up as a pilot site in 2004, and active and continuous injection has been in operation from June 2008 until August 2013.

6.2 Processes of Retaining CO_2 in Porous Reservoir Rocks

Suitable reservoir rocks are predominantly porous sedimentary rocks in the subsurface. The most important rocks for geological CO_2 storage are sandstones with sufficient porosity and permeability, allowing the CO_2 to be injected efficiently into these formations. CO_2 is injected into the reservoir via wells with the aid of pumps that ensure injection pressure high enough to overcome the flow resistance in the rock, which depends on permeability and other rock properties but also on the flow resistance of the displaced formation fluid in case of saline aquifers.

Different physical and chemical processes ensure that the injected CO_2 is retained in the reservoir rocks (Fig. 6.1). The relative importance and contribution of these different processes on the overall reservoir′s retention potential vary over a logarithmic time scale (IPCC 2005). On the shortest time scale of years, during injection and directly afterwards, the injected CO_2 migrates upwards because it is less dense than the formation fluid initially contained in the geological formation. The CO_2 accumulates and is physically concentrated below the impermeable caprock, which is usually clay or salt rock (Fig. 6.1).

Within decades, parts of the CO_2 are retained by capillary forces (Figs. 6.1 and 6.2) if the pore necks have such a small diameter that the CO_2 can no longer migrate

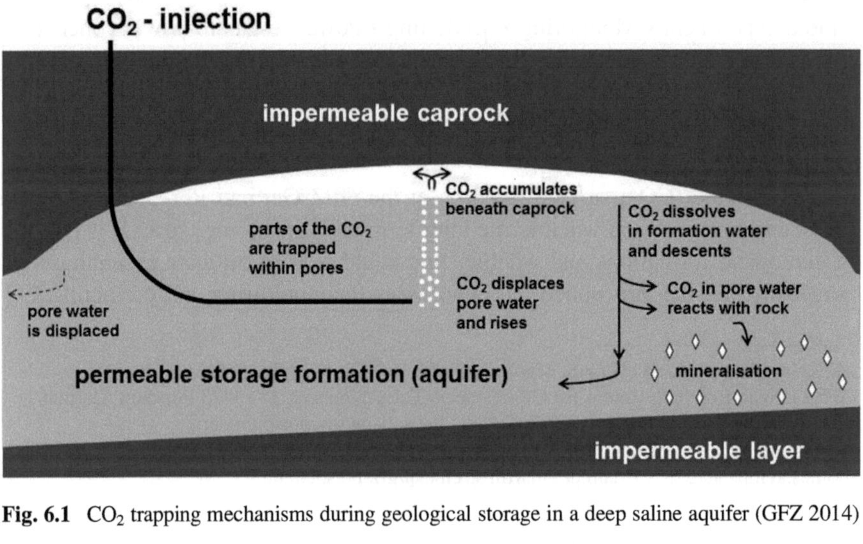

Fig. 6.1 CO_2 trapping mechanisms during geological storage in a deep saline aquifer (GFZ 2014)

upwards despite the density difference compared to the ambient formation fluid. The gas can then only be displaced by other fluids if they flow into the storage formation under elevated pressure.

Over a period of centuries, the major fraction of the CO_2 dissolves in the formation fluid, and carbonic acid is formed. The binding of CO_2 to the water remains stable as long as the pressure on the solution does not decrease and/or the temperature does not rise. This CO_2-enriched water has a slightly higher density than the original formation fluid and tends to migrate downwards due to gravity (Fig. 6.1).

In the long term, on a time scale of some thousand years, the process of mineralization binds fractions of the carbon dioxide in the form of carbonates. Carbonization is the chemical neutralization reaction between the earth alkalines of the rock and the carbonic acid. Thus, mineralization of the CO_2 leads to permanent trapping in the rock in the form of calcite, dolomite or siderite for example.

Overall, the four trapping mechanisms in the storage formation facilitate permanent and safe storage. Only the fraction of CO_2 that exists as a free gas phase is driven upwards by buoyancy forces and could escape from the storage complex. The increasing effect of CO_2 trapping over time via the four trapping mechanisms continuously reduces the fraction of the free gas phase in the storage formation (Fig. 6.2), which has been verified, for example, by studies of natural CO_2 reservoirs. These studies show that around 18 % of the CO_2 mineralizes over a long period of time, and that the major fraction of the CO_2 is found dissolved in the formation water (Gilfillan et al. 2009).

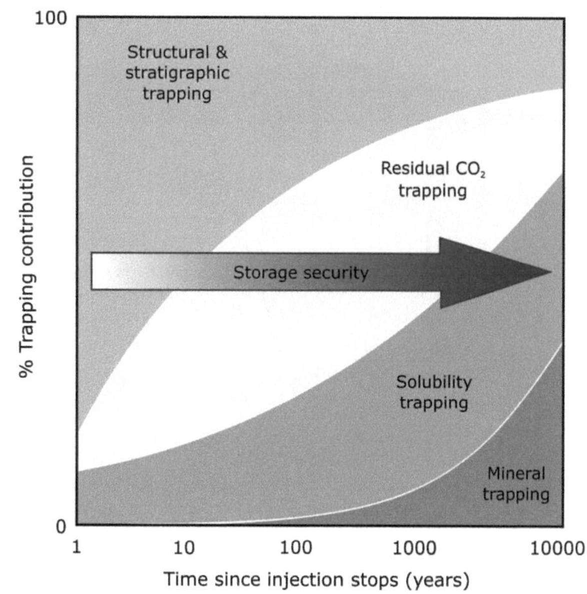

Fig. 6.2 Trapping mechanisms increase the safety of geological CO_2 storage over time (After IPCC 2005. GFZ 2014)

6.3 Potential Leakage from CO_2 Storage

Figure 6.3 shows the schematic principle of geological CO_2 storage, as well as potential risks associated with the technology. The CO_2 is injected into the storage formation underneath an impermeable caprock (Fig. 6.1). A multibarrier system above the storage complex, as shown in Fig. 6.3, comprises alternating layers of potential reservoir rock and caprock. The largest risk of leakage is location specific but in most cases probably posed by existing wells. Both active and abandoned wells are potential migration pathways because firstly they provide a direct connection between the surface of the Earth and the storage formation, and secondly they contain man-made materials (piping and cementing), which can corrode in the long term.

Considering a multibarrier system containing a large number of passive wells (old wells and observation wells), statistical methods can be applied together with an analytical solution in order to estimate the potential leakage rate (Nordbotten et al. 2004). The more barrier units that are present, the smaller the cumulative amount of CO_2 that can migrate along corroded wells towards the Earth's surface because major fractions of the leaking CO_2 could be taken up by the formations lying above the storage complex if corrosion connected those layers as well. Calculations show that 10 % of the total amount stored would leak from the storage complex if one caprock layer was present, 1 % if two caprock layers were present (Fig. 6.3), and 0.1 % if three were present, etc. (Nordbotten et al. 2004). If there is only one well 100 m away from the injection well and if this is leaking, then between 0.1 and 0.2 % of the total amount stored in the aquifer is expected above

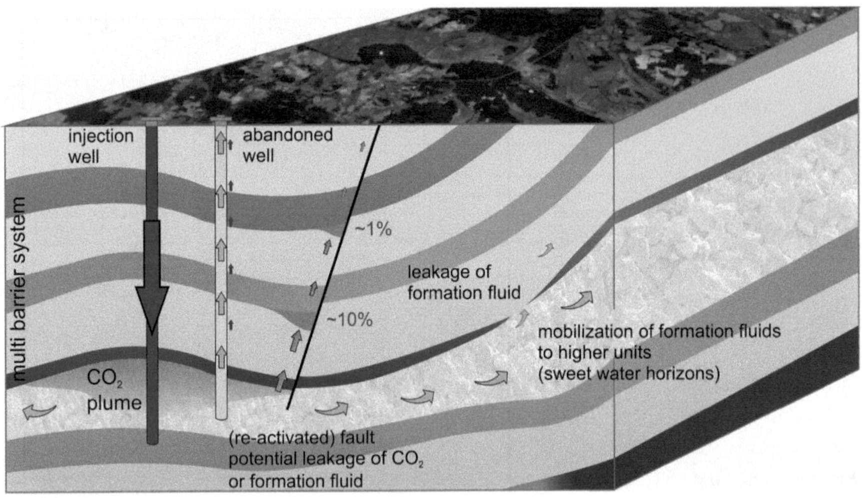

Fig. 6.3 Schematic principle of the geological storage of CO_2 with a multibarrier system. Potential anthropogenic and natural leakage pathways for CO_2 and the mobilization of saline water are also shown (GFZ 2014)

this level (Ebigbo et al. 2007). The potential leakage rate essentially depends on the number of (leaking) wells in the vicinity of the storage facility. During the injection of CO_2 into a depleted oil field in America (West Pearl Queen, New Mexico), the real leakage rate near the injection well was measured using tracers. These studies estimated a leakage rate of about 0.0085 % per year of the total CO_2 injected (Wells et al. 2007). In this particular experiment, approx. 2,000 tCO_2 were injected over a period of 2 months. In addition to direct measurements of CO_2 at the surface and in the wells, seismic monitoring can also be used in order to detect CO_2 migration and potential leakage at an early stage (Bohnhoff et al. 2010) and facilitate countermeasures. As discussed, leaking wells can occur and need to be observed, which makes monitoring of CO_2 storage sites essential.

In addition to wells, potential natural leakage pathways exist. These are flow pathways along fractures and faults (Fig. 6.3). They may be present in the reservoir rock and caprock, as well as in the overlying rock layers, and are more complex than wells because they comprise non-uniform surfaces with variable permeability. Geological faults can be impermeable to fluids, but as natural CO_2 seeps have shown, they can also be permeable to gases. One of the largest measured degassings at a natural CO_2 source was in Italy, where an emission flux of 2,000 tCO_2 per day was measured across an area of approx. 0.5 km^2 (Chiodini et al. 2010). The degassing system in this case, however, is located in a mountainous region, and is thus characterized by a very different geological structure than potential CO_2 storage sites. This region in Italy is tectonically highly active as evidenced by several earthquakes and thus not a prime target for CO_2 storage. Many fracture and fault systems in sedimentary basins (e.g. North German basin), in contrast, are impermeable to fluids, as verified by the discovery of natural gas and crude oil fields millions of years old. If these systems were not predominantly impermeable, then no hydrocarbon deposits would be found in them.

In case of CO_2 storage in saline aquifers, another effect that must be investigated in detail for every site is the displacement of saline water (Fig. 6.3). The CO_2 injected into the reservoir rock displaces the saline water initially present in the pore space. It must be ensured that the saline water does not flow along migration pathways into the drinking water reservoirs of shallow aquifers, and contaminate the drinking water with so much salt that it would be unusable for the drinking water supply. This necessitates a comprehensive and thorough exploration of each potential site. Such data acquisition then provides the basis for precautionary safety analyses. A theoretical study showed that for undisturbed systems (no fractures or faults or other direct fluid flow conduits), there is no risk of saline water migrating upward into drinking water reservoirs despite pressure increases at a distance of tens of kilometres away from the injection zone of the geological storage formation (Birkholzer et al. 2009). Similar investigations of disturbed systems show a very low tendency towards possible salinization of near-surface aquifers even for highly permeable fault systems (Tillner et al. 2013).

A general evaluation of the suitability of sites in advance cannot be performed effectively. It is therefore essential that a comprehensive exploratory investigation will be performed. This investigation is site-specific, and is the most important and indeed the only possible method of evaluating risks in detail and deciding whether a site is suitable for geological CO_2 storage in general. To answer the question if long-term and safe geological CO_2 storage can be realized requires CO_2 injection accompanied by extensive monitoring.

6.4 Safety of the Geological Storage of CO_2

The most important question that has to be answered about CO_2 storage technology is how safe it is for humans and the environment. In order to predict the safety of geological CO_2 storage at the present time, two 'analogues' are taken into consideration. These are (i) naturally occurring CO_2 reservoirs and/or sources and (ii) other sites where gas has been stored in porous rocks (Kühn 2011).

The underground geological storage of CO_2 is not a human invention but rather a natural phenomenon. Numerous naturally occurring CO_2 reservoirs have existed throughout the world for thousands or even millions of years, e.g. the Rhön region in Germany, the south of France, and Italy. These naturally occurring reservoirs prove that rocks can store CO_2 for geologically long periods of time and that caprocks can efficiently retain the gas. If future CO_2 storage facilities are chosen accordingly and investigated using state-of-the-art methods, then long-term storage of most of the CO_2 will be possible. Naturally occurring reserves of CO_2 help us to understand the conditions under which gas can be retained. In contrast, natural CO_2 seeps show what consequences are to be expected when CO_2 escapes. With their study of 286 natural CO_2 sources in Italy, Roberts et al. demonstrate that an appropriate risk management in advance of industrial CO_2 storage can minimize the health risk associated with the unintended leakage of CO_2 (Roberts et al. 2011). The calculated risk of death in regions surrounding the natural seeps in Italy is 10^{-8} per year, which is much lower than other everyday risks to human life which are accepted by society. For example, the probability of being killed in a car crash is $1.8 \cdot 10^{-4}$ or of being struck by lightning in America $2.3 \cdot 10^{-5}$ (Roberts et al. 2011).

On the shorter technological time scale, experience in gas storage technology provides insights for the geological storage of CO_2, such as on the diffusion behaviour of gases in porous rocks. The technology of storing large quantities of natural gas in deep underground rock formations to compensate for seasonal fluctuations in demand has proven its worth over decades in many places in the world. The storage volume of the 23 porous gas storage reservoirs in Germany is around 12.5 billion Nm^3 natural gas (Sedlacek 2009). An example of successful and safe natural gas storage is the underground storage facility in Berlin, which has been in operation since 1992 at a depth of 800 m. This storage facility stretches underneath a protected natural area as well as directly underneath residential areas, sports grounds, and recreation areas.

These two analogues provide important information for the geological storage of CO_2 and also demonstrate that it is possible to control the process technically and to operate it safely. They also confirm that large amounts of CO_2 can be stored for long periods of time in reservoir rocks. Despite this, possible leakage pathways and the risks associated with the technology must be identified in order to ensure that geological CO_2 storage will not pose any danger to humans or the environment.

Even if CO_2 would escape to the Earth's surface in spite of all of the safety measures, the dangers associated with it are relatively small compared to other gases (e.g. natural gas) because CO_2 is non-toxic and neither combustible nor explosive. Depending on the framework conditions such as flow rate, topography, wind speed, and wind direction, fugitive CO_2 mixes quickly with the ambient air and is diluted to a harmless level. However, it should be noted that continuously inhaling high concentrations of CO_2 (TLV = threshold limit value = 5,000 ppm) poses a health hazard for humans.

An overview of findings from natural CO_2 sources and observed leakages from gas storage sites helps us to estimate the hazard potential (Lewicki et al. 2007):

- Carbon dioxide can accumulate in primary and secondary reservoir rocks underneath impermeable caprock layers but it can also permeate these layers under certain conditions and escape at the Earth's surface.
- Many natural releases are directly connected to an event (e.g. earthquakes).
- Permeable fracture and fault systems can act as migration pathways for CO_2, allowing the gas to escape at the Earth's surface.
- Wells with construction defects represent the main leakage pathways for CO_2.
- The way in which CO_2 is released via leakage pathways and the amount released is always a site-specific phenomenon.
- The hazard potential for humans is mostly small as the population affected has usually been informed and monitoring systems have been installed. Different naturally occurring events, however, have also led to fatalities, e.g. in residential blocks due to elevated concentrations of CO_2 (Lewicki et al. 2007; Chiodini et al. 2010).
- Changes in groundwater quality associated with the release of CO_2 have also been observed, although the respective limits for drinking water were not exceeded in most cases.

As is the case for every technology, there is also a technological risk associated with the geological storage of CO_2. However, this appears to be manageable (Roberts et al. 2011), particularly when modern monitoring systems are used.

6.5 Monitoring of CO_2 Storage

In assessing the surveillance of CO_2 injected into a geological formation for storage, it must be noted that this is a new technology still in the technical and scientific demonstration phase. Empirical values are currently only available for a few pilot sites. Furthermore, most of these sites are purely scientific projects with comparatively low amounts of injected CO_2. Findings from CO_2 storage projects on a commercial scale come from the Sleipner project (Norway), where more than 1 $MtCO_2$ per year has been injected since 1996, the In Salah project (Algeria), where around 0.7 $MtCO_2$ per year has been injected since 2004, the Snohvit project (Norway), where about 0.5 $MtCO_2$ per year is injected, and the Weyburn Midale project (Canada), which has injected around 2.8 $MtCO_2$ per year since 2000 within the framework of enhanced oil recovery (EOR). In addition to these projects, recourse can be taken to experience with gas storage technology, as already mentioned. However, the differences between storing natural gas and CO_2 in terms of chemical and physical properties must be taken into account here. The volumes and objectives are also different. While natural gas storage aims at the best possible recoverability, i.e. a high fraction of working gas, CO_2 storage aims at the best possible trapping of the injected CO_2 in the reservoir. This leads to different requirements on geological characteristics in the respective reservoirs. Experience can therefore only be transferred to a limited extent.

A range of direct and indirect methods can be used to monitor the injected CO_2. Most of these methods are based on established geophysical and geochemical techniques, which may have to be modified depending on the requirements of CO_2 storage. Each storage site has very specific monitoring requirements for the injected CO_2, which must be precisely defined before start of storage operation within a needs and risks assessment. This definition requires the creation of a comprehensive geological, hydraulic, and geomechanical model of the storage reservoir. Based on the requirements, a monitoring concept must then be developed under surveillance of the responsible authorities in order to combine different monitoring methods in the most appropriate manner.

Based on current knowledge, such a monitoring concept, which is tailored to the individual storage reservoir, will make it possible to reliably monitor and control the injected CO_2. It should be noted that the described methods allow a qualitative description of CO_2 distribution, but that the amount of CO_2 is difficult to quantify.

A key point for predicting the behaviour of the injected CO_2, and for the related risk analysis, is the use of numerical simulations. Modelling and storage site monitoring are iterative processes, and both the modelling results and monitoring concepts can be adjusted accordingly during the storage process and validated. Further, results from monitoring will guide the injection operation in order to allow for a safe storage procedure. While monitoring methods can only reflect the actual situation, numerical modelling can be used to predict long-term behaviour of the storage formation. These predictions can then be used to develop an adaptive monitoring system and to optimize the operation of the storage facility.

In terms of optimal delineation of the storage reservoir and the related determination of the spatial distribution of potential irregularities during storage operation, the spreading of the injected CO_2 itself must be differentiated from spreading of the pressure increase in the reservoir caused by the injection process. The latter can cover a significantly larger spatial area than the CO_2 itself. While the distribution of CO_2 is decisive for the possible leakage of CO_2 from the storage facility, the spatial distribution of the pressure increase is decisive for displaced saline formation water potentially migrating upwards. Furthermore, it is important to distinguish between spatial distribution or spreading at reservoir depth and spatial spreading of CO_2 at the surface. Whereas CO_2 or displaced saline formation water migrates upwards, horizontal migration can occur in overlying rock units, which may extend far beyond the spreading at the depth of the reservoir. Therefore, the underground distribution of the CO_2 and the pressure increase cannot be directly transferred to the corresponding distribution at the surface. The spatial distribution of the CO_2 and of the pressure increase at depths of the reservoir can be determined using existing monitoring and simulation methods. The potential spatial spread at the surface resulting from this spatial spread at reservoir level must be individually determined for each storage site as part of a risk analysis.

Not only must the monitoring concept and the spatial distribution and/or limitation be specifically designed or defined for each storage site, but a risk analysis is also only possible on a case-by-case basis. Each storage site is characterized by very specific geological conditions, which often differ from each other considerably. Due to these different geological settings, no generalizable criteria for risk assessments can be defined or applied to all storage sites. Exclusion criteria for a storage site include caprock that has not formed fully, as well as migration pathways, and thus hydraulic connections between the storage formation and the aquifers found above the caprock.

6.6 Experience from the Pilot Site in Ketzin

The underground geological storage of CO_2 is being studied near the town of Ketzin/Havel (Brandenburg) around 40 km west of Berlin (Martens et al. 2012). The geological target horizons for CO_2 storage at the pilot site in Ketzin are porous sandstone layers at a depth of 630–650 m (Förster et al. 2006; Norden et al. 2010) (Fig. 6.4).

Above the sandstone storage formation are layers of clay, which act as a seal and are more than 165 m thick. From the 1960s until 2000, the Ketzin site was initially used to store town gas and then natural gas in a shallower sandstone formation at a depth of around 280 m. For this reason, the site is well explored. Based on existing knowledge and additional exploratory investigations, in 2007 three new wells were drilled for the geological storage of CO_2 with final depths of up to around 800 m each. One of these wells (Ktzi 201) is used to monitor and inject the CO_2, while the other two (Ktzi 200 and Ktzi 202) are used to monitor the injection and distribution of the CO_2 (Prevedel et al. 2009). In summer 2011, another well (P300) was drilled at the pilot site to a depth of 446 m in order to observe the geochemical and hydrogeological conditions in the first aquifer above the storage formation. The final well (Ktzi 203) was drilled in 2012 into the reservoir especially to retrieve rock samples which were in contact with CO_2 for 4 years (Fig. 6.4).

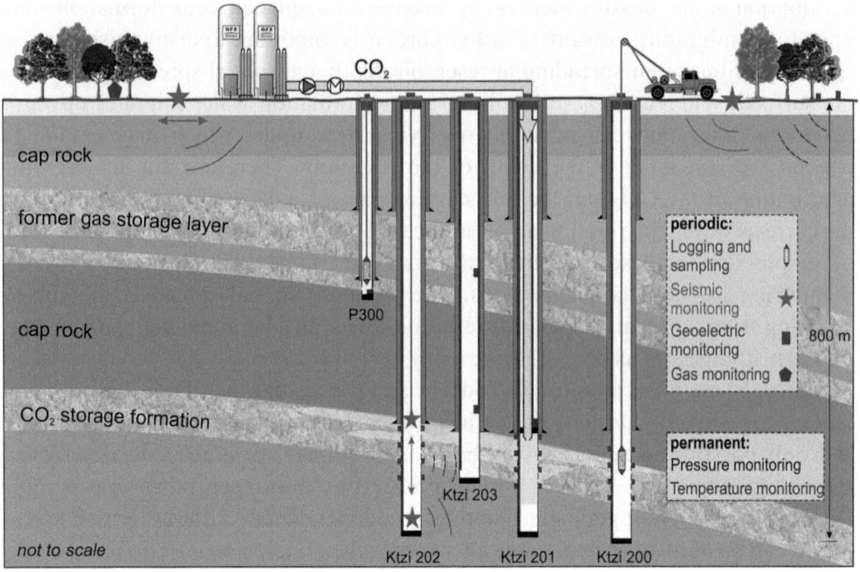

Fig. 6.4 Schematic profile cross-section of the pilot site in Ketzin with all five wells and an overview of monitoring methods (GFZ 2014)

6.6.1 Storage of CO_2 Is Safe and Reliable

From June 2008 until August 2013, mainly food-grade CO_2 has been fed into the underground formation via the injection well. Over the entire period, the injection has been safe and reliable. The injection conditions at the injection wellhead are approx. 62 bar and 35 °C. The respective injection regime is determined in accordance with the scientific tasks and requirements. Injection began on 30 June 2008, and 67,271 tCO_2 had been injected by August 2013.

The injection led to an initial pressure rise in the reservoir from originally 62 bar to 76 bar. Continuous pressure measurements show that pressure conditions in the reservoir have remained stable between 72 bar and 76 bar since spring 2009 (Möller et al. 2012). The allowed maximal reservoir pressure of 85 bar defined in the storage permission, which was approved by the Brandenburg state agency for mining, geology and raw materials (Landesamt für Bergbau, Geologie und Rohstoffe, LBGR), was neither reached nor exceeded at any time during injection. Overall, the measurements verify a stable and reliable storage operation (Liebscher et al. 2012, 2013).

6.6.2 Combination of Geochemical and Geophysical Monitoring Methods for Detecting Small Amounts of CO_2

At the pilot site in Ketzin, the primary objective is to develop, test and apply geophysical and geochemical monitoring methods. These will provide general information on monitoring of CO_2 storage reservoirs, and thus facilitate the monitoring of the spatial distribution of CO_2 injected underground. In this context, the most comprehensive monitoring programme in the world is in place at the pilot site in Ketzin (Giese et al. 2009; Fig. 6.4). It comprises permanent monitoring methods, such as pressure and temperature measurements (Möller et al. 2012), as well as periodic measurements, such as surface measurements of CO_2 flows in the upper soil layers (Zimmer et al. 2011), borehole measurements (Henninges et al. 2011), deep fluid sampling (Morozova et al. 2011), geoelectric (Kiessling et al. 2010; Labitzke et al. 2012; Schmidt-Hattenberger et al. 2011), and active and passive seismic monitoring (Bergmann et al. 2011; Kazemeini et al. 2009; Lüth et al. 2011; Yordkayhun et al. 2009a, b).

The results of geoelectric and seismic monitoring, in particular, show that even very small amounts of CO_2 can be determined indirectly in the subsurface with sufficient precision. Geoelectric methods reliably detect approx. 5,000 tCO_2, and seismic methods image the spatial distribution of CO_2 for an injected volume of around 22,000 tCO_2. The results of both methods also show good agreement with each other.

6.6.3 Fluid Rock Interactions Do Not Impact the Storage Integrity

Sandstone samples from the Ketzin storage formation were treated in the laboratory with CO_2 and saline water with near in situ conditions (55 bar and 40 °C). For comparison, samples were studied with and without CO_2 in contact with saline water. Overall, the dissolution of calcium-rich plagioclase, K-feldspar, and anhydrite was observed, while albite appears to be stable (Fischer et al. 2011). The petrophysical properties of the sandstone samples also show changes with a slightly increased porosity (Zemke et al. 2010). The observed chemical reactions occurred on such a small scale that the integrity of reservoir and caprocks is not affected.

6.6.4 Numerical Simulations Depict the Temporal and Spatial Behaviour of Injected CO_2

Static and dynamic modelling complement the monitoring methods at the Ketzin site and provide support to the operational management by delivering predictions.

Dynamic modelling is the only method for predicting the long-term behaviour of a storage site based on the known hydraulic, thermal, chemical, and mechanical processes (Bergmann et al. 2010; Kempka et al. 2010; Lengler et al. 2010).

Based on new findings obtained during injection operation so far, the underlying geological model was and will be continuously further developed and adapted (Liebscher et al. 2012; Martens et al. 2012). Numerical simulations performed to date on the basis of this geological model reveal good agreement between simulation results and monitoring measurements. It can therefore be assumed that site-specific predictions derived for further spreading of the CO_2 after stop of injection are reliable (Fig. 6.5).

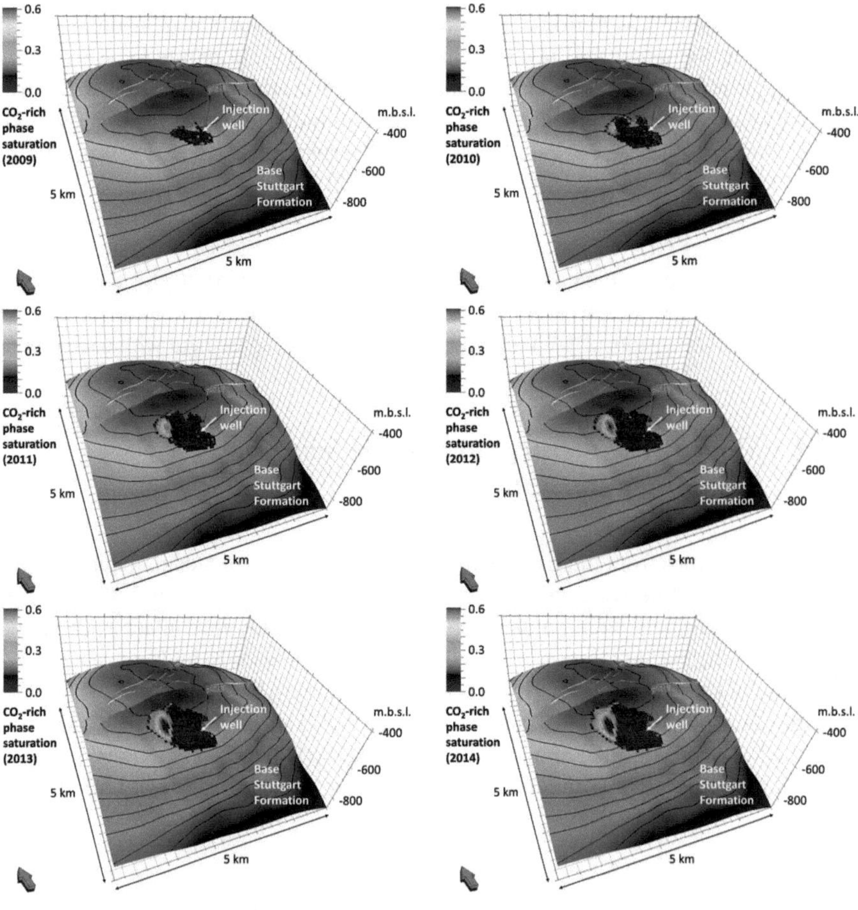

Fig. 6.5 Simulated distribution of gaseous CO_2 in the storage formation (model di-mensions: 5 km × 5 km) after 1, 2, 3, and 4 years (from *top left* to *bottom right*) and after 5 and 6 years (*bottom left* and *right*, prediction) (GFZ 2014)

6.7 CO_2 Storage as a Component of Energy Storage for a Closed Carbon Cycle

The geological storage of CO_2, however, does not only play a key role for the mitigation of CO_2 emissions in the atmosphere in the long term, but it could also become a central component in the hydrogen economy as 'dynamic' storage, as proposed in the power-to-gas concept.

If excess electricity is to be converted into methane and stored, a CO_2 source will be required. According to the German federal government's climate change mitigation targets, CO_2 emissions are to be cut by at least 80 % by 2050 in regard to levels of 1990. In order to achieve this target, CO_2 from biogas production must be used for any power-to-gas concept because it is not considered to be additional CO_2, and is assigned to the natural carbon cycle. Another option is to use process-related CO_2 produced in industrial processes.

If we were to go a step further, the power-to-gas concept could be extended to include the separation, storage, recycling and reuse of CO_2 produced during the energy generation process (e.g. via combined cycle power plants = CCGT; Fig. 6.6). If all components are integrated in one site, then a local closed carbon cycle is the result (Streibel et al. 2013). This would safeguard the advantages of fossil fuels – the ability to be stored in large quantities and thus supply very

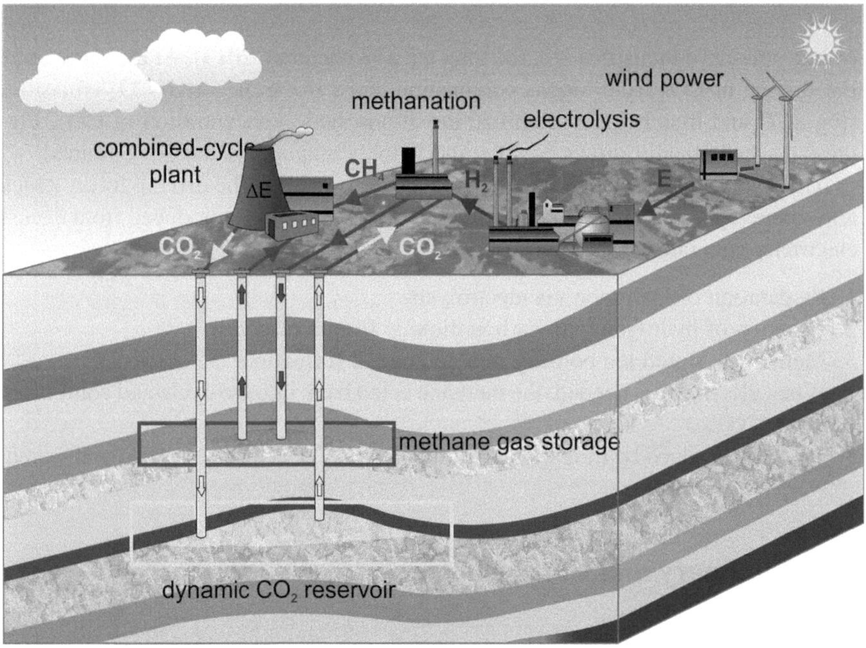

Fig. 6.6 Closed carbon cycle achieved by coupling CO_2 storage with methane gas storage to store excess renewable wind and solar energy (GFZ 2014)

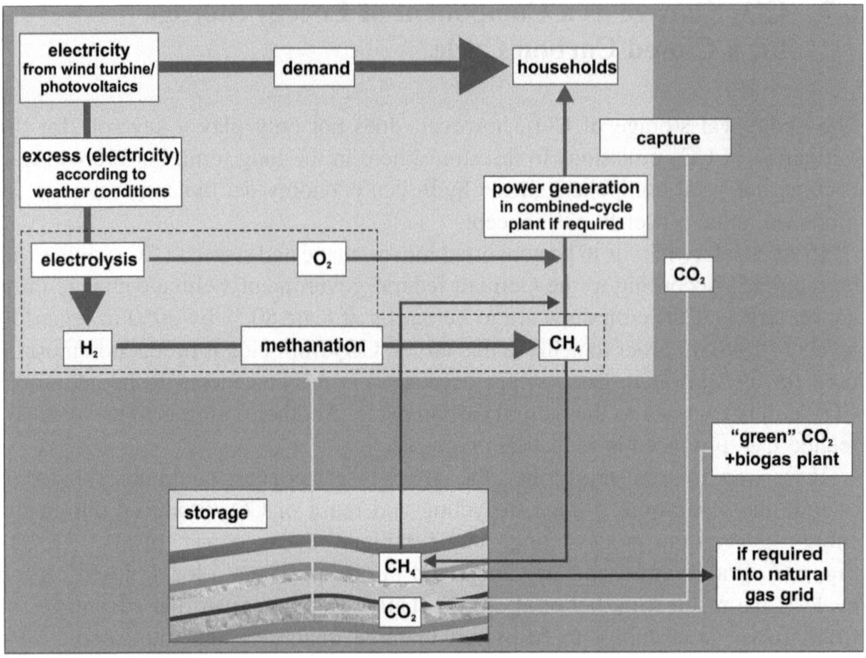

Fig. 6.7 The closed carbon cycle (GFZ 2014)

energy-intensive industries – in the long term (Kühn et al. 2013). At the same time, the storage of CO_2 from biogas combustion gives rise to negative CO_2 emissions (Fig. 6.7) and thus helps to stabilize the atmospheric concentration of CO_2. Figure 6.6 schematically shows how the individual components can be combined.

This cycle comprises five stages (Kühn 2012, Fig. 6.7). The driving force, which allows the repeated conversion of CO_2, is hydrogen, which is produced from excess electricity generated from renewables:

1. Preparation of hydrogen via electrolysis.
2. Reaction of hydrogen with carbon dioxide from a reservoir.
3. Methane is stored temporarily in a geological formation.
4. When electricity is needed, the methane is fed back into the cycle and combusted in a CCGT.
5. The CO_2 produced during combustion in the CCGT is then separated and stored.

6.8 Summary and Conclusions

Reservoir rocks with the potential for storing CO_2 are mainly sandstones, as they are characterized by sufficient porosities and permeabilities allowing CO_2 to be injected efficiently into these formations. Overall, four trapping mechanisms in the layers of the storage formation facilitate permanent and safe storage: (i) structural trapping below an impermeable caprock, (ii) immobilization via capillary forces in the pore space, (iii) dissolution of CO_2 in the formation water, and (iv) mineral trapping via carbonization.

As demonstrated, leaks can occur and must be detected, which makes monitoring of CO_2 storage sites essential. A general evaluation of the suitability of sites in advance cannot be performed effectively without a geological site characterization. It is therefore essential that a comprehensive exploration will be performed before a project begins. As is the case for every technology, there is also a technological risk associated with the geological storage of CO_2. However, this appears to be manageable, particularly when modern monitoring systems are used.

Near the town of Ketzin/Havel in Brandenburg, the first continental European field laboratory for CO_2 storage was set up as a pilot site in 2004, and it is in operation until today with active and continuous injection from June 2008 until August 2013. During that period 67,271 tCO_2 have been stored. The pilot site in Ketzin was thus the first and is still the only active CO_2 storage project in Germany. The injection of CO_2 has been accompanied by one of the most extensive scientific research and development programmes in the world. The results show that: (i) the geological storage of CO_2 at the pilot site in Ketzin is safe and reliable, and poses no danger to humans or the environment, (ii) a well-thought-out combination of different geochemical and geophysical monitoring methods can detect small amounts of CO_2 and image its spatial distribution, (iii) the interactions between fluid and rock induced by CO_2 injection at the pilot site in Ketzin have no significant impacts and do not influence the integrity of the reservoir or the caprock, and (iv) numerical simulations can depict the temporal and spatial behaviour of injected CO_2.

Work at the pilot site in Ketzin demonstrates the safety and reliability of CO_2 storage on a research scale, and is thus an important milestone on the way to decarbonizing society and making an important contribution to using underground geological formations in an environmentally friendly manner. In addition, results from studies at Ketzin provide basic and transferable knowledge which is of value for a new integrated concept of CO_2 mitigation and utilization in combination with the power-to-gas concept based on a closed carbon cycle approach.

References

Bergmann P, Lengler U, Schmidt-Hattenberger C, Giese R, Norden B (2010) Modelling the geoelectric and seismic reservoir response caused by carbon dioxide injection based on multiphase flow simulation: results from the CO2SINK project. Chemie der Erde 70:173–183

Bergmann P, Yang C, Lüth S, Juhlin C, Cosma C (2011) Time-lapse processing of 2D seismic profiles with testing of static correction methods at the CO_2 injection site Ketzin (Germany). J Appl Geophys 75:124–139

Birkholzer JT, Zhou Q, Tsang C-F (2009) Large-scale impact of CO_2 storage in deep saline aquifers: a sensitivity study on pressure response in stratified systems. Int J Greenh Gas Control 3, 181–194

Bohnhoff M, Zoback MD, Chiaramonte L, Gerst JL, Gupta N (2010) Seismic detection of CO_2 leakage along monitoring wellbores. Int J Greenh Gas Control 4:687–697

Chiodini G, Granieri D, Avino R, Caliro S, Costa A, Minopoli C, Vilardo G (2010) Non-volcanic CO_2 earth degassing: case of Mefite d'Ansanto (southern Apennines), Italy. Geophys Res Lett 37, L11303

Ebigbo A, Class H, Helmig R (2007) CO_2 leakage through an abandoned well: problem-oriented benchmarks. Comput Geosci. doi:10.1007/s10596-006-9033-7

Fischer S, Zemke K, Liebscher A, Wandrey M, CO2SINK Group (2011) Petrophysical and petrochemical effects of long-term CO_2-exposure experiments on brine-saturated reservoir sandstone. Energy Procedia 4:4487–4494

Förster A, Norden B, Zinck-Jørgensen K, Frykman P, Kulenkampff J, Spangenberg E, Erzinger J, Zimmer M, Kopp J, Borm G, Juhlin C-G, Cosma C, Hurter S (2006) Baseline characterization of the CO2SINK geological storage site at Ketzin, Germany. Environ Geosci 133:145

Giese R, Henninges J, Lüth S, Morozova D, Schmidt-Hattenberger C, Würdemann H, Zimmer M, Cosma C, Juhlin C, CO2SINK Group (2009) Monitoring at the CO2SINK site: a concept integrating geophysics, geochemistry and microbiology. Energy Procedia 1:2251–2259

Gilfillan SMV, Sherwood Lollar B, Holland G, Blagburn D, Stevens S, Schoell M, Cassidy M, Ding Z, Zhou Z, Lacrampe-Couloume G, Ballentine CJ (2009) Solubility trapping in formation water as dominant CO_2 sink in natural gas fields. Nature 458:614–618

Henninges J, Liebscher A, Bannach A, Brandt W, Hurter S, Köhler S, Möller F, CO2SINK Group (2011) P-T-ρ and two-phase fluid conditions with inverted density profile in observation wells at the CO_2 storage site at Ketzin (Germany). Energy Procedia 4:6085–6090

IPCC (2005) Special report on carbon dioxide capture and storage. http://www.ipcc.ch/activity/srccs/index.htm. IPCC, 15 Aug 2012

Kazemeini H, Juhlin C, Zinck-Jorgensen K, Norden B (2009) Application of the continuous wavelet transform on seismic data for mapping of channel deposits and gas detection at the CO2SINK site, Ketzin, Germany. Geophys Prospect 57:111–123

Kempka T, Kühn M, Class H, Frykman P, Kopp A, Nielsen CM, Probst P (2010) Modelling of CO_2 arrival time at Ketzin – Part I. Int J Greenh Gas Control 4:1007–1015

Kiessling D, Schmidt-Hattenberger C, Schuett H, Schilling F, Krüger K, Schöbel B, Danckwardt E, Kummerow J, CO2SINK Group (2010) Geoelectrical methods for monitoring geological CO_2 storage: first results from cross-hole and surface-downhole measurements from the CO2SINK test site at Ketzin (Germany). Int J Greenh Gas Control 4:816–826

Kühn M (2011) Chancen und Risiken – CO_2-Speicherung. Chemie in unserer Zeit 45:126–138

Kühn M (2012) Patent WO 2013156611 A1. http://www.google.com/patents/WO2013156611A1?cl=en&hl=de. Last access 12 Oct 2014

Kühn M, Nakaten NC, Streibel M, Kempka T (2013) Carbon neutral and flexible underground storage of renewable excess energy. Erdöl Erdgas Kohle 129(10):348–352 (in German)

Labitzke T, Bergmann P, Kießling D, Schmidt-Hattenberger C (2012) 3D surface-downhole electrical resistivity tomography data sets of the Ketzin CO_2 storage pilot from the CO2SINK project phase. Scientific technical report, Data: 12/05. doi:10.2312/GFZ.b103-12051

Lengler U, de Lucia M, Kühn M (2010) The impact of heterogeneity on the distribution of CO_2: numerical simulation of CO_2 storage at Ketzin. Int J Greenh Gas Control 4:1016–1025

Lewicki JL, Birkholzer J, Tsang C-F (2007) Natural and industrial analogues for leakage of CO_2 from storage reservoirs: identification of features, events, and processes and lessons learned. Environ Geol 52:457–467

Liebscher A, Martens S, Möller F, Lüth S, Schmidt-Hattenberger C, Kempka T, Szizybalski A, Kühn M (2012). Überwachung und Modellierung der geologischen CO_2-Speicherung – Erfahrungen vom Pilotstandort Ketzin, Brandenburg (Deutschland). Geotechnik 35. doi:10.1002/gete.201200005

Liebscher A, Möller F, Bannach A, Köhler S, Wiebach J, Schmidt-Hattenberger C, Weiner M, Pretschner C, Ebert K, Zemke J (2013) Injection operation and operational pressure-temperature monitoring at the CO2 storage pilot site Ketzin, Germany – design, results, recommendations. Int J Greenhouse Gas Control 15:163–173

Lüth S, Bergmann P, Cosma C, Enescu N, Giese R, Götz J, Ivanova A, Juhlin C, Kashubin A, Yang C, Zhang F (2011) Time-lapse seismic surface and down-hole measurements for monitoring CO_2 storage in the CO2SINK project (Ketzin, Germany). Energy Procedia 4:3435–3442

Martens S, Kempka T, Liebscher A, Lüth S, Möller F, Myrttinen A, Norden B, Schmidt-Hattenberger C, Zimmer M, Kühn M, Ketzin Group (2012) Europe's longest-operating on-shore CO_2 storage site at Ketzin, Germany – Progress report after three years of injection. Environ Earth Sci. doi:10.1007/s12665-012-1672-5

Möller F, Liebscher A, Martens S, Schmidt-Hattenberger C, Kühn M (2012) Yearly operational datasets of the CO_2 storage pilot site Ketzin, Germany. Scientific technical report, Data: 12/06. doi:10.2312/GFZ.b103-12066

Morozova D, Zettlitzer M, Let D, Würdemann H, CO2SINK Group (2011) Monitoring of the microbial community composition in deep subsurface saline aquifers during CO_2 storage in Ketzin, Germany. Energy Procedia 4:4362–4370

Nordbotten JM, Celia MA, Bachu S (2004) Analytical solutions for leakage rates through abandoned wells. Water Resour Res 40, W04204. doi:10.1029/2003WR002997

Norden B, Förster A, Vu-Hoang D, Marcelis F, Springer N, Le Nir I (2010) Lithological and petrophysical core-log interpretation in CO2SINK, the European onshore research storage and verification project. SPE Reservoir Eval Eng 13:179–192

Prevedel B, Wohlgemuth L, Legarth B, Henninges J, Schütt H, Schmidt-Hattenberger C, Norden B, Förster A, Hurter. S (2009) The CO2SINK boreholes for geological CO_2-storage testing. Energy Procedia 1:2087–2094

Roberts JJ, Wood RA, Haszeldine RS (2011) Assessing the health risks of natural CO_2 seeps in Italy. PNAS early edition: www.pnas.org/cgi/doi/10.1073/pnas.1018590108

Schmidt-Hattenberger C, Bergmann P, Kießling D, Krüger K, Rücker C, Schütt H, Ketzin Group (2011) Application of a Vertical Electrical Resistivity Array (VERA) for monitoring CO_2 migration at the Ketzin site: first performance evaluation. Energy Procedia 4:3363–3370

Sedlacek R (2009) Untertage-Gasspeicherung in Deutschland. Erdgas Erdöl Kohle 125:412–426

Streibel M, Nakaten NC, Kempka T, Kühn M (2013) Analysis of an integrated carbon cycle for storage of renewables. Energy Procedia 40:202–211

Tillner E, Kempka T, Nakaten B, Kühn M (2013) Brine migration through fault zones studied in 3D numerical simulations for a prospective CO_2 storage site in Northeast Germany. Int J Greenh Gas Control, 19, 689–703. doi:10.1016/j.ijggc.2013.03.012

Wells AW, Diehl JR, Bromhal G, Strazisar BR, Wilson TH, White CM (2007) The use of tracers to assess leakage from the sequestration of CO_2 in a depleted oil reservoir, New Mexico, USA. Appl Geochem 22:996–1016

Yordkayhun S, Ivanova A, Giese R, Juhlin C, Cosma C (2009a) Comparison of surface seismic sources at the CO2SINK site, Ketzin, Germany. Geophys Prospect 57:125–139

Yordkayhun S, Juhlin C, Norden B (2009b) 3D seismic reflection surveying at the CO2SINK project site, Ketzin, Germany: a study for extracting shallow subsurface information. Near Surface Geophys 7:75–91

Zemke K, Liebscher A, Wandrey M, CO2SINK Group (2010) Petrophysical analysis to investigate the effects of carbon dioxide storage in a subsurface saline aquifer at Ketzin, Germany (CO2SINK). Int J Greenh Gas Control 4:990–999

Zimmer M, Pilz P, Erzinger J (2011) Long-term surface carbon dioxide flux monitoring at the Ketzin carbon dioxide storage test site. Environ Geosci 18:119–130

[illegible]

Part II
Economic and Social Perspectives

Chapter 7
Economic Analysis of Carbon Capture in the Energy Sector

Wilhelm Kuckshinrichs and Stefan Vögele

Abstract The cost of carbon capture is a crucial factor for the deployment of the technologies in the electricity sector. In general, much higher electricity generation costs arise in case of carbon capture. With an increase of approximately 80 %, lignite-based CCS plants are particularly affected. The CO_2 avoidance costs are € 34–38/tCO_2 for lignite plants, € 41–48/tCO_2 for hard coal plants, and with approx. € 67/tCO_2 highest for natural gas plants. This depends on the lower level of CO_2 avoided in case of gas-fired power plants. Only when the price of allowances rises to these levels will the use of CCS power plants be cost-effective.

However, capture plants must be refinanced through the electricity market, as long as other market design options, e.g. capacity market or feed-in-tariffs, don't render possible returns. In general, the question arises as to the degree to which higher revenues due to merit order effects can cover the additional investment costs for capture plants and the subsequent transport and storage of CO_2. With further increase of renewable energy, there is a danger that the power plant capacities of an existing fleet will be potentially underused. As a result, there would be a short-term cost recovery problem for fossil power plants. Regardless of the possible development of capacity markets, the comparatively high refinancing needs compared to conventional power plants will remain if capacity revenues are to be incorporated.

Keywords Levelized cost of electricity • CO_2 avoidance cost • Merit order effect • Capacity market

7.1 Introduction and Motivation

This chapter concentrates on the economic analysis of the use of CCS in the electricity industry. The focus is on cost projections for carbon capture technologies, because the costs for CCS arise mainly from the capture of CO_2 and only to a lesser extent from its transportation and storage. For economic considerations, the

W. Kuckshinrichs (✉) • S. Vögele
Institute of Energy and Climate Research – Systems Analysis and Technology Evaluation (IEK-STE), Forschungszentrum Jülich GmbH, D-52425 Jülich, Germany
e-mail: w.kuckshinrichs@fz-juelich.de; s.voegele@fz-juelich.de

W. Kuckshinrichs, J.-F. Hake (eds.), *Carbon Capture, Storage and Use*,
DOI 10.1007/978-3-319-11943-4_7

price of CO_2 plays a key role because it alters the relative costs of the individual technologies depending on the technical specifications, and this is the only way of economically assessing the use of CCS technologies. In addition to investment costs and efficiency, the number of full-load hours is another relevant parameter.

Carbon capture technologies have not yet reached a level of maturity that can satisfy the requirements of commercial use in the electricity industry. Refining the technologies, building demonstration plants, and constructing commercial plants are tasks associated with a series of uncertainties of a technical and scientific nature, such as integration in the power plant process, level of investment costs, and economies of scale. Consequently, no empirical cost data are available for commercial usage yet. A series of investigations deals with expert estimates on the costs of technologies for demonstration plants and for the first commercial power plants. Within the framework of commercialization, the potential to cut the investment costs in particular is discussed using learning curves.

In the current electricity market design (energy only), a merit order or supply function covering the variable costs of power plants is relevant for pricing on the wholesale market. Here, it is of central significance whether and how CCS power plants contribute to the pricing process on the wholesale market, and what implications this could have for the refinancing of investments in CCS power plants.

This chapter is structured on the basis of the approach described. Following the introduction, Sect. 7.2 will discuss the process and cost parameters, as well as learning curve projections. Section 7.3 will examine the costs of electricity generation and CO_2 avoidance, and present cost projections for Germany. Using sensitivity calculations, key parameters such as the CO_2 price and number of full-load hours will be analysed in terms of their economic significance. Against the background of the discussion on the future structure of electricity supply, Sect. 7.4 will classify CCS technologies from an energy economics point of view within the framework of the electricity market design. Section 7.5 will conclude with a summary.

7.2 Demonstration Plants

7.2.1 Demonstration Plants for Electricity Generation

Against the background of EU efforts aiming at the demonstration of CCS technologies, an information system for energy technologies (SETIS[1]) was created as part of the SET-Plan Initiative.[2] In a study, Tzimas (Tzimas 2009) analysed and harmonized cost data from 13 different sources (Table 7.1). These sources comprise

[1] SETIS: SET-Plan Information System.

[2] SET-Plan: European Strategic Energy Technology Plan.

Table 7.1 Reference values for process and cost parameters for CCS demonstration power plants (~2,015)

Fluctuation margins for cost data according to Tzimas ±30 %.		Reference plants		CCS demonstration plants			
		PF	NGCC	IGCC-CCS	PF-CCS	Oxyfuel	NGCC-CCS
Specifications							
Net capacity	MW_e	400	400	400	400	400	400
Efficiency	%	46	58	35	35	35	46
Carbon capture efficiency	%	–	–	85	85	85	85
Cost data							
Investment	€$_{2011}$/kW	1,546	776	2,833	2,614	3,032	1,359
FOM	€$_{2011}$/kW	67	28	78	68	94	40
VOM	€$_{2011}$/MWh	0.9	0.05	2.2	4.7	0.9	0.9

Source: Adapted from Tzimas (2009)
FOM fixed operating and maintenance costs, *VOM* variable operating and maintenance costs, *PF* pulverized fuel, *NGCC* natural gas combined cycle, *IGCC* integrated gasification combined cycle. Availability: 85 %. Cost data are extrapolated using 1.5 %/year from €$_{2008}$ to €$_{2011}$

pre-feasibility studies, cost models, literature reviews, and expert opinions from industry and other organizations, and thus provide cost projections for CCS plants in different ways and with varying degrees of detail. Tzimas identifies pre-feasibility studies for concrete plant planning as the sources with the highest degree of data robustness.

With respect to technical and economic parameters, a difference must be made between demonstration projects and the commercial use of CCS power plants. Demonstration projects prove that a plant can be used on an industrial scale, and they explicitly aim to acquire experience on the application of the technology and to induce the first learning effects. Other learning effects will arise during the commercial operation of plants.

For coal-fired power plants, no significant difference in the investment costs could be identified for the various technology lines, even though the absolute difference of the values presented was €$_{2008}$ 400/kW. Assuming a determined reference value €$_{2011}$ 2,823/kW, the deviation is only ±7 %. Based on a data inaccuracy of approx. 30 % (Tzimas 2009), this is a relatively low value.

Little information is available on the fixed (FOM) and variable (VOM) operating and maintenance costs. Nonetheless, the literature data for IGCC, PF, NGCC plants with CCS are in good agreement with each other. The costs for transportation and storage are project-specific and depend on the location of the conversion plant, the storage facility, the pipeline routes, and the type of storage formation. They vary between €$_{2011}$ 5/t and €$_{2011}$ 42/t. The average cost of transportation and storage is around €$_{2008}$ 20/tCO_2. Cost projections by IPPC (2005) and MIT (2007), in contrast, are much lower. On average, investment costs of around €$_{2011}$ 1,900/kW$_e$ are incurred for CCS plants (cf. overview in Lohwasser and Madlener 2009). The low costs must be interpreted against the background of the originally more optimistic cost estimates.

7.2.2 *Learning Rates*

An empirical analysis shows that the specific capital investment for energy conversion plants decreases at a considerable rate with the total installed capacity over long periods of time (Rubin et al. 2004; McDonald and Schrattenholzer 2001). Compared to technologies such as wind energy or photovoltaics there are relatively few publications on learning curves for fossil-fired power plants with carbon capture. In analogy to technology developments such as those for the desulfurization and denitrification of power plants, some empirical values are also taken into account. Experience has shown that learning rates of 10–12 % are also expected for plants with carbon capture. Considerably higher rates are quoted only for photovoltaics (approx. 20 %) and much lower, partially negative, rates are quoted for nuclear energy (Al-Juaied and Whitmore 2009).

A detailed overview in Van den Broek et al. (2009) distinguishes learning rates for capture technologies (without transportation and storage) according to the plant components and parameters such as efficiency and availability. For plant components, the mean learning rates range between 0 % for CO_2 compression to 11 % for carbon capture using amine scrubbing and 12 % for Selexol concepts in IGCC plants. The mean learning rates cover a range with upward and downward deviations of 50 %.

In Neij (2008), learning rates of 5 % are assumed for all types of coal-fired power plants including CCS technologies. For gas-fired plants, a learning rate of 10 % is assumed. Furthermore, it is suggested that sensitivities of ±2 % be calculated in order to account for uncertainties. A more pessimistic assessment is provided by the GCCS Institute (Global CCS Institute 2011) with a predicted cost reduction of less than 5 % for the transition from FOAK (first of a kind) to NOAK (n^{th} of a kind), which is explained by the fact that a series of key components for carbon capture are already tried-and-tested technologies, and the level of maturity will therefore only increase minimally with increasing capacity, which means that no significant potential cost reduction is expected. However, greater potential for reducing costs is expected after the introduction of the next generation of technologies (Rubin et al. 2007a).

Rubin et al. (2007b) derive learning rates of 2.1 % (PC), 5.0 % (IGCC), 2.8 % (oxyfuel), and 2.2 % (NGCC) for investment costs. Expanding capacity to 100 GW (which is equivalent to doubling the capacities around 3.5–4.5 times) results in percentage reductions in the investment costs totalling 15 % (PC + IGCC), 13 % (oxyfuel), and 20 % (NGCC). For operating and maintenance costs, the corresponding learning rates are 5.7 %, 4.8 %, 3.5 %, and 3.9 %, respectively.

Preliminary Conclusions

- The degree of uncertainty with regard to costs is high for demonstration power plants.
- For demonstration plants, the following reference values are proposed for investment costs (calculated from Tzimas 2009):
  - IGCC-CCS: $€_{2011}$ 2,833/kW
  - PF-CCS: $€_{2011}$ 2,614/kW
  - Oxyfuel: $€_{2011}$ 3,032/kW
  - NGCC-CCS: $€_{2011}$ 1,359/kW
- The cost uncertainty is estimated as ± 30 %. In relation to a mean value of $€_{2011}$ 2,823/kW, coal-based technology lines therefore show no significant differences in investment costs.
- For carbon capture, there is only very little experience and knowledge on learning curves. Initial analyses quote learning rates of 2.1 % (PC), 5.0 % (IGCC), 2.8 % (oxyfuel), and 2.2 % (NGCC) (Rubin et al. 2007b).

7.3 Commercial Use of CCS

7.3.1 Cost and Process Parameters

Over the past few years, a series of cost analyses have been published on power plants with carbon capture (e.g. ETP ZEP 2011; Global CCS Institute 2011; IEA NEA OECD 2010; IPCC 2005; McKinsey 2008; MIT 2007). In the context of an increasing number of pilot and demonstration projects, knowledge of technical processes is improving, and higher costs are now being assumed, particularly for investments. However, there are still numerous uncertainty factors, which must be taken into account for cost and process parameters. Expectations are therefore often still very different.

The results of individual cost analyses cannot be compared directly due to the different assumptions regarding reference year, plant sizes and configurations, fuel prices, interest rates, etc. (Rubin 2012). The most recent ZEP study (ETP ZEP 2011) compares the results using harmonized reference parameters. Figure 7.1 shows the results for hard coal plants with carbon capture as an example.

For post-combustion plants, no clear trend can be discerned for the costs of generating electricity. Costs range between € 65/MWh to much more than € 80/MWh. For IGCC pre-combustion and oxyfuel power plants, the costs of generating electricity are similarly high, ranging from € 60/MWh to almost € 80/MWh.

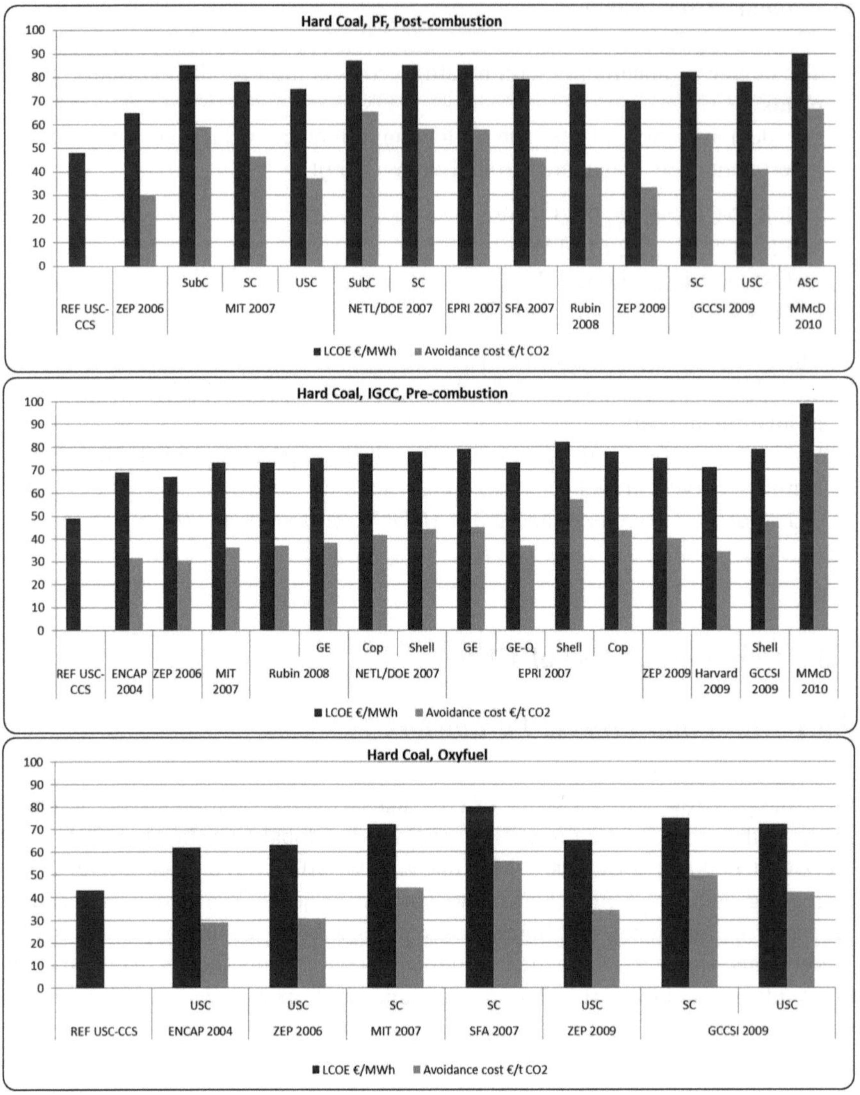

Fig. 7.1 Electricity generation and CO_2 avoidance costs for hard coal plants with carbon capture from different studies
Fuel price: € 2.42/GJ
For IGCCs, the power plant concepts of the companies are depicted
The following studies were incorporated: (Al-Juaied and Whitmore 2009; ENCAP 2008, 2009; Global CCS Institute 2009; Holt and Booras 2007; McKinsey 2008; MIT 2007; MMcD 2010; NETL 2007; Rubin et al. 2007a; SFA Pacific Inc. 2007; ZEP 2008)
(Source: Adapted from ETP ZEP 2011)

The CO_2 avoidance costs in post-combustion plants have a large range from € 30/tCO_2 to approx. € 65/tCO_2. For IGCC pre-combustion and oxyfuel plants, this range is smaller but still appreciable at approx. € 30–55/tCO_2. These figures reveal no clear advantage for any of the technologies discussed. The cost values for lignite plants are lower, and for natural gas plants, they are higher, as expected.

Tzimas and Georgakaki[3] (2010) and ETP ZEP (2011) compared the demonstration plants (Table 7.1) with the first commercial plants in 2020 and revealed a clear increase in the expected investment costs. This increase for PF-CCS and IGCC-CCS is particularly high for lignite and not quite as high for hard coal. ETP ZEP is much more optimistic with its expectations for oxyfuel plants: the expected specific investment costs decrease. For natural gas plants, in contrast, an increase is expected in the specific investment costs.

Compared to earlier analyses for commercial application in Germany (Hake et al. 2009), considerable changes emerge for individual cost components. The specific investment costs tend to be much higher both for the respective reference plants as well as for plants with carbon capture: for lignite plants between 50 and 75 %, for hard coal plants between 40 and 55 %, and for natural gas between 60 and 100 %. A notable exception here is oxyfuel technology, for which the specific investment costs for hard coal are estimated as 25 % higher and for lignite merely 9 % higher than in Hake et al. (2009). Fixed operating costs with the exception of the natural gas reference plant are consistently higher, while estimates for variable operating costs – again with the exception of the natural gas reference plant – tend to be much lower. Overall, a clear transition in the cost structure towards fixed costs is visible (see Lohwasser and Madlener 2009). The capital cost share increases considerably, and the share of fixed operating costs also increases. This is accompanied by a loss of economic flexibility of the plants, which are reliant on high annual utilization because of the high share of fixed costs. In a future energy supply system with a high proportion of fluctuating renewable energy and increasing demands on the flexible use of fossil-fired power plants, this situation is not very beneficial.

Table 7.2 shows important process and cost parameters based on Hake et al. (2009) and ETP ZEP (2011), which are assumed for power plant concepts in Germany and provide the basis for the subsequent calculations.

The investment costs of lignite power plants tend to be higher than those of hard coal power plants. For natural gas plants, the investment costs are much lower but the fuel costs are much higher. Compared to the respective reference plants, the investment costs of power plants with carbon capture are approx. 70–90 % higher.

[3] The cost breakdown in Tzimas & Georgakaki reflects the situation up to 2007 (Tzimas and Georgakaki 2010).

Table 7.2 Process and cost parameters for the first commercial CCS power plants

		Lignite				Hard coal				Natural gas	
		REF	PF-CCS	IGCC-CCS	Oxyfuel	REF	PF-CCS	IGCC-CCS	Oxyfuel	REF NGCC	NGCC-CCS
Specifications											
Capacity	MW_{el}	1,100	1,100	900	900	1,100	1,100	900	900	400	400
Lifetime	a	40	40	40	40	40	40	40	40	25	25
Efficiency	%	46.5	34.9	39.8	38.5	49.5	37.8	41.8	41.5	61.0	52.5
Carbon capture efficiency	%	–	90	90	90	–	90	90	90	–	90
Cost data											
Investment	$€_{2011}$/kW	1,700	2,950	3,200	2,900	1,600	2,800	3,100	2,800	700	1,250
Fixed costs	$€_{2011}$/kW	29	58	70	58	28	54	66	54	19	37
Variable costs	$€_{2011}$/MWh	3	5	5	5	3	5	5	5	1	3

Source: Calculation based on Hake et al. (2009) and ETP ZEP (2011)

7.3.2 Electricity Generation and CO_2 Avoidance Costs

In order to predict the costs of CCS technologies, the electricity generation costs [€/MWh] must be taken into account for production as must the CO_2 avoidance costs [€/tCO_2] for the reduction of CO_2 emissions.

Electricity generation costs reflect the costs of converting primary energy carriers and/or fuels into electricity along the process chain of a power plant. The costs of capturing CO_2 increase the total plant costs in three ways: (1) costs for additional plant components, e.g. CO_2 compression; (2) costs for additional plant capacity required to compensate for efficiency losses if net capacity is to be kept at the same level; (3) costs for additional fuel due to efficiency losses. In the system, additional costs are also incurred for the transportation and storage of CO_2.

In contrast, the CO_2 avoidance costs reflect the costs incurred for the CO_2 emissions that are not released into the atmosphere and are thus 'avoided'. The avoidance costs are calculated by comparing a reference technology without CCS with the corresponding technology with CCS. The CO_2 avoidance costs must be distinguished from the carbon capture costs. These are based on the corresponding amount of CO_2 captured at the power plant. The CO_2 avoidance costs are always higher than the carbon capture costs due to efficiency losses and the necessary compensation by means of a higher output and a higher fuel input.

For the following economic analysis, the concept of the levelized cost of electricity (LCOE) is used. This approach makes a financial analysis possible that focuses on estimating the electricity generation costs by taking into account the most important cost components, such as capital costs, fuel costs, and operating costs. The approach calculates the cash value of investments and the operation of a plant over the lifetime of the plant converted into uniform periodic payments.[4] The electricity generation costs can be calculated in this way for plants with and without carbon capture.

In addition to the process and cost parameters of power plants (Table 7.2), a number of other parameters also play a key role for the economic analysis. Of particular interest are the investment costs and the efficiency of the plants, as well as the transportation and storage costs. Berry (2008) also refers to the fuel costs, which have been taken into account in the approach here with price escalation (Table 7.3). For further analyses, the fundamental data for the energy sector shown in Table 7.3 are included in the calculations.

Fuel prices increase in real terms by approx. 1.2 %/a in accordance with Lindenberger et al. (2010). Transportation and storage costs vary widely depending on the transport distance, the amount transported, spatial conditions for a pipeline route, etc. In the literature, the costs quoted vary depending on the pipeline length, the terrain, and the gas volume, as well as on the storage medium (e.g. onshore/offshore, depleted gas field/saline aquifer) (Tzimas 2009). They range between € 5/tCO_2 (Global CCS Institute 2011) and € 9–18/tCO_2 (McKinsey 2008).

[4] For the mathematical principles, see Appendix.

Table 7.3 Economic data for the energy sector

Fuel price	$€_{2011}$/GJ	
Lignite		1.52
Hard coal		2.63[a]
Natural gas		6.39[a]
Transportation and storage costs	$€_{2011}/tCO_2$	5.00
Escalation	%/a	
Fuel price		1.20[a]
Operating costs		1.50
Transportation and storage costs		1.50
Full-load hours	h/a	7,500
Interest rate	%	5.00

[a]Calculated based on Lindenberger et al. (2010)

The infrastructural aspect is also important because a future transport network depends on the geographical distribution of the storage facilities and the CO_2 sources. For the following analysis, simplified transportation and storage costs of € $5/tCO_2$ are assumed.[5] The plants are assumed to operate with a high number of full-load hours.

The electricity generation costs at high utilization are shown in Fig. 7.2. A clear increase is visible for the electricity generation costs (LCOE) of CCS plants compared to the reference plants: for lignite from € 36/MWh to up to € 65/MWh (IGCC-CCS), for hard coal from € 44/MWh to up to approx. € 73/MWh (IGCC-CCS), and for natural gas from € 54/MWh to € 73/MWh (CCGT-CCS).

The CO_2 avoidance costs at high utilization are lowest for lignite plants at € 34–38/tCO_2 and highest for natural gas plants at € 67/tCO_2. Of the coal-fired power plants, the oxyfuel CCS plant is the most advantageous (Fig. 7.3).

7.3.3 Sensitivity Calculations

Against the background of uncertainties regarding the process and cost parameters as well as the economic data for the energy sector, sensitivities are calculated in the following in order to portray the impacts on the electricity generation costs of CCS technologies. The focus here is on monetary parameters (investment, fuel, CO_2 allowances) and process parameters (efficiency, full-load hours).

The investment costs represent the highest share of fixed costs. Higher investment costs cause a direct increase in the generation costs.

In order to account for different development opportunities on the electricity market, the number of full-load hours is modified in the following. The number of full-load hours is significant for offsetting other fixed costs and the investment

[5] For a differentiated analysis of transportation and infrastructure costs, see Chap. 9.

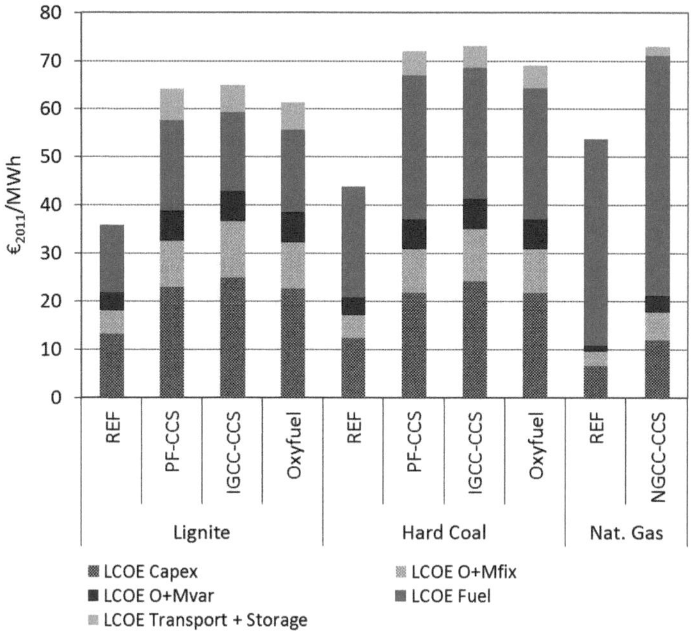

Fig. 7.2 LCOE based on fundamental economic data for the energy sector (With no learning rate effect)

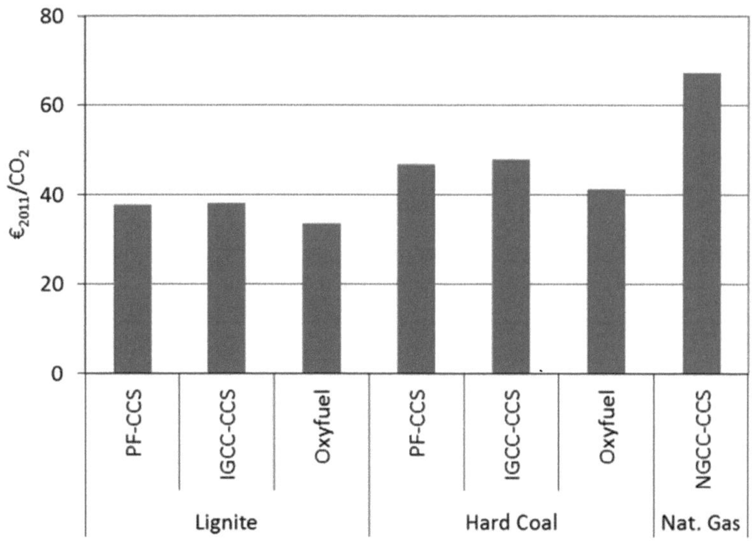

Fig. 7.3 CO_2 avoidance costs based on fundamental economic data for the energy sector (With no learning rate effect)

costs. The number of full-load hours has an impact on the apportionment of the fixed costs so that a higher number of full-load hours facilitate a degression of the generation costs.

The prices of fuel and emissions allowances, on the other hand, affect the variable costs, so that if they are increased, the variable costs also increase.

Changes in efficiency, in contrast, affect both the fixed and variable costs. A higher efficiency allows more electricity to be generated, and thus improves the apportionment of fixed costs. At the same time, improved efficiency leads to improved fuel utilization, which is accompanied by decreasing CO_2 intensity, and thus a lower specific need for CO_2 allowances. The variable costs therefore decrease.

The trading of allowances is a market-based solution for pricing CO_2. In IEA NEA OECD (2010), the price of CO_2 is explicitly incorporated as a variable cost factor in calculating the LCOE. This increases by a value that is calculated based on the price of allowances [€/tCO_2] multiplied by the CO_2 intensity [tCO_2/MWh]. The degree to which these technologies are affected by the level of the CO_2 price depends on their CO_2 intensity. In general, the higher the CO_2 intensity (after capture for CCS), the more the CO_2 price affects the generation costs.

For coal-fired power plants without CCS, this aspect is particularly relevant. The high CO_2 emissions must be covered by CO_2 allowances, while CCS plants only require allowances to cover the remaining CO_2 emissions. However in this case, the lower allowance costs must be balanced against the much higher investment costs. Ultimately, this alters the relative competitiveness of the conversion technologies (Nicholson et al. 2011). Only above a certain emissions allowance price level can the use of CCS power plants be justified. Conversely, it may be cost-effective for (permanently) low allowance prices to operate power plants without carbon capture.

Figure 7.4 shows the results of sensitivity calculations for CCS power plants. The calculations are based on a ± 10 % variation in the starting value of the parameters (Table 7.2 and Table 7.3). The base CO_2 price is assumed to be € 30/tCO_2.

The reaction patterns are very similar for all coal technologies. Variations of ± 10 % in full-load hours, efficiency, and fuel and purchase prices lead to changes in the generation costs of approx. € 2–4/MWh. These parameters are extremely important for the technologies discussed here, although sometimes in a different sequence. For lignite, the number of full-load hours tends to be pivotal, while for hard coal, efficiency tends to be more important. In both cases, the purchase price is not of overriding importance.

A very low number of full-load hours (~2,500 h) tends to cause the CO_2 avoidance costs to double. These costs then tend to be highest for natural gas plants (€ 123/tCO_2), and lowest for lignite plants (€ 71–78/tCO_2). For a low number of full-load hours, the avoidance costs for hard coal plants are very high at € 86–107/tCO_2. Overall, it appears that a relatively high CO_2 price is necessary to justify the implementation of the CCS technologies described here, particularly for a low number of full-load hours.

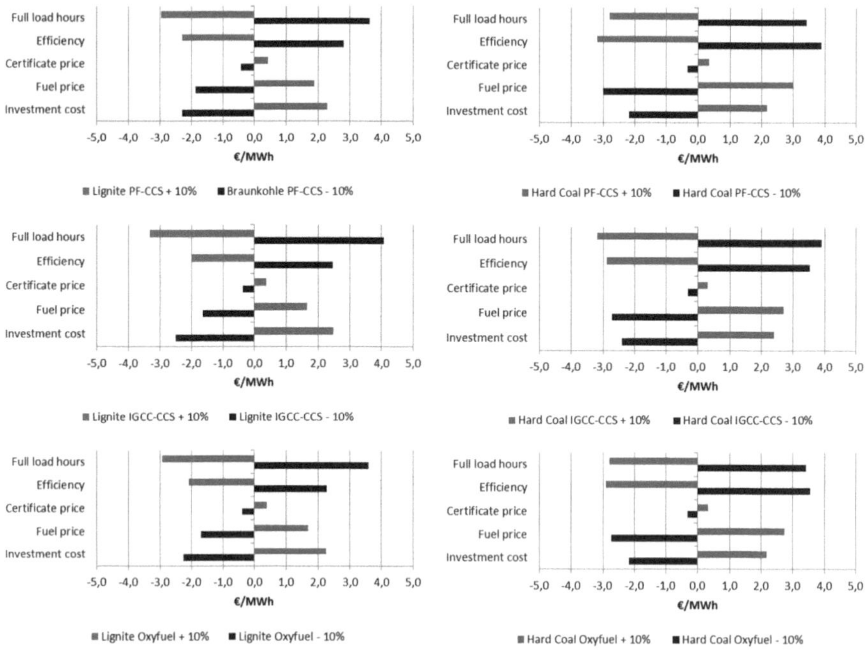

Fig. 7.4 Sensitivity of electricity generation costs for parameter variation. Parameters modified by ± 10 % respectively; € 30/tCO_2 is assumed for CO_2 allowances

Variations in the CO_2 price play a comparatively small role. A variation of ±10 % in the initial price of € 30/tCO_2 results in a change in the generation costs of approx. € 0.5/MWh.

Preliminary Conclusions

- The costs of capturing CO_2 increase the plant costs in three ways: (1) costs for additional plant components, e.g. CO_2 compression; (2) costs for additional plant capacity to compensate for efficiency losses if net capacity is to be kept at the same level; (3) costs for additional fuel due to efficiency losses.
- In the system, additional costs are also incurred for the transportation and storage of CO_2. They range between € 5/tCO_2 (Global CCS Institute 2011) and € 9–18/tCO_2 (McKinsey 2008). Depending on technical parameters the costs may be even higher in special cases.
- For the first commercial plants, considerably higher investment costs are expected than in earlier studies. Investment costs for CCS power plants tend to be around 70–90 % higher than for conventional power plants without CCS.
- The electricity generation costs for CCS power plants at a high utilization are approx. 26 % (natural gas) to 80 % (lignite) higher than the reference plants. The

CO_2 avoidance costs are € 34–38/tCO_2 for lignite power plants, approx. € 41–48/tCO_2 for hard coal plants, and approx. € 67/tCO_2 for natural gas plants.
- The utilization of lignite CCS power plants is only cost-effective if the price of allowances is at least approx. € 34–38/tCO_2. Cost effectiveness demands higher emissions allowance prices for hard coal, and considerably higher allowance prices for natural gas.
- Variations of ± 10 % in full-load hours, efficiency, and fuel and purchase prices lead to changes in the generation costs of approx. € 2–4/MWh. A corresponding variation of the CO_2 price, on the other hand, is not significant.
- A very low number of full-load hours (~2,500 h) tends to cause the CO_2 avoidance costs to double. As a result, a relatively high CO_2 price would be necessary to justify the implementation of the CCS technologies described here.

7.4 Electricity Production and Power Exchange Price for CCS Power Plant Usage in Germany

While the last section focused on the technical costs, this section will concentrate on the issue of whether and how the use of CCS power plants can influence the pricing process on the electricity wholesale market as well as producer surplus. Producer surplus is calculated for the electricity sold from the difference between the price of electricity and the variable power plant costs. Producer surplus allows the electricity suppliers to refinance their power plant investments. C.p. the price level of CO_2 allowances plays an important role here.

7.4.1 Pricing on the Electricity Market

The merit order is significant for pricing on the electricity market. It is a supply function, and ranks the power plant capacities in order of merit based on their variable costs. Power plants with low variable costs (e.g. wind energy, photovoltaics, and lignite power plants) are utilized first, followed by power plants with the next-highest variable costs (e.g. hard coal power plants, then gas power plants).[6]

The variable costs of fossil-fired power plants are influenced by CO_2 emissions allowance trading. The decisive factors here are the price of allowances (€/tCO_2), which is set on the electricity market, and the specific CO_2 emissions (tCO_2/MWh) of each individual power plant. Low CO_2 prices mean that the use of conventional power plants without CCS is more advantageous than the use of CCS power plants despite relatively high specific CO_2 emissions if the fuel costs of CCS power plants are correspondingly high. This depends on the efficiency losses of the CCS power

[6] The prioritization of feed-in from renewables is anchored in the legislation.

plant. This situation can change for higher allowance prices, which would then make CCS power plants relatively more advantageous in relation to variable costs, and the merit order would then alter accordingly. A quantity effect can also emerge as an additional effect if the electricity export also changes as a result of the price effects. Taking account of the electricity price and production volume, altered producer surpluses could result as a consequence.

7.4.2 Use of CCS Power Plants

For the analysis, different scenarios were created for different CO_2 allowance prices and for additional conventional and/or renewable energy capacity. An overview is shown in Table 7.4.

The model calculations were specified based on studies by EURELECTRIC and ENTSO-E on the development of future electricity generation in Europe (EURELECTRIC 2010; ENTSO-E 2011a, b). Data on installed power plant capacity in 2030 were taken from the EURELECTRIC study. The study was also used to derive specific annual efficiencies for the selected power plant types. Furthermore,

Table 7.4 Scenario overview for merit order analysis

Name of scenario	Description
Reference scenario	Situation in 2030
REF20	CO_2 allowance price of € 20/tCO_2
REF30	CO_2 allowance price of € 30/tCO_2
REF40	CO_2 allowance price of € 40/tCO_2
REF30 + RE	REF30 scenario with increased renewable energy capacity (+10 GW wind, +20 GW PV, +10 GW gas as backup)
REF30red +RE	REF30 scenario with reduced coal-fired power plant capacity (−10 GW) and increased renewable energy capacity (+10 GW wind, +20 GW PV, +10 GW gas as backup)
CCS scenarios	In contrast to the reference situations, 14.6 GW hard coal power plants and 13.7 GW lignite power plants in Germany are equipped with CCS
CCS20	CO_2 allowance price of € 20/tCO_2
CCS30	CO_2 allowance price of € 30/tCO_2
CCS40	CO_2 allowance price of € 40/tCO_2
RE-CCS scenarios	
CCS30 + RE	CCS30 scenario with an additional 10 GW wind and 20 GW PV (+10 GW gas as backup)
CCSred30 +RE	CCS30 scenario with reduced CCS capacity (instead of 28 GW only 18 GW) but with an additional 10 GW wind and 20 GW PV (10 GW gas as backup)

data on the development of the demand for electricity were taken from the study. As the average end-use price for electricity is considerably higher than the prices on the electricity market, and thus (at least in the short term) no reaction is expected from the end users to compensate for the price changes on the spot market, an inelastic electricity demand was assumed.

The data on existing and expected future power exchange capacities in Europe were taken mainly from ENTSO-E (2010, 2012). Data on the specific operating costs of power plants were taken from the IEA Outlook (2011), and calculated on the basis of data on the power-plant-specific efficiencies and fuel prices (IEA 2011).

In the reference scenarios, the situation is described without CCS for different allowance prices. The reference scenarios were compared to scenarios which assume the use of CCS in Germany. Here, it was assumed that 14.6 GW hard coal power plants and 13.7 GW lignite plants are operated with CCS. Of these, 19 GW are old power plants retrofitted with CCS.

In order to analyse the effects of the availability or non-availability of renewables, the calculations were performed on an hourly basis, i.e. for every hour in a year, the optimal usage of the power plants in the individual countries in Europe was calculated. The calculations are based on a model of power plant use described in Rübbelke and Vögele (2012).

In the scenarios investigated, the use of CCS led to a decrease in the average annual prices on the electricity market. This can be explained by the fact that the use of CCS means that the specially equipped coal-fired power plants are less affected by the price of CO_2 allowances, and thus maintain a competitive edge over conventional coal-fired power plants. An example of the development of electricity prices on the electricity market is shown in Fig. 7.5.

This graph shows the annual distribution of the calculated price of electricity in the scenarios REF30 and CCS30. At 4,400 h, the electricity price in scenario CCS30 is below that of scenario REF30 due to differences in the operating costs

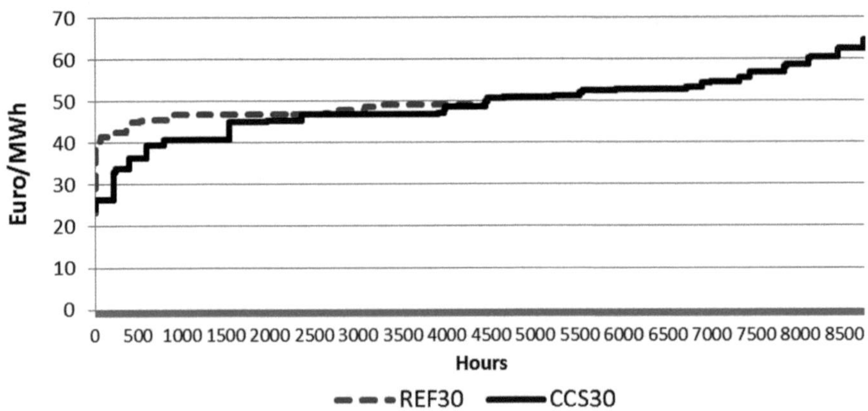

Fig. 7.5 Distribution of electricity prices over a year according to price level

Table 7.5 Impacts of CCS usage on electricity price and production for different CO_2 prices

		CCS20 – REF20	CCS30 – REF30	CCS40 – REF40
Changes in prices (Germany)		−2.2 %	−3.7 %	−6.4 %
Changes in electricity generation (Germany)		+0.9 %	+1.8 %	+4.7 %
Specific contribution margin (Germany)		€ 32/kW	€ 66/kW	€ 89/kW
Amortization period for additional investments	Post-combustion	14–38	7–18	5–13
	Pre-combustion	19–47	9–22	7–16
	Oxyfuel	22–38	10–18	8–13

Note: Estimation of the required years based on IEA (2011), Hake et al. (2009) and ETP ZEP (2011)

of CCS and conventional coal-fired power plants. At these points in time, the electricity price is determined by coal-fired power plants. At other points in time, gas power plants usually determine power plant marginal costs.

Due to changes in operating costs, electricity imports into Germany decrease in the model calculations and domestic production increases. The changes in production coupled with the price changes and the shape of the merit order curve lead to an additional producer surplus. From the additional revenues, specific contribution margins of € 32 per kW installed CCS capacity (comparison of CCS20 with REF20), € 66 (comparison of CCS30 with REF30), and € 89 (comparison of CCS40 with REF40) are obtained.

At a contribution margin of € 32, it takes approx. 14 years of operation to amortize additional investments for post-combustion plants under favourable conditions and about 38 years under unfavourable conditions (i.e. in the case of higher investment costs). In the case of higher investment costs for pre-combustion and oxyfuel plants, the amortization period increases. For higher allowance prices, the refinancing period decreases accordingly (Table 7.5).

Table 7.6 shows the results of a comparison of CCS scenarios with an increased renewable energy capacity. In the scenarios in which the share of renewables has been expanded further than in the reference case, the average annual price of electricity on the spot market is lower than without increased capacity. If the end-users pay the additional investment costs for renewables, as has been the case to date, and the additional revenues generated are then available to cover the additional investment costs for CCS, the specific contribution margins for CCS are € 44/kW (comparison of CCS30 + RE with REF30 + RE) and € 80/kW (comparison of CCS30 + RE with REF30). The use of CCS in the scenarios with an increased renewable energy capacity and unchanged output in the coal area leads to a clear reduction in the prices on the spot market and to an increase in domestic production. A reduction in the use of coal-fired power plants and a simultaneous increase in renewable energy capacity, basically leads to higher average electricity prices because in this case the comparatively more expensive backup capacities must be

Table 7.6 Impacts of CCS usage on electricity price and production for different CO_2 prices and increased renewable energy capacity

	CCS30 + RE		CCSred30 + RE	
Comparative scenario	REF30	REF30 + RE	REF30	REF30red + RE
Changes in prices (Germany)	−9.4 %	−5.5 %	1.2 %	−2.3 %
Changes in electricity generation (Germany)	5.6 %	2.7 %	−3.8 %	0.7 %
Changes in revenues (Germany)	+ € 2.3 billion	+ € 1.3 billion	+ € 2.4 billion	+ € 1.4 billion

used in order to compensate for fluctuations in electricity generation. In the case investigated, the higher costs give rise to a growth in electricity imports. The use of CCS dampens the rising energy carrier prices and the growing electricity imports. As a result, the electricity prices in scenario CCSred30 + RE are lower than those in scenario REF30red + RE. At the same time, domestic production increases in this CCS scenario.

If renewable energy is further integrated into the current electricity market design ('energy only'), there is a danger that the power plant capacities of an existing fleet will be potentially underused. In addition to the generation cost effect caused by a low number of full-load hours, the drop in residual demand would lead to a merit order effect. As a result, there would be a short-term cost recovery problem for fossil plants in the installed power plant fleet.

The long-term effect involves a decreased investment incentive for new builds. This applies even more so to CCS plants with comparatively high investment costs. Capacity markets attempt to provide a framework for paying compensation for making capacities available. The capital costs of power plants would therefore no longer be exclusively covered by revenues from the energy actually generated, but rather from the revenues of providing capacity as well. In addition, steady capital returns would also be made available for periods during which the power plant produces no electricity (Brunekreeft et al. 2011).

For Germany and Europe, the introduction of capacity markets is a topic of heated debate (Achner et al. 2011; Böckers et al. 2012; Bode and Groscurth 2011; Brunekreeft and Meyer 2011; Cramton and Ockenfels 2012; EWI 2012; Siegmeier and von Hirschhausen 2011). Regardless of the possible concrete development of capacity markets, the comparatively high refinancing needs compared to conventional power plants will remain if capacity revenues are to be incorporated. Against the background of the cost structure of CCS power plants, lignite CCS plants will tend to be used for base load, as shown by analyses within the framework of the discussion of possible capacity markets (EWI 2012). Security of supply will be ensured by gas turbines as backup power plants. The very few full-load hours tend to make the amortization of investment costs in these cases difficult, which mean that revenues from a possible capacity market are more important here.

Preliminary Conclusions

- Low loading because of low prices for CO_2 allowances mean that the operating costs of CCS power plants (despite efficiency losses) are below those of comparable coal-fired power plants. The use of CCS power plants therefore leads to a situation where periods of low electricity demand or periods when wind and photovoltaic plants can only be used to a limited extent give rise to lower electricity prices than in the reference situation. In addition, a merit order effect emerges, i.e. additional revenues are generated.
- Considering the existing uncertainties with regard to the additional investment costs for CCS plants, it can be assumed that CCS plants will only become

interesting to investors when the allowance price is at least € 40/tCO_2. Only then will the potential revenues be sufficient to offset the additional investment costs for CCS power plants in an acceptable time frame.

- Increasing the share of renewable energy will cause price effects, which will make refinancing more difficult for CCS power plants. On the other hand, price effects of CCS power plants will decrease the revenues for renewable energy, which in turn will have impacts on the level of renewable energy surcharges (EEG surcharges).

7.5 Summary and Conclusions

The costs of capturing CO_2 increase the plant costs in three ways: (1) costs for additional plant components, e.g. CO_2 compression; (2) costs for additional plant capacity to compensate for efficiency losses if net capacity is to be kept at the same level; (3) costs for additional fuel due to efficiency losses.

The plant costs for CCS power plants still involve uncertainties, despite the continuing development of demonstration facilities. An increasing understanding and ongoing technological development means that the investment costs of the first commercial CCS plants are predicted to be much higher than in previous studies. The investment costs for CCS power plants tend to be around 70–90 % higher than for conventional plants without CCS. The costs for the transportation and storage of CO_2 depend on the quantities to be transported, the transport distance, and the type and location of the geological storage facility, and they vary considerably. In all cases, the costs of capturing CO_2 dominate.

Even for high plant utilization, much higher electricity generation costs arise. With an increase of up to 80 % (lignite), coal-based CCS plants are particularly affected. The CO_2 avoidance costs are € 34–38/tCO_2 for lignite plants, € 41–48/tCO_2 for hard coal plants, and approx. € 67/tCO_2 for natural gas plants. Only when the price of allowances rises to the same level will the use of CCS power plants be cost-effective.

Variations of ± 10 % in the individual process and cost parameters (full-load hours, efficiency, fuel, and purchase prices) lead to moderate changes in the generation costs for CCS plants of approx. € 2–4/MWh. A corresponding variation in the price of CO_2, in contrast, is not significant due to the low specific CO_2 emissions. A very low number of full-load hours (~2,500 h) tends to cause the CO_2 avoidance costs to double. As a result, a relatively high CO_2 price would be necessary to justify the implementation of the CCS technologies described here for a low number of full-load hours.

CCS power plants must be refinanced through the electricity market. Furthermore, the use of CCS power plants can have an effect on the price of electricity on the wholesale market under certain conditions. Assuming perfect competition, the price on the electricity market is determined by the costs of the last power plant used, whereby the order in which the power plants are used is based on their marginal costs (merit order) and the costs for electricity imports must be considered. For high allowance prices (e.g. € 30/tCO_2), the operating costs of CCS power plants are below those of a comparable coal-fired power plant. The price of electricity, particularly during periods of low electricity demand, is determined by the costs of coal power plants. During these periods, the use of CCS dampens the price of electricity. Lower electricity generation costs domestically lead to a drop in electricity imports and a rise in electricity exports.

In general, the question arises as to the degree to which higher revenues due to merit order effects can cover the additional investment costs for CCS power plants. Owing to the high uncertainties with respect to the additional investment costs for CCS plants, it can be assumed that they will only become interesting to investors when the allowance price is at least € 40/tCO_2. The development in the area of renewable energy must also be considered here. As long as sufficient 'cheap' backup capacities, i.e. power plants with low operating costs, are available, the increased use of renewables will lead to a decrease in the average annual price on the electricity market. In addition, merit order effects occur. The use of CCS also dampens the price of electricity and thus boosts the level of domestic production. It must be noted that price effects caused by the increased use of renewable energy will make refinancing for CCS power plants more difficult, and the price effects of CCS power plants will decrease the revenues for renewable energy, which in turn will have impacts on the level of EEG surcharges.

If renewable energy is further integrated into the current electricity market design ('energy only'), there is a danger that the power plant capacities of an existing fleet will be potentially underused. In addition to the generation cost effect caused by a low number of full-load hours, the drop in residual demand would lead to a merit order effect. As a result, there would be a short-term cost recovery problem for fossil plants in the installed power plant fleet. Regardless of the possible concrete development of capacity markets, the comparatively high refinancing needs compared to conventional power plants will remain if capacity revenues are to be incorporated.

Appendix

LCOE

LCOE according to Global CCS Institute (2009), supplemented with a cost term for CO_2 allowances (IEA NEA OECD 2010):

$$LCOE\,[€/MWh] = \frac{CRF \cdot I + F_{FOM} \cdot C_{FOM}}{CF \cdot E_{Annual}} + F_{VOM} \cdot C_{VOM} + F_{Fuel} \cdot C_{Fuel} + F_{Carb} \cdot C_{Carb}$$

$$CRF = \frac{i}{1-(1+i)^{-n}} \text{ (Capital Recovery Factor)}$$

$$F_j = \frac{K_j\left(1-K_j^n\right)}{A\left(1-K_j\right)} \text{ (Levelisation Factor)};$$

$$A = \frac{(1+i)^n-1}{i(1+i)^n} \text{ (Present Value Factor)}; \quad K_j = \frac{1+R_j}{1+i} \text{ (Escalation Factor)}$$

with i = interest rate

and

F_{FOM} = Levelisation Factor fix O + M

F_{VOM} = Levelisation Factor variable O + M

F_{Fuel} = Levelisation Factor Fuel

F_{Carb} = Levelisation Factor CO_2

R_j = Escalation rate for cost j (excluding inflation)

CAC

$$CAC\,[€/tCO_2] = \frac{EGC_{CCS} - EGC_{REF}}{CO_{2,\,REF} - CO_{2,\,CCS}} + C_{Carb}$$

Where

EGC_{CCS}: energy generation costs of a plant with carbon capture,
EGC_{REF}: energy generation costs of the plant without carbon capture,
$CO_{2,REF}$: specific CO_2 emissions without carbon capture,
$CO_{2,CCS}$: specific CO_2 emissions with carbon capture

Learning Curves

Learning Curve

$K = K_0 \cdot X^{-E}$ with E : Learning index; X : cumulative capacity

Progress Rate

$PR = 2^{-E}$ Cost development with doubling capacity

Learning Rate

$LR = 1 - PR$

$$\Rightarrow E = -\frac{\ln(1-LR)}{\ln 2}$$

Methodological Approach for Merit Order Analyses

The methodological approach is based on the assumption of full competition on the electricity market. The price of electricity is regulated there depending on supply and demand. The price of electricity is determined by the marginal costs of the most expensive power plant needed to cover demand. The target function of the optimization formulation is thus:

$$\min Z_t = \sum_n \sum_i c_i \cdot s_{i,n,t} \cdot X_{i,n} + \sum_n \sum_m c_l imp_{n,m,t}$$

where

t: time index []
n, m: country index
i: index for power plant type
c_i: electricity generation costs of power plant type i [€/MWh]
$s_{i,n,t}$: utilization of power plant type i in country n at time t, where $0 \leq s_{i,n,t} \leq 1$ []
$X_{i,n}$: installed capacity of power plant type i in country n [MW]
c_l: costs for exchange of electricity [€/MWh]
$imp_{n,m,t}$: net imports of electricity from country n to country m [€/MWh]

A secondary condition here is that demand must always be covered.

$$\sum_i s_{i,n,t} \cdot X_{i,n} + \sum_m imp_{n,m,t} \geq d_{n,t} \qquad \forall n$$

In addition, electricity import and export capacities must be considered.

$$imp_{n,m,t} \leq NTC_{n,m} \qquad \forall (n, m)$$

with

$NTC_{n,m}$: net transfer capacities.

References

Achner S, Michels A, Nailis D, Ritzau M, Schuffelen L (2011) Kapazitätsmarkt. Rahmenannahmen, Notwendigkeit und Eckpunkte einer Ausgestaltung. Studie im Auftrag des Bundesverband Neuer Energieanbieter e.V. (bne). Aachen, http://www.neue-energieanbieter.de/data/uploads/20110907_bne_bet_studie_kapazitaetsmarkt_final.pdf. 15 Aug 2012: BET Aachen

Al-Juaied M, Whitmore A (2009) Realistic costs of carbon capture. Harvard Kennedy School, http://belfercenter.ksg.harvard.edu/publication/19185/realistic_costs_of_carbon_capture.html. 15 Aug 2012

Berry D (2008) Investment risk of new coal-fired power plants. http://iowa.sierraclub.org/Energy. 9 Aug 2011
Böckers V, Giessing L, Haucap J, Heimeshoff U, Rösch J (2012) Braucht Deutschland einen Kapazitätsmarkt für Kraftwerke? Eine Analyse des deutschen Marktes für Stromerzeugung DICE Ordnungspolitische Perspektiven, http://www.dice.hhu.de/fileadmin/redaktion/Fakultaeten/Wirtschaftswissenschaftliche_Fakultaet/DICE/Ordnungspolitische_Perspektiven/024_OP_Bo__ckers_Giessing_Haucap-Heimeshoff_Ro__sch.pdf. 15 Aug 2012, 24
Bode S, Groscurth H-M (2011) Die künftige Rolle von Gaskraftwerken in Deutschland. Hamburg, http://www.die-klima-allianz.de/wp-content/uploads/2011/10/Klima-Allianz-Studie-Gaskraftwerke-Okt-2011.pdf. 15 Aug 2012: arrhenius Institut für Energie- und Klimapolitik
Brunekreeft G, Meyer R (2011) Kapitalkosten und Kraftwerksinvestitionen bei zunehmender Einspeisung aus erneuerbaren Energie – Die Diskussion um Kapazitätsmärkte. Zeitschrift für Wirtschaftspolitik 60:62–73
Brunekreeft G, Damsgaard N, De Vries L, Fritz P, Meyer R (2011) A raw model for a North European capacity market. Elforsk rapport. http://www.elforsk.se/Programomraden/Anvandning/MarketDesign/Publications/2011/. 15 Aug 2012
Cramton P, Ockenfels A (2012) Economics and design of capacity markets for the power sector. Zeitschrift für Energiewirtschaft 36(Jg): 113–134
ENCAP (2008) Reference cases and guidelines for technology concepts. refman.et-model.com. 15 Aug 2012
ENCAP (2009) Power systems evaluation and benchmarking. Public version. www.encapco2.org. 15 Aug 2012
ENTSO-E (2010) Ten-year network development plan 2010–2020, Brussels, https://www.entsoe.eu/system-development/tyndp/. 15 Aug 2012
ENTSO-E (2011a) Data portal CE – production. https://www.entsoe.eu/data-portal/production/Pages/default.aspx
ENTSO-E (2011b) NTC values [Online]. Available: https://www.entsoe.eu/resources/ntc-values/ntc-matrix. Accessed 2 Mar 2011
ENTSO-E (2012) Ten-year network development plan 2012 project for consultation, Brussels, https://www.entsoe.eu/system-development/tyndp/. 15 Aug 2012
ETP ZEP (2011) The costs of CO_2 capture. http://www.zeroemissionsplatform.eu/library/publication/165-zep-cost-report-summary.html. 15 Aug 2012: European Technology Platform for Zero Emission Fossil Fuel Power Plants (ZEP)
EURELECTRIC (2010) Power statistics 2010 edition, Brussels, http://www.eurelectric.org/PowerStats2010/. 15 Aug 2012
EWI (2012) Untersuchungen zu einem zukunftsfähigen Strommarktdesign. In: Energiewirtschaftliches Institut UK (ed) Köln, www.ewi.uni-koeln.de. 15 Aug 2012: Energiewirtschaftliches Institut, Universität Köln
Global CCS Institute (2009) Strategic analysis of the global status of carbon capture and storage. Report 2: Economic assessment of carbon capture and storage technologies http://www.globalccsinstitute.com/. 15 Aug 2012: Global CCS Institute
Global CCS Institute (2011) Economic assessment of carbon capture and storage technologies. 2011 update. http://www.globalccsinstitute.com/. 15 Aug 2012: Global CCS Institute
Hake J-F, Hansen P, Heckler R, Linßen J, Markewitz P, Martinsen D, Weber K (2009) Projektionsrechnungen bis 2050 für das Energiesystem von Deutschland im Rahmen des VDI-Projektes "Future Climate Engineering Solutions". STE research report, 03/2009
Holt N, Booras G (2007) Updated cost and performance estimates for clean coal technologies including CO_2 capture – 2006. http://my.epri.com. 15 Aug 2012: EPRI
IEA (2011) World energy outlook – investment costs. Available: http://www.worldenergyoutlook.org/investments.asp. Accessed 7 Feb 2012
IEA NEA OECD (2010) Projected costs of generating electricity. Paris, www.iea.org/speech/2010/Tanaka/iea_nea.pdf. 15 Aug 2012, OECD

IPCC (2005) Special report on carbon dioxide capture and storage. www.ipcc.ch/special-reports/srccs/srccs_wholereport.pdf

Lindenberger D, Lutz C, Schlesinger M (2010) Energieszenarien für ein Energiekonzept der Bundesregierung Köln/Osnabrück/Basel, http://www.bmu.de/files/pdfs/allgemein/application/pdf/energieszenarien_2010.pdf. 15 Aug 2012: EWI/GWS/PROGNOS

Lohwasser R, Madlener R (2009) Impact of CCS on the economics of coal-fired power plants – why investment costs do and efficiency doesn't matter. FCN Working Paper, E.ON Research Center, Aachen, http://www.eonerc.rwth-aachen.de/aw/cms/website/themen/home/~sxv/publications_eon/?scol=1&cat=aaaaaaaaaaaasqz&sasc=0&pl=24&pn=7&lang=en. 15 Aug 2012, 7/2009

Mcdonald A, Schrattenholzer L (2001) Learning rates for energy technologies. Energy Policy 29:255–261

McKinsey (2008) Carbon capture and storage: assessing the economics. www.mckinsey.it/idee/practice_news/carbon-capture-and-storage-assessing-the-economics.views

MIT (2007) The future of coal. Cambridge, http://web.mit.edu. 15 Aug 2012: Massachusetts Institute of Technology

MMCD (2010) UK electricity generation costs update. http://www.decc.gov.uk/assets/decc/statistics/projections/71-uk-electricity-generation-costs-update-.pdf. 15 Aug 2012: Mott MacDonald

Neij L (2008) Cost development of future technologies for power generation – a study based on experience curves and complementary bottom-up assessments. Energy Policy 36:2200–2211

NETL (2007) Cost and performance baseline for fossil energy plants http://www.netl.doe.gov/energy-analyses/baseline_studies.html. 15 Aug 2012: DOE/NETL

Nicholson M, Biegler T, Brook BW (2011) How carbon pricing changes the relative competitiveness of low-carbon baseload generating technologies. Energy 36:305–313

Rübbelke D, Vögele S (2012) Short-term distributional consequences of climate change impacts on the power sector: who gains and who loses? Climatic Change, Online first

Rubin E (2012) Understanding the pitfalls of CCS cost estimates. Int J Greenh Gas Control 10:181–190

Rubin E, Taylor M, Yeh S, Hounshell D (2004) Learning curves for environmental technology and their importance for climate policy analysis. Energy 29:1551–1559

Rubin ES, Chen C, Rao AB (2007a) Cost and performance of fossil fuel power plants with CO_2 capture and storage. Energy Policy 35:4444–4454

Rubin ES, Yeh S, Antes M, Berkenpas M, Davison J (2007b) Use of experience curves to estimate the future cost of power plants with CO_2 capture. Int J Greenh Gas Control I:188–197

SFA Pacific Inc (2007) Gasification – critical analysis of technology, economics, and markets. http://www.sfapacific.com/gasification.shtml. 15 Aug 2012

Siegmeier J, von Hirschhausen C (2011) Energiewende: Brauchen wir noch "Kapazitätsmärkte" für konventionelle Kraftwerke? In: Klusmann B, Schütz D (eds) Die Zukunft des Strommarktes. Ponte Press, Bochum

Tzimas E (2009) The cost of carbon capture and storage demonstration projects in Europe. JRC Scientific and Technical Reports. Luxembourg, http://publications.jrc.ec.europa.eu/repository/browse?type=author&value=TZIMAS+EVANGELOS. 15 Aug 2012: European Commission

Tzimas E, Georgakaki A (2010) A long-term view of fossil-fuelled power generation in Europe. Energy Policy 38:4252–4264

van den Broek M, Hoefnagels R, Rubin E, Turkenburg W, Faaij A (2009) Effects of technological learning on future cost and performance of power plants with CO_2 capture. Prog Energy Combust Sci 35:457–480

ZEP (2008) EU demonstration programme for CO_2 capture and storage (CCS) – ZEP's proposal. In: European Technology Platform for Zero Emission Fossil Fuel Power Plants (ZEP) (ed) Brussels, http://www.zero-emissionplatform.eu/website/docs/ETP%20ZEP/EU%20Demonstration%20Programme%20for%20CCS%20-%20ZEP's%20Proposal.pdf. 15 Aug 2012

Chapter 8
Cost Analysis for CCS in Selected Carbon-Intensive Industries

Johannes Fleer and Wilhelm Kuckshinrichs

Abstract The cost of CCS is a crucial factor for the deployment of the technologies in industry. In order to analyze the potential of the options the levelized cost of electricity (LCOE) approach has been modified to be suitable for industrial processes. In order to allow for a cross-sector comparison, the CO_2 avoidance costs (CAC) are assessed as well. Based on a case-specific model plant, the costs for CCS in two carbon-intensive industries, cement production and refineries, are investigated.

For the cement works, the levelized production costs (LPC) are calculated for the reference plant without CCS, and for two CCS options (post-combustion, oxyfuel). For refineries, LPC are calculated for the reference plant without CCS and for a retrofit with oxyfuel technology.

The results show a major increase in LPC for the cement sector from € 84 per tonne cement in the baseline case to € 170/t for the post-combustion option (+102 %) and to € 112/t for the oxyfuel variant (+32 %) respectively. The LPC in the refinery case rise from € 70 per tonne crude oil in the baseline case to € 81/t for the oxyfuel option (+15 %). The CO_2 avoidance costs add up to € 143/tCO_2 for the cement works with post-combustion capture, € 55/tCO_2 for the oxyfuel option and € 62/tCO_2 for the oxyfuel refinery.

Keywords CCS in industry • Cement • Refinery • Production costs • CO_2 avoidance costs

8.1 Introduction and Motivation

In addition to fossil-fired power plants, industrial facilities such as cement works, steel works, and refineries contribute a significant share to global CO_2 emissions. While all CO_2 emissions in fossil-fired power plants are energy-related, many industrial facilities also release process-related CO_2 emissions. In contrast to energy-related emissions, these cannot be reduced by switching to energy carriers

J. Fleer (✉) • W. Kuckshinrichs
Institute of Energy and Climate Research – Systems Analysis and Technology Evaluation (IEK-STE), Forschungszentrum Jülich GmbH, D-52425 Jülich, Germany
e-mail: j.fleer@fz-juelich.de; w.kuckshinrichs@fz-juelich.de

W. Kuckshinrichs, J.-F. Hake (eds.), *Carbon Capture, Storage and Use*,
DOI 10.1007/978-3-319-11943-4_8

that contain less carbon. The use of carbon capture and storage (CCS) is a feasible industrial-scale mitigation measure because numerous industrial facilities – like fossil-fired power plants – can be characterized as large CO_2 point sources.

This chapter concentrates on an economic analysis of CCS deployment in selected carbon-intensive industrial sectors. The cement and refinery sectors, which are analysed here, are among the largest industrial emitters. As in the previous chapter, the analysis focuses on cost projections for carbon capture technologies, because the proportion of transport and storage in overall CCS costs is relatively small.

The maturity of carbon capture technology has not yet reached a level that can satisfy the requirements of commercial use in industry (see Chap. 2). Empirical cost data are therefore not available. Instead, the cost analysis is based on expert estimates and projections of energy and commodity prices, which are subject to uncertainty in terms of technological advances and developments on the energy and commodity markets.

The methodology for the cost analysis is taken from the previous chapter (Chap. 7) and has been adapted for application to industrial processes (see Fleer 2011). While in the electricity sector, comparability of different types of power plants is given because they generate an identical product (a megawatt hour of electricity), this is not the case in industry. Products vary from one industry sector to another, in some cases even within sectors (refineries). In addition, parameters relevant for the energy economy (e.g. efficiency, full-load hours) play a secondary role in industrial facilities, if they apply at all.

The cost analysis is founded on studies based on model facilities with defined boundary conditions for the individual industrial processes. These model plants do not represent average plants for the respective industrial process, but rather possible plant configurations based on real European plants. The objective of this chapter is to assess CCS costs for selected industrial sectors. The structure of this chapter corresponds to the approach taken in the analysis. Following this introduction, Section 8.2 will discuss the methodology of cost analysis and present the model plants. Section 8.3 will outline the results of the analysis and examine the economic significance of key parameters by means of a sensitivity analysis. Section 8.4 summarizes the findings.

8.2 Methodology of Cost Analysis

8.2.1 Methodological Approach

The methodology of cost analysis follows the approach taken in Chap. 7 ‘Economic Analysis of Carbon Capture in the Energy Sector’. As part of the cost projections for CCS technologies, two key figures are calculated: levelized cost of electricity (LCOE) – i.e. electricity generation cost – and CO_2 avoidance costs.

Since there is no uniform end product in industry that is comparable to a megawatt hour of electricity in the electricity generation sector, the electricity generation costs are replaced with the more general concept of production or processing costs. This is referred to here as levelized production costs (LPC) based on the concept of LCOE. LPC is defined analogously to LCOE, but is normalized to a product that is specific for the respective industrial sector instead of a megawatt hour of electricity. A cross-sector comparison cannot be performed with this approach; however, the key figure calculated makes it possible to compare the economics of different mitigation options within an industrial sector to each other and to describe the impact of CCS on the cost of goods production. While the electricity generation costs/LCOE reflect the costs of converting primary energy carriers into the end product of electricity along the process chain of a power plant, the LPC indicates the costs of converting raw materials and energy carriers into a process-specific end product along the process chain of an industrial plant.

For the cement industry, 1 tonne of cement is defined as the end product. In the refinery sector, it is difficult to specify an end product. Oil refineries produce a number of different products (e.g. petrol, diesel, lubricants) in coupled manufacturing processes instead of a uniform product. The crude oil used as a raw material in the refineries, however, is a uniform starting material. In order to avoid having to allocate the production costs to different outputs, a tonne of crude oil is chosen here as a reference quantity for the LPC.

The CO_2 avoidance costs (CAC) are relevant for a cross-sector comparison. They represent the costs of avoiding the release of 1 tonne of CO_2 into the atmosphere and can be calculated by comparing the reference technology without CCS to the corresponding plant with CCS technology. In this chapter, only the CO_2 emissions avoided directly at the plant are accounted for in the calculation of CO_2 avoidance costs. Possible mitigation effects or additional emissions caused by feeding electricity into the grid or consuming electrical energy from the grid are not taken into consideration.

Of particular interest in the economic analysis are the investment costs, fixed operating costs, and the variable costs for energy carriers and commodities. In the LPC approach, the present value of investment costs is converted into equal periodic instalments over the lifetime of a plant. The expected increase in fixed operating costs as well as commodity and fuel prices during plant lifetime are taken into account using specific escalation factors.

All available cost data are recalculated for the reference year 2020 from their respective base year, assuming an average inflation rate of 1.5 %, and expressed in real values for the year 2011 ($€_{2011}$). The cost of energy carriers (coal, electricity) is taken from Lindenberger et al.'s projection (Lindenberger et al. 2010). The figures for the cost of transporting and storing the captured CO_2 vary between approx. € 5/tCO_2 (Global CCS Institute 2011) and € 18/tCO_2 (McKinsey 2008). As in the previous chapter, simplified transportation and storage costs of € 5/tCO_2 are assumed.

Due to the uncertainties related to the process and cost parameters as well as the economic data, sensitivities are calculated. The impact of any change on conversion costs is calculated by varying the parameters. The parameters varied include those that are part of the fixed costs (investment costs, interest rate, and fixed operating costs) as well as those contributing to the variable costs (energy prices, commodity prices, transportation and storage costs, CO_2 allowances).

Allowances for CO_2 emissions from large industrial plants must be allocated or purchased under the European Union's emission trading scheme (EU ETS). Emission trading puts a price tag on every tonne of carbon dioxide emitted. In the baseline calculation for this chapter, a CO_2 price of € 0/tCO_2 is included in order to calculate the CO_2 avoidance costs. In the sensitivity calculations, however, the impact of the CO_2 price is also part of the analysis. For this purpose, an allowance price of € 30/tCO_2 is assumed.

8.2.2 Model Plants and Baseline Data for Cost Analysis

This economic analysis of CCS technology in industrial applications is based predominantly on two studies. The analysis of the cement industry is founded on a report published by the IEA Greenhouse Gas R&D Programme (IEA GHG 2008). For the refinery sector, a study exists on retrofitting the process heaters and steam boilers in a refinery with oxyfuel technology (Allam et al. 2005). This study constitutes the basis for the analysis of the refinery sector.

Both studies are based on a respective model plant that defines the key technical parameters. In the IEA GHG report, the model plant is a European cement works with a capacity of 1 million tonnes per year. This cement works represents the best available technique in accordance with the European IPPC Directive (European Commission 2010). The refinery study is based on the BP refinery in Grangemouth (Scotland), which has a crude oil processing capacity of 10 million tonnes per year. Important process and cost parameters for these two plants are listed in Table 8.1. The costs of commodities, energy carriers, and CO_2 transportation and storage are summarized under variable operating and maintenance costs.

In addition, the cost analysis makes assumptions regarding the energy prices, transportation and storage costs, and the interest rate on investments. The baseline data are listed in Table 8.2. The escalation factors, which allow for price trends during the operating life of the plants, are also listed in Table 8.2.

8.3 Results

8.3.1 Levelized Production Costs and CO_2 Avoidance Costs

Figure 8.1 shows the levelized production costs (LPC) for the cement works (per tonne of cement) and for the refinery (per tonne of crude oil). The processed crude

Table 8.1 Process and cost parameters for the model plants

		Reference plants		CCS model plants		
		Cement	Refinery	Cement post-combustion	Cement oxyfuel	Refinery oxyfuel
Specifications						
Capacity	million t/a	1	10	1	1	10
Production	million t/a	0.9	9.5	0.9	0.9	9.5
CO_2 emissions	million t/a	0.73	2.17	0.19	0.28	0.52
Capture efficiency	%	–	–	85	62	78
Cost data						
Investment	million €$_{2011}$	276	2,536	585	341	3,030
FOM[a]	€$_{2011}$/a	20.1	91.5	37.2	23.9	97.4
VOM[a]	€$_{2011}$/a	28.9	337.8	60.1	42.5	391.6

Source: Figure adapted from IEA GHG (2008), Allam et al. (2005), Favennec and Pigeyre (2001) and Fleer (2011)
[a]*FOM* fixed operating and maintenance costs, *VOM* variable operating and maintenance costs; cost data are extrapolated to €$_{2011}$ based on an inflation rate of 1.5 %/a

Table 8.2 Baseline data for cost calculation

Fuel price	Hard coal	€$_{2011}$/t	147[a]
	Petroleum coke		114[b]
Electricity price	Industry feed-out	€$_{2011}$/MWh	109[a]
	CHP feed-in		40.39[c]
Transport and storage costs		€$_{2011}$/tCO_2	5.00
Escalation	Fuel price	%/a	1.20[a]
	Commodity price		1.50
	Operating cost		1.50
	Transport and storage cost		1.50
Interest rate		%	5.00

[a]From Lindenberger et al. (2010)
[b]Recalculation based on Energy Argus (2011)
[c]From EEX (2012)

oil itself is not included in the calculations as a cost factor. In all cases, the cost of capital (CAPEX), the fixed operating and maintenance costs (O&M$_{fix}$), and the costs of energy carriers represent the largest proportion in overall LPC. The costs of commodities as well as transportation and storage (T&S) represent considerably smaller shares. In the case of the cement works, a use of CCS technology leads to a major increase in production costs, from € 84/t in the baseline case to € 170/t for the post-combustion option. This represents a 102 % increase. The deployment of

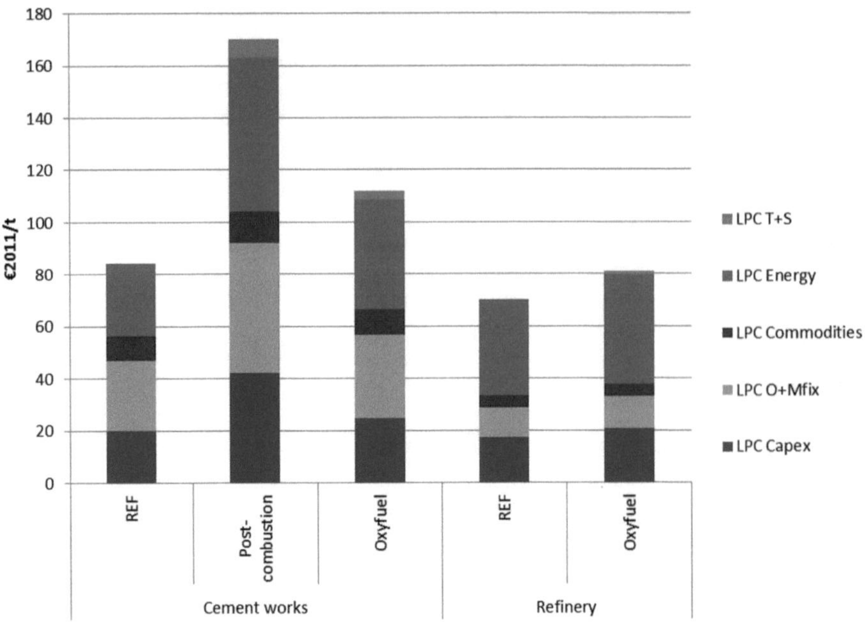

Fig. 8.1 Levelized production costs of cement and crude oil with/without CCS

oxyfuel technology entails a 32 % increase to € 112/t. The increase in processing costs is less pronounced in the refinery studied. The LPC increases from € 70/t in the baseline case to € 81/t for the oxyfuel variant, which is an increase of 15 %.

Figure 8.2 shows the CO_2 avoidance costs for the three CCS options. CO_2 avoidance costs are highest for the cement works with post-combustion capture at € 143/t$CO_{2,avoid}$ but much lower in the case of the oxyfuel option (€ 55/t$CO_{2,avoid}$) and the oxyfuel refinery (€ 62/tCO_2).

In comparison to CCS for power plants (see Chap. 7), the CO_2 avoidance costs of the oxyfuel cement works and the oxyfuel refinery are higher than for lignite and hard-coal power plants, but lower than for the natural gas plant (CCGT-CCS) studied there. The avoidance costs for the cement works with post-combustion technology are considerably higher than for power plants.

8.3.2 Sensitivity Calculations

Both the process and cost parameters of CCS technology and the underlying economic data are subject to uncertainties. In order to map the effects of changes in these parameters or data on the production or processing costs, sensitivity calculations were performed. The following parameters were varied (Fig. 8.3):

- Investment costs
- Interest rate

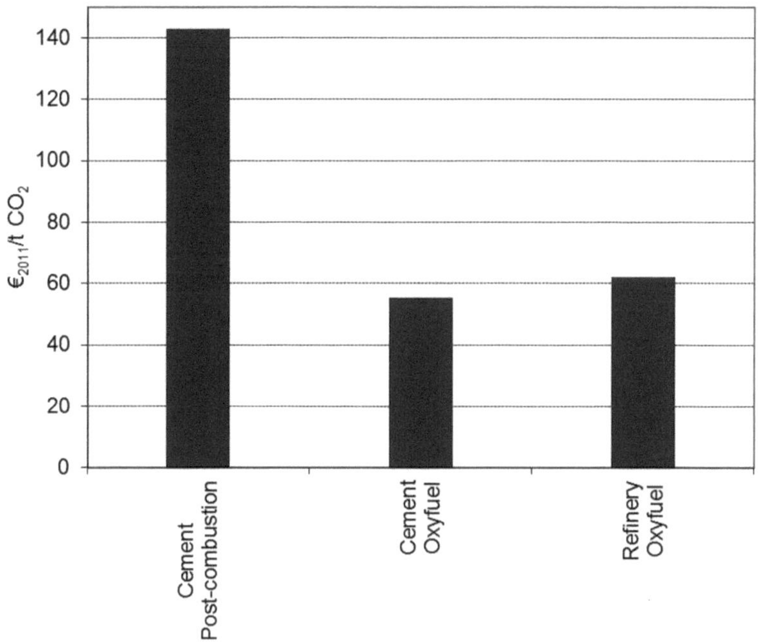

Fig. 8.2 CO_2 avoidance costs

- Fixed operating and maintenance costs ($O\&M_{fix}$)
- Commodity prices
- Energy prices
- Costs of CO_2 transportation and storage
- Allowance prices (assuming a baseline price of € 30/tCO_2)

The sum of investment costs and fixed operating and maintenance costs are the fixed costs. They are independent of the production volume of the plant. An increase in these cost parameters directly causes an increase in LPC. A change in the interest rate has an impact on the costs for interest on investments and therefore on the fixed costs.

The costs of commodities, energy, transportation, and CO_2 storage are variable costs. They depend on the output or processed quantity of the respective plant. If the produced/processed amount is constant, an increase in commodity or energy prices directly causes an increase in the LPC. The costs incurred for the transportation and storage of CO_2 are determined by the amount of captured CO_2.

EU ETS Emissions trading puts a price on a tonne of CO_2 emissions. This creates economic incentives for reducing CO_2 emissions. While a price of € 0/t was assumed for the calculation of the LPC and CO_2 avoidance costs in the previous chapter, an allowance price of € 30/tCO_2 was assumed when calculating the sensitivity to parameter variations. The costs incurred for CO_2 allowances can be determined by multiplying the price per allowance with the amount of carbon emitted into the atmosphere (after capture). These are part of the variable costs and

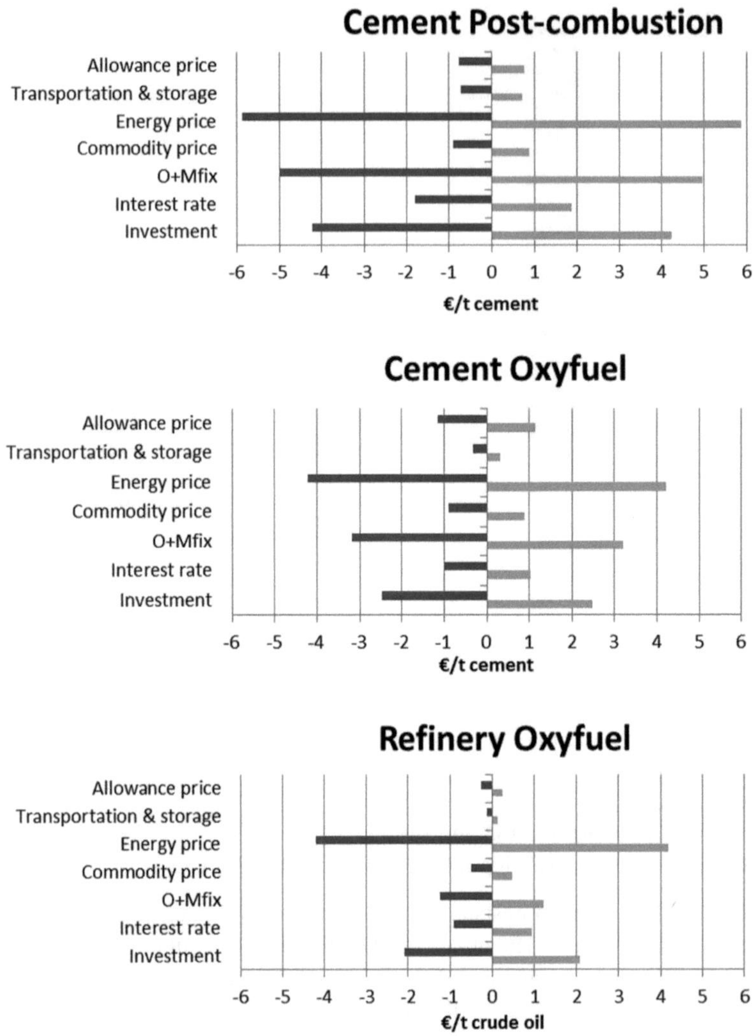

Fig. 8.3 Sensitivity of LPC for parameter variation. Parameters varied by ±10 % respectively; € 30/tCO_2 is assumed for CO_2 emission certificates

increase with the amount of emitted CO_2. This in turn is correlated with plant output and the carbon intensity of the process.

As no information is available on the impact of the process parameters of utilization and efficiency on the variable costs, no variation calculation is possible for these parameters. The sensitivity analysis is restricted to the monetary parameters listed above.

Figure 8.3 shows the results of the sensitivity calculations for industrial plants with CCS. The calculations are based on a ±10 % variation of the starting value of the parameters.

The reaction patterns are very similar for all three model plants. It is particularly obvious that in all three cases, the energy price plays a central role. A variation in the energy price results in a € 4–6/t change in the LPC. The investment costs are another important factor, which causes a change in LPC of € 2–2.5/t for the oxyfuel option and more than € 4/t for the post-combustion cement works. In the two cement works, the fixed operating and maintenance costs are also significant. They cause a change in LPC of about € 3–5/t. A change in any of the other parameters has relatively little impact on production or processing costs.

8.4 Summary

The deployment of CCS as a CO_2 mitigation measure for industrial plants seems to be technically feasible, but neither demonstration nor commercial CCS plants are currently being operated on the industrial scale. For this reason, the estimates for plant costs currently continue to be subject to great uncertainty. The results outlined here are closely linked to the reference plant data and should be treated as reference points that need to be specified further for future studies, for example for a review of the German plants (see study of the Institute for Applied Ecology (Öko-Institut 2012)).

The cost analysis shows an increase of approx. 32 % in production costs if oxyfuel technology is used in a cement works with a capacity of 1 million tonnes of cement per year. In the case of carbon capture with post-combustion technology, production costs rise by about 102 %. Retrofitting an oil refinery with a capacity of 10 million tonnes of crude oil per year with oxyfuel technology results in a 15 % increase in processing costs.

The cost analysis results in CO_2 avoidance costs of about € 55/t$CO_{2,avoid}$ for the oxyfuel cement works and about € 62/t$CO_{2,avoid}$ for the oxyfuel refinery. Avoidance costs are much higher for the cement works with post-combustion capture (€ 143/t $CO_{2,avoid}$).

References

Allam R, White V, Ivens N, Simmonds M (2005) The oxyfuel baseline: revamping heaters and boilers to oxyfiring by cryogenic air separation and flue gas recycle. In: Thomas DC, Benson SM (eds) Carbon dioxide capture for storage in deep geologic formations. Elsevier, Amsterdam, pp 451–475

EEX (2012) European Energy Exchange: Strom Spotmarkt – EPEX SPOT. http://www.eex.com/de/Downloads. 9 Aug 2012

Energy Argus (2011) Petroleum coke. Petroleum coke market prices, news and analysis. Argus Media. http://www.argusmedia.com/Petroleum/Petroleum-Products/~/media/Files/PDFs/Samples/Energy-Argus-Petroleum-Coke.ashx

European Commission (2010) Reference document on best available techniques in the cement, lime and magnesium oxide manufacturing industries. European Commission. http://eippcb.jrc.es/reference/cl.html

Favennec J-P, Pigeyre A (2001) Refining: a technical summary. Investments, margins, costs. Probable future developments. In: Favennec J-P (ed) Refinery operation and management. Editions Technip, Paris, pp 117–170

Fleer J (2011) Technische und ökonomische Analyse von Carbon Capture and Storage (CCS) für CO_2-intensive Industrieprozesse. Lehrstuhl für Energiesysteme und Energiewirtschaft. Ruhr-Universität Bochum, Bochum

Global CCS Institute (2011) Economic assessment of carbon capture and storage technologies. 2011 update. Global CCS Institute, Canberra

IEA GHG (2008) CO_2 capture in the cement industry. IEA Greenhouse Gas R&D Programme. http://ieaghg.org/docs/General_Docs/Reports/2008-3.pdf

Lindenberger D, Schlesinger M, Lutz C (2010) Energieszenarien für ein Energiekonzept der Bundesregierung. EWI/GWS/PROGNOS, Köln/Osnabrück/Basel

McKinsey (2008) Carbon capture & storage: assessing the economics. http://assets.wwf.ch/downloads/mckinsey2008.pdf

Öko-Institut (2012) Potenziale und Chancen der Technologie zur CO_2-Abtrennung und -Ablagerung (CCS) für industrielle Prozessemissionen. WWF, Öko-Institut, Berlin

Chapter 9
CCS Transportation Infrastructures: Technologies, Costs, and Regulation

Joachim Geske

Abstract The deployment of CCS technology requires CO_2 to be transported from capture- to designated storage-sites. For this purpose the creation of a CO_2 pipeline transportation infrastructure is considered advantageous. In this chapter it is analyzed how the provision of a CO_2 pipeline transportation infrastructure should be organized from an economic point of view. It is shown that a regulated or a private provision of the infrastructure is preferable depending on returns to scale property of the infrastructure on the system level. This property on its part depends on the spatial distribution of capture and storage sites, and of the volume transported.

Keywords CO_2 pipeline transportation infrastructure • Returns to scale

9.1 Introduction

CCS technology comprises not only the "capture" process, but also the additional component of CO_2 storage. In many countries, including Germany, most of the geological formations suitable for storage are located at some distance from the plants where CO_2 can be captured (e.g. GeoCapacity 2009). The deployment of CCS technology therefore requires CO_2 to be transported from the plants where it is produced to the designated storage sites. CO_2 transportation cannot be realized using existing infrastructures for the transportation of gas, water, or vehicles because they are either technically unsuitable, cannot provide the necessary capacities, or the underlying technologies are not cost optimal (e.g. Odenberger and Svensson 2003, Fig. 9.1). Instead, pipeline transportation and the creation of a CO_2 pipeline transportation infrastructure are considered advantageous (GeoCapacity 2009).[1,2]

[1] Where waterways are accessible, the transportation of CO_2 by ship can provide an alternative to pipeline transportation, in particular because this makes the construction of a pipeline infrastructure unnecessary.

[2] For more information on the technical aspects of CO_2 pipelines, see Chap. 3.

J. Geske (✉)
Institute of Energy and Climate Research – Systems Analysis and Technology Evaluation (IEK-STE), Forschungszentrum Jülich GmbH, D-52425 Jülich, Germany
e-mail: j.geske@fz-juelich.de

W. Kuckshinrichs, J.-F. Hake (eds.), *Carbon Capture, Storage and Use*,
DOI 10.1007/978-3-319-11943-4_9

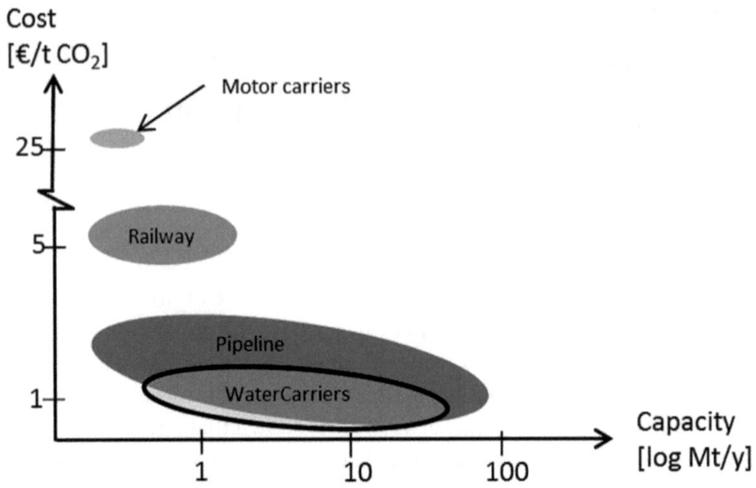

Fig. 9.1 Costs and capacities of transportation alternatives for a distance of 250 km (Source: Odenberger and Svensson 2003.)

From an economic point of view,[3] the returns-to-scale property plays a key role when it comes to building and operating a CO_2 transportation infrastructure.[4] A decentralized transportation infrastructure with constant returns to scale can be operated efficiently in a market economy. In comparison, an infrastructure with increasing returns to scale can, under perfect competition, only be operated at a loss. As businesses strive to maximize their profits, this type of infrastructure cannot be implemented as a market-driven system. Increasing returns to scale can only be exploited if individual pipeline sections are made available by a central provider (natural monopoly, e.g. Knieps 2002). The resulting market power and associated deadweight loss can be counteracted by means of regulation. These cases, which require different economic policy measures, can be distinguished by their returns-to-scale properties.

In order to calculate the returns-to-scale property of a CCS infrastructure, the returns to scale can be determined for individual pipelines and these figures can then be aggregated, resulting in the infrastructure returns-to-scale property. From a technical point of view, a pipeline infrastructure consists of individual pipelines that contain the goods to be transported within a cylinder, so that transportation takes place along one-dimensional "paths". As the surface-area-to-volume ratio of cylinders decreases with volume, a smaller volume means that less material is required to build the pipeline, reducing the specific construction costs as a function

[3] Legal issues including their economic implications for building a CCS infrastructure are discussed in Parfomak and Folger (2008).

[4] Decreasing/constant/increasing returns to scale: if the factor input is increased by the factor a, the transportation capacity increases by less than/equal to/greater than the factor a. Increasing returns to scale are equivalent to a decrease in average costs.

of the volume transported. This is one reason why individual pipelines of an infrastructure have increasing returns to scale.[5]

However, the pipeline level is not directly indicative for the returns-to-scale property of the infrastructure as a system, because in extreme cases – which could occur in a pipeline infrastructure for the transportation of CO_2 from industrial sources of similar size to nearby saline aquifers – short, direct connections from capture to sink can be the cost optimal solution (e.g. Oei et al. 2011). In such a case, the infrastructure costs correspond to a linear scaling of the pipeline costs, and the returns-to-scale property is transferred from the pipeline level. However, if integrating additional connections requires numerous new pipelines to be built, the returns to scale relative to the volume transported through the infrastructure may even decrease. This means that on the system level, the returns to scale depend on the spatial distribution of capture and storage sites, and of the volume transported.[6]

This situation makes it difficult to derive generally valid statements, which would be required to estimate the need for regulation associated with different infrastructure development strategies. In addition, only 'relatively little experience with the combination of CO_2 capture, transport and storage in a fully integrated CCS system' is available (IPCC 2005), which makes it impossible to empirically determine the spatial structure of a CCS infrastructure. It is therefore unavoidable to draw conclusions based on normative principles. The dominating and most plausible normative principle[7] for deducing the spatial structure of an envisaged new infrastructure is the minimization of cost (Middleton and Bielicki 2009a). In Sect. 9.2 cost optimal infrastructures and, using these as a basis, relevant analyses (Middleton and Bielicki 2009a, b) for two model regions are presented, including the spatial structures and returns-to-scale property of the optimal infrastructure.

Their analyses show that the returns to scale of optimal transportation technologies do indeed depend on the regions studied, as well as numerous technical details. Generally valid statements for evaluating infrastructure development strategies and the associated need for regulation therefore cannot be deduced directly from an analytical approach. However, in order to nevertheless supplement the case studies for optimal infrastructures with an analytical approach, a model is presented in Sect. 9.3 that proposes a concept for an infrastructure with a one-dimensional topography. This makes the analysis less complex without eliminating relevant properties of the infrastructure, and allows the returns-to-scale property to be derived from cost parameters and spatial distribution parameters. Model predictions

[5] This correlation applies to non-compressible liquids. See also Erich (1980), Fritsch (2014, p. 185 f.) and Knieps (2008, p. 22). The case of CO_2 is discussed in the next section.

[6] Bakken and von Streng Velken (2008) or Poiencot and Brown (2011), for example. Dooley et al. (2005) also analyze system costs, but restrict themselves to a simple source-sink matching heuristic. Fischedick et al. (2006) provide a technology-focused overview of infrastructure options for Germany in Chap. 9.

[7] Mendelevitch et al. (2010) present an optimal European infrastructure. Oei et al. (2010, 2011) focus on an infrastructure for CO_2 from industrial sources.

on the returns-to-scale properties were validated for the two optimized infrastructures (Middleton and Bielicki 2009a, b) discussed above.

To date, the analysis of optimal CCS transportation infrastructures has been restricted to given connection points. In the long term, however, infrastructures already established will play a role in the site selection considerations of carbon-emitting industries. In order to consider these decisions, in Sect. 9.4 the model for deriving a welfare-maximizing infrastructure is generalized based on the cost structure described in Sect. 9.3. The results show that transmission costs must be taken into account when planning a welfare-maximizing infrastructure. These interdependencies make it more difficult to implement the social optimum by regulating a natural monopoly compared to conventional monopolies.

In Sect. 9.5 the effects of 'non-linear tariffs' and of the regulation instrument of Ramsey pricing are analyzed for the model in Sect. 9.3. The analysis shows that the welfare maximum can only be implemented through cross-subsidies if the construction of the infrastructure is considered in conjunction with its operation.

9.2 Optimal CCS Infrastructures and Costs

It is not difficult to define what an optimal infrastructure is. However, determining an infrastructure that fulfils these optimality requirements is a demanding task due to the wide range of infrastructure options. The basic idea of (Middleton and Bielicki 2009b) involves breaking down all potential infrastructure networks into pipeline sections and then piecing them together again in a cost minimizing manner. This requires information on the investment and operating costs for individual point-to-point pipelines. The costs of these connections are determined in Baufumé et al. (2011), Odenberger and Svensson (2003) and IEA (2008), for example. Methodologically, conclusions on the costs of envisaged new CCS pipelines are drawn from the costs of pipelines already in operation. These have shown that the onshore pipeline investment costs C_{Inv} are linearly dependent on the pipeline length L and quadratically dependent on the pipeline diameter D (A and B are constants):

$$C_{Inv}(D, L) = \left(A + BD^2\right)L. \tag{9.1}$$

The investment costs therefore have a linear relationship with the connection length L. This course of the function can be made plausible by the fact that the material used for building pipelines represents an important share in construction costs and that the surface of these pipelines grows quadratically with the radius of the pipe and linearly with the length of the pipeline.

Taking into account an energy balance for the flowing CO_2 results in the following relationship between the pipeline diameter, pipeline length, and mass or volume flow $\dot{Q}$ (C is a constant; see also McCoy and Rubin 2008):

$$\dot{Q} = C\sqrt{D^5/L}. \tag{9.2}$$

This means that by doubling the diameter D, the volume flow $\dot{Q}$ can be increased almost sixfold if the length L remains constant. If the length of the pipeline is increased while keeping the diameter constant, then the volume flow will decrease. Therefore, in accordance with (9.2), due to the reduced flow, the pipeline length causes additional investment costs:

$$C_{Inv}\left(\dot{Q},L\right) = \left(A + \overline{B}L^{2/5}\dot{Q}^{4/5}\right)L. \tag{9.3}$$

According to Odenberger and Svensson (2003), the operating and maintenance costs C_{var} (E, F, and G are constants) are:

$$C_{var}\left(\dot{Q},L\right) = E + FL^{1/5}\dot{Q}^{2/5} + GL. \tag{9.4}$$

If the investment costs (9.3) and the operating and maintenance costs (9.4) are combined – as in Baufumé et al. (2011) – with an interest rate of 5 % and a depreciation period of 20 years, and if the operating and maintenance costs for intermediate compression are added (but not for liquefaction/compression at the power plant), the resulting onshore transportation price is € 1–4/tCO_2 per 100 km. For offshore transportation, it is € 1–5.5/tCO_2 per 100 km.

The exact investment and variable cost functions (9.3) and (9.4) are unsuitable for infrastructure optimization, because breaking down all potential infrastructure networks into pipeline sections and putting them together in a cost optimal manner using a linear program means that the transportation paths for CO_2 cannot be tracked individually. In this case, the length of the total transportation pathway of CO_2 L, inside parentheses in Eq. (9.2), is no longer known ex ante. This is why a constant correction factor is introduced for the average transportation pathway length $\overline{L}$ in parentheses in (9.3), which is contained in the constant b (step '$\approx_1$ in (9.5)). Linearization with reference to $\dot{Q}$ (step '$\approx_{LinQ}$' in (9.5)) makes it possible to handle the optimization problem as a linear program.[8] In addition, variable length-specific costs related or unrelated to the amounts of CO_2 transported can also be integrated into the cost function. The approximated investment costs $C_{MB}\left(\dot{Q},L\right)$ are:

$$C_{Inv}\left(\dot{Q},L\right)\approx_1\left(a + b\dot{Q}^{4/5}\right)L \approx_{LinQ}\left(c_{fix} + c_{var}\dot{Q}\right)L =: C_{MB}\left(\dot{Q},L\right). \tag{9.5}$$

In the following calculations, this cost function is used as a cost function for point-to-point connections which can be combined to form complex infrastructure architectures.

[8] One of the reasons why this approximation is sufficient as the exponent of Q (4/5) is close to 1.

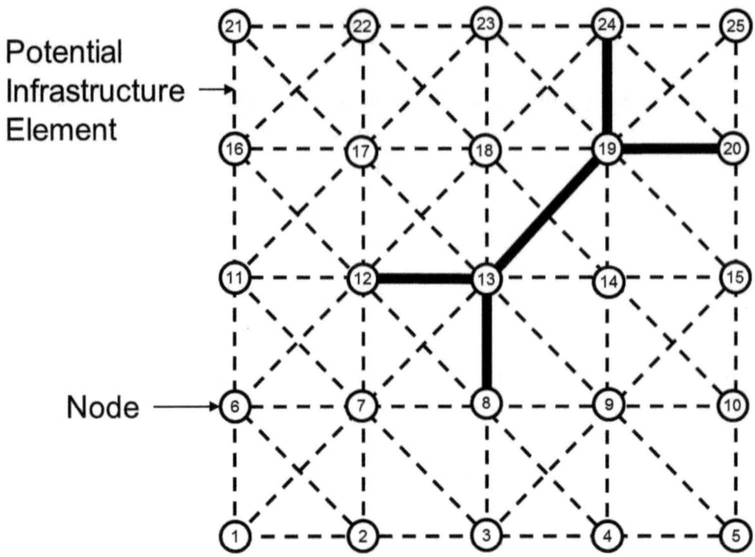

Fig. 9.2 Network topology
Infrastructure (black connections) $\boldsymbol{y}_{12,13}$, $\boldsymbol{y}_{8,13}$, $\boldsymbol{y}_{13,19}$, $\boldsymbol{y}_{19,20}$, $\boldsymbol{y}_{12,13}$, $\boldsymbol{y}_{19,24} > 0$ with connection points $\boldsymbol{a}_{24}$, $\boldsymbol{a}_{20} > 0$ and storage sites $\boldsymbol{b}_{12}$, $\boldsymbol{b}_8 > 0$

In order to derive the infrastructure costs from these point-to-point connection costs, it is necessary to define connection points (inlets and outlets) as well as the pipelines available for transportation. To this end, the surface on which an infrastructure is to be built is covered with a network of nodes $I = \{1, \ldots, N\}$ (Fig. 9.2). An amount $a_{i \in C}$ of CO_2 is captured at nodes i that form a set $C \subseteq I$. Storage of a volume $b_{j \in S}$ occurs at nodes j of a set $S \subseteq I$. The set of all possible infrastructures is then described by defining for each node a set of other nodes with which a connection can be established.

In the case of the topology shown in Fig. 9.2, each node can be connected to any of its eight adjacent nodes. According to Eq. (9.5), for each route from node i to node j, fixed costs c_{ij}^{fix} and variable costs $c_{ij}^{var} m_{ij}$ will be generated when a pipeline ij ($y_{ij} = 1$; without pipeline $y_{ij} = 0$) is built. c_{ij}^{fix} and c_{ij}^{var} may include site-specific characteristics such as topographic features in addition to investment, operating, and maintenance costs.

When describing an infrastructure, the amount of CO_2 entering the pipeline at each potential node must be equal to the amount of CO_2 transported away from that node (9.7). The transportation of positive amounts m_{ij} is only possible if the pipeline has been built (9.8):

$$\min_{m_{ij}, y_{ij}} c^{fix} \sum_{i,j} y_{ij} c_{ij}^{fix} + c^{var} \sum_{i,j} c_{ij}^{var} m_{ij} \tag{9.6}$$

$$s.t.: \sum_j m_{ij} - \sum_j m_{ji} - a_i + b_i = 0 \tag{9.7}$$

$$s.t.: 0 \leq m_{ij} \leq y_{ij} \sum_i a_i. \tag{9.8}$$

By minimizing the costs (9.6) across all possible connections y_{ij} and amounts of CO_2 transported m_{ij} under the boundary conditions of (9.7) and (9.8), the 'network' is chosen from all possible pipeline connections that fulfils the transportation requirements in the cost minimizing manner. The set of pipelines $y_{ij} > 0$ that solves the mixed integer problem (9.6), (9.7) and (9.8), is currently the best definition[9] of an optimal CCS infrastructure,[10] .[11] The model can be extended to include pipeline capacity limits, regional cost characteristics,[12] and capture and storage costs, and even in terms of spatial resolution (number of nodes), it is only limited by the time required to calculate a solution to the problem.

Using the example of California, (Middleton and Bielicki 2009b) derive optimal CCS systems[13] for the 37 largest CO_2 sources[14] and 14 storage sites (depleted oil fields) with a maximum total capacity of 56 $MtCO_2$/a for different amounts of captured CO_2. The resulting distribution of costs between the system components is shown in Fig. 9.3. The curves can thus also be interpreted as component-specific average cost functions.

The average system costs show a slight linear increase as a function of the stored amount, with the slope starting to increase at about 40 Mt/a. The capture costs represent about 90 % of the system costs.[15] While the capture costs are almost constant up to 40 Mt/a, they increase linearly at stored amounts of >40 Mt/a. This is due to the fact that sources with above-average CO_2 avoidance costs are included. The average costs of the system components of transportation and storage, in

[9] Less satisfactory approaches: e.g. Bumb et al. (2009) and Benson and Odgen (2002).

[10] These methods were developed for the area of water network design. A comprehensive overview can be found in Jezowski (2010).

[11] Infrastructures can be classified according to their spatial structure. Connected infrastructures, of which at least one infrastructure element is used as a production factor for several independent transports, is referred to as a network infrastructure.

[12] These can include significant cost reductions achieved by reducing planning efforts by building a new infrastructure close to existing infrastructures.

[13] As capture and storage costs were taken into account in this example in addition to the infrastructures, costs are not referred to here as infrastructure costs but as system costs.

[14] Including 21 gas power plants, 1 coal power plant, 10 oil refineries, and 5 cement works.

[15] Although it is expected that infrastructure costs only make up a small share of CCS system costs in Germany and the USA (e.g. McKinsey 2008), they can have a decisive impact on the system cost curve.

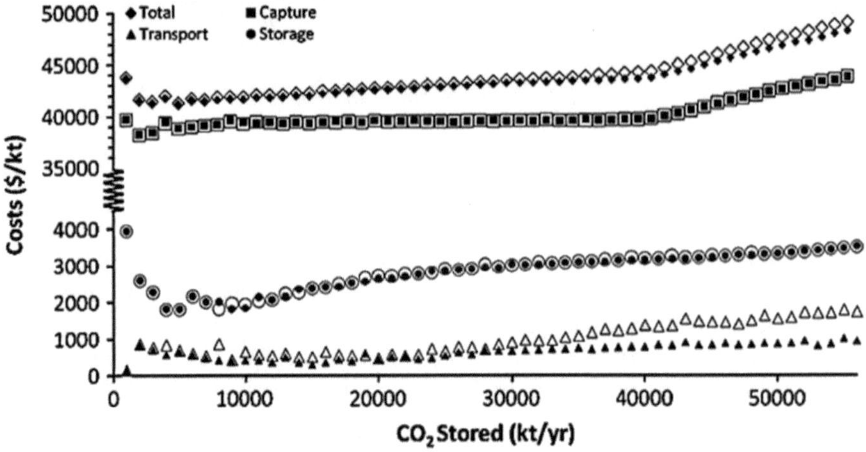

Fig. 9.3 CCS average system and component costs – Example 1 (Source: Fig. 5 from Middleton and Bielicki 2009b (Non-filled symbols represent the costs of direct connections from sources to storage sites. In this case, potential cost reductions through the construction of an optimized infrastructure are not exploited)

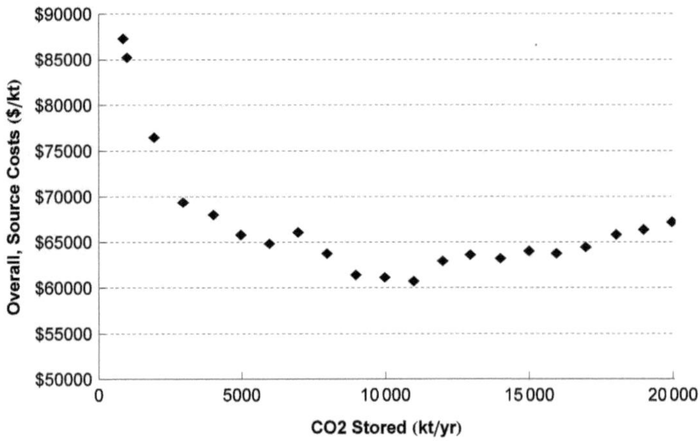

Fig. 9.4 CCS average system costs – Example 2 (Source: Fig. 2 from Middleton and Bielicki 2009a)

contrast, fall slightly with increasing system capacity, before increasing linearly with the amounts stored.

In Example 2, (Middleton and Bielicki 2009a) identify an optimal CCS system for the US Pacific Northwest region with 12 CO_2 sources[16] and a system capacity of up to

[16] Comprising 3 refineries, 3 cement works, 6 power plants (4 gas power plants and 2 coal power plants).

20.8 $MtCO_2$/a for storage in two deep-sea basalt formations 300 km west of the coast. The average system costs as a function of the stored amount of CO_2 are shown in Fig. 9.4.

In both examples, the authors explain in detail that the spatial structure of the optimal infrastructure changes significantly depending on the total capacity. This shows that an analysis of individual cases is needed to derive the infrastructure costs. It is not possible to evaluate an abstract infrastructure strategy directly using this approach. In order to supplement this analysis technique, a one-dimensional model is presented in the following section that includes the key characteristics of this analysis, and derives its cost function.

9.3 One-Dimensional Infrastructure Model

Firstly, the cost function of a one-dimensional infrastructure is established (see Fig. 9.5). This is a special case of the general network topology presented above, in which each node is only connected with two of its neighbouring nodes. At each connection point A_i, a CO_2 amount m_i is fed into the infrastructure. The distance between adjacent connection points A_i and A_{i+1} is L_{i+1}. The infrastructure is designed in such a way that source A_N located at the greatest distance from the storage site is connected to source A_{N-1} at the second greatest distance via a pipeline with length L_N and capacity m_N. The CO_2 amount m_{N-1} enters the pipeline at point A_{N-1}, so that the pipeline capacity between A_{N-1} and A_{N-2} is $m_{N-1} + m_{N-2}$, and so on up to source A_1.

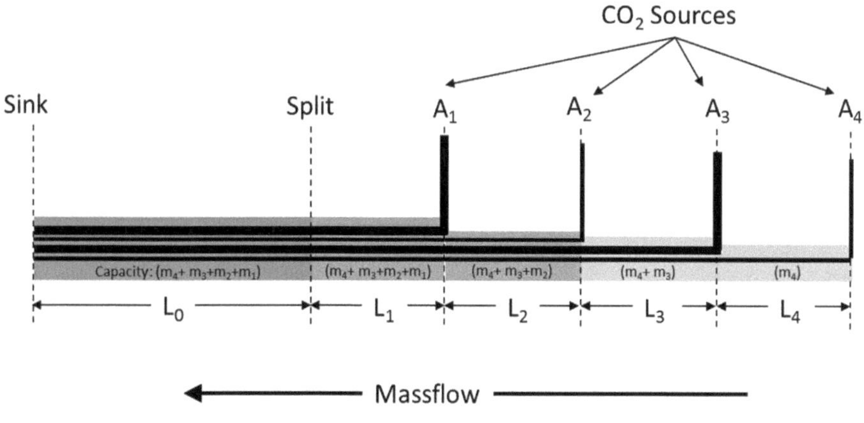

Fig. 9.5 Four sources in one-dimensional configuration

The total costs of building and operating an infrastructure for N sources with $m_0 = 0$ are:

$$C\Big((m_i, L_i)_{i=0,\ldots,N}\Big) = \sum_{i=0}^{N}\left(c_{fix} + c_{var}\sum_{j=i}^{N} m_j\right)L_i. \tag{9.9}$$

This equation can be used to derive the costs $\Delta C(m_k)$ for integrating an additional source into the infrastructure at point $k \in \{1,\ldots,N\}$ using the amount of injected CO_2 m_k and the construction of the new pipeline $L(m_k)$:

$$\Delta C(m_k) = C\Big((m_i, L_i)_{i=0,\ldots,N}\Big) + c_{fix}L(m_k) + c_{var}m_k\sum_{j=0}^{k} L_j. \tag{9.10}$$

Increasing the pipeline capacity by m_k gives rise to additional costs $c_{fix}L(m_k)$ for building the new pipeline, which causes the variable costs to increase by $c_{var}\, m_k \sum_{j=0}^{k} L_j$ on all pipeline sections through which the additional amount is transported. The cost increase is determined by the index k – the point where the additional CO_2 enters the pipeline – and by the new pipeline which is required to integrate the new source into the network $L(m_k)$. The cost increase for an extension of the pipeline at the point k by L_k can be determined in the same manner:

$$\Delta C(L_k) = C\Big((m_i, L_i)_{i=0,\ldots,N}\Big) + L_k\left(c_{fix} + c_{var}\sum_{j=i}^{N} m_k\right). \tag{9.11}$$

The extension of the infrastructure by the section L_k causes additional construction costs $L_k c_{fix}$ and transmission costs $L_k c_{var} \sum_{j=i}^{N} m_k$. When approaching the limit for m_k and L_k respectively to zero the marginal costs can be deduced from (9.10) and (9.11):

$$C_m(i) = c_{var}\sum_{j=0}^{i} L_j \tag{9.12}$$

$$C_L(i) = c_{fix} + c_{var}\sum_{j=i}^{N} m_j. \tag{9.13}$$

Whether the average costs rise or fall as a result of the cost increase depends on whether the cost increase is higher or lower than the average costs. This, in turn, depends on the distribution of connection points and the amounts of CO_2 entering the pipeline; in a special case where $m_{i>0} = m$ and $L_{i>0} = L$, $M = \sum_{i=1}^{N} m_i$ and $N = M/m$, however, the average costs according to (9.9) are:

$$\frac{C\Big(\left(0,L_0\right),\left(m,L\right)_{i=1,\dots,N}\Big)}{M}=\underbrace{\frac{c_{fix}L_0}{M}}_{\text{decreasing term}}+c_{fix}\frac{L}{m}+\underbrace{c_{var}\left(L_0+\frac{L}{2}\right)}_{\text{constant term}}+\underbrace{c_{var}\frac{L}{2m}M}_{\text{increasing term}}. \quad (9.14)$$

With respect to the capacity M the average costs are determined by a decreasing, a constant, and a linearly increasing term. The decreasing term results from fixed-cost degression on the initial pipeline section L_0, while the linear term comprises the total costs arising from the integration of additional sources. When the capacity is small, the decreasing term determines the average cost curve, while the linearly increasing term is dominant when the capacity is high. This is due to the fact that pipeline sections whose utilization increases due to the increasing amounts of CO_2 transported contribute to a reduction in the average costs (utilization effect, fixed-cost degression). However, as soon as supplying these amounts to be transported becomes more complex (connection effect), the average costs increase. Therefore (9.9) has a U-shaped curve with a local minimum at M^*:

$$M^*=\sqrt{2\frac{L_0\ c_{fix}}{L\ mc_{var}}} \quad (9.15)$$

$$\frac{C\Big(\left(0,L_0\right),\left(m,L\right)_{i=1,\dots,N}\Big)}{M^*}=\sqrt{\frac{L_0\ c_{fix}}{L\ c_{var}}2m}. \quad (9.16)$$

In addition to the ratio of variable costs to fixed costs, the density of connection points and their distance to the storage site are also important for the average cost curve. The discussion of the system cost curves for the model infrastructures 1 and 2 in Sect. (9.2) illustrates the relevance of these considerations.[17]

In Example 1, utilization effects at low capacities lead to a slight decrease in infrastructure average costs before the integration of more remote sources dominates the utilization effects and the average costs increase linearly. In Example 2, in contrast, it is expected that the infrastructure is subject to pronounced utilization effects due to the long offshore pipeline leading to the storage site (high value of L_0). This is indeed the case. In contrast to the previous example, the average system costs fall significantly down to a total amount of 10 Mt/a (half the maximum capacity). This shows that the cost functions of CCS systems display not only minor quantitative, but also qualitative differences depending on their scope and local circumstances, and that these differences can be understood using the simple model presented here.

[17] These findings also explain the results of Wildenborg et al. (2004): 'The backbone transport infrastructure becomes cost optimal when storage is restricted to offshore hydrocarbon fields. The costs amount to 9.74 €/t CO_2 without backbone and equal 4.48 €/t CO_2 with backbone.' However, their method for determining the backbone infrastructure is not transparent.

9.4 A Welfare-Maximizing Infrastructure Taking into Account Long-Term Business Decisions

So far, only the costs of building and operating a CCS infrastructure and the impact of given geographic connection and storage structures have been considered. Beyond this simplified and static spatial analysis, there are also dynamic elements when it comes to practical application. For example, a CCS infrastructure, once established, will be a factor in the decision of where to locate CO_2 sources, such as fossil-fired power plants and other CO_2-emitting industries. In this sense, it is necessary to differentiate between a short-term infrastructure analysis for existing pipeline structures and a long-term analysis. In the long term, the connection structures can no longer be considered a given. Instead, businesses will choose the location that is most favourable for them, so that in addition to the amounts of CO_2 fed into the pipeline system, the connection structure is also variable. What should the infrastructure capacity be set for the long term?

One approach to answering this question is based on maximizing social welfare W It is assumed here that it consists of the sum of the homogeneous good of electricity generated at N locations minus the costs for building and operating an infrastructure for the removal of the CO_2 emitted during the production process (no factor procurement costs) (Fig. 9.6). The production at one of these locations i is described by the production function $f^i(m_i, L_i)$ and a factor leading to an amount of CO_2 emissions m_i. The production site, which can be chosen in one dimension, is defined as the distance to the next producer L_i.

Modelling the distance L_i as a production factor is based on the notion that a greater distance to the next electricity producer increases the local market power of the provider on a market i with homogeneously distributed consumers, because they must take transportation costs into account when purchasing from competing providers. On the local markets i, the potential surcharge increases with the distance to the next production site.

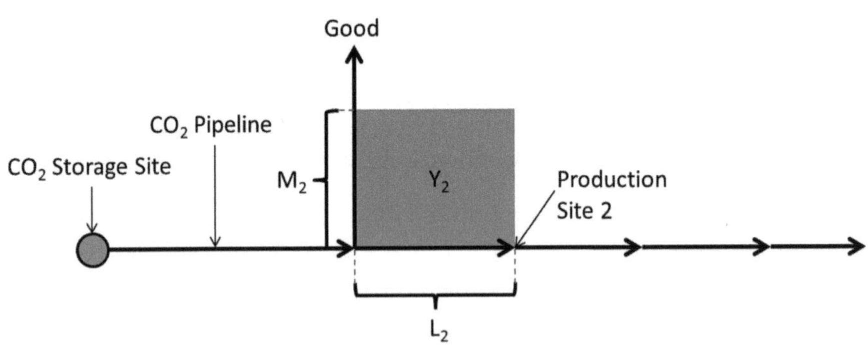

Fig. 9.6 Model of the production structure and spatial arrangement for long-term business decisions

As an example, the real revenue could be L_i^γ. At the same time, the expenses for operating a small-scale electricity distribution infrastructure of $L_i^{-\beta}$ increase with market size. The electricity generation Y_i on the market i for the special case where $\gamma = 1$ and $\alpha = \beta$ can be described as:

$$Y_i = m_i^\alpha L_i^{\gamma-\beta} = m_i^\alpha L_i^{1-\alpha}.$$

With this in mind, the electricity generation function $f^i(m_i, L_i)$ is now used with decreasing marginal revenues for the different factors. The welfare calculation of a social planner is:

$$\max_{m_i, L_i} W = \sum_{i=1}^{N} p^i f^i(m_i, L_i) - C\Big((m_i, L_i)_{i=0,\ldots,N}\Big). \tag{9.17}$$

From this equation, the first-order conditions with the marginal costs (9.12) and (9.13) can be derived:

$$\frac{\partial W}{\partial m_i} = p^i f^i{}_m - c_{var} \sum_{j=0}^{i} L_j = 0 \tag{9.18}$$

$$\frac{\partial W}{\partial L_i} = p^i f^i{}_L - c_{fix} - c_{var} \sum_{j=i}^{N} m_j = 0. \tag{9.19}$$

As a consequence, welfare is maximized when the production f^i is increased by introducing m_i only as long as the marginal revenue of the production increase $p^i f^i{}_m$ exceeds the variable cost increases caused by transporting additional CO_2 through all 'transmission sections' $c_{var} \sum_{j=0}^{i} L_j$ (9.18). If the connection points can also be selected, the distance L_i to the next production site $i-1$ must only be increased as long as the resulting revenue $p^i f^i{}_L$ exceeds the fixed costs and the variable costs $c_{fix} + c_{var} \sum_{j=i}^{N} m_j$ of transmission through pipeline section i (19).

Choosing the optimal amount of CO_2 to be transported m_i requires that the costs incurred between the storage site and the connection point $\sum_{j=0}^{i} L_j$ outside pipeline section L_i be taken into account. If the individual producers (owners of the production site) are equipped with ownership rights of the respective pipeline sections L_i, the costs incurred outside their 'private' pipeline section L_i can be interpreted as external effects for those sending CO_2 through that section. In addition, the amounts of CO_2 to be transported that are not caused by the owner of an individual pipeline section must also be taken into account when choosing the site. This means that the decision on where to locate is also subject to external effects.

9.5 Regulation

In Sect. 9.3 it is shown that the transportation infrastructure can have increasing returns to scale depending on the spatial distribution of the amounts of CO_2 transported. If the transportation capacities are low and the density of connection points is high, or if utilization effects play a significant role, returns to scale are likely to increase.

The relevance of this statement is illustrated by the examples in Sect. 9.2. In Example 2, the average system costs are at least \$ 60/tCO_2 (which corresponds to € 42/tCO_2)[18] and account for a 50 % share of the CO_2 identified as storable. Within the EU ETS (Emissions Trading Scheme), this means that the system would only produce a profit if emissions allowances cost more than \$ 60/tCO_2 (€ 42/tCO_2). If the price of emissions allowances is lower or if less than 10 MtCO_2 can be avoided per year, regulation would be required to make operation profitable for a business. In Example 1, in contrast, the average system costs are almost constant. In this case, an efficient allocation including the CCS system can be implemented at an allowance price of approx. \$ 41–47/tCO_2 (€ 28.7–29.9/tCO_2), depending on system capacity.

The economic significance of increasing returns to scale of the aggregated cost function is that it makes increases in capacity attractive which would inevitably lead to losses for the providers under perfect competition (cut-throat competition). Unless these losses are compensated for by subsidies, an infrastructure that takes full advantage of increasing returns to scale requires a natural monopoly. However, if the supply is unregulated, the market power of the monopolist will lead to deadweight loss, which can be avoided or at least reduced if the infrastructure is state-owned or if the natural monopoly is regulated.

Numerous proposals have been put forward for the regulation of a natural monopoly. In this section, only the well-known options of non-linear tariffs and Ramsey pricing are applied to the cost function (9.9), and the efficient tariff and pricing rule are discussed. It is important to bear in mind here that the marginal costs of the infrastructure do not equal those of the pipeline.

Non-linear pricing: if the monopolist can set a non-linear price according to:

$$V^*(m_i, L_i) = C_m(i)\, m_i + C_L(i)\, L_i + c_i \tag{9.20}$$

$$Nc_i = \left(c_{fix} + c_{var}M\right)L_0 - c_{var}\sum_{i=1}^{N} m_i \sum_{j=0}^{i} L_j. \tag{9.21}$$

then the welfare-maximizing infrastructure based on Eq. (9.18) and (9.19) can be implemented. A tariff V^* that implements a welfare-maximizing infrastructure comprises a term for the input $C_m(i)\ m_i$, a term for pipeline construction $C_L(i)\ L_i$, and a lump sum c_i. The prices for the input and pipeline construction correspond to

[18] At an average exchange rate of \$ 1 = € 0.7 in 2009.

the marginal costs $C_m(i)$ and $C_L(i)$ (9.12) and (9.13). The lump sum consists of a charge for pipeline investment and operating costs for the pipeline section L_0, which is levied on all infrastructure users, and a repayment of variable costs paid twice amounting to $-c_{var}\sum_{i=1}^{N} m_i \sum_{j=0}^{i} L_j$. The repayment function of this term suggests that an efficient infrastructure supply requires cross-subsidies between the construction of new pipelines and the operation of the infrastructure.

Ramsey pricing: In contrast, if only linear pricing is possible, then the prices for constructing and operating the infrastructure can be selected in such a way that they are welfare-maximizing in the sense of (9.17), but that the profits of the two infrastructure elements (construction and operation) do not become negative. In this case, cross-subsidies are also explicitly allowed. This includes regional cross-subsidies. In this case, the Ramsey rules are:

$$\frac{p_m - C_m(i)}{p_m} = \frac{k}{\varepsilon_{m,i}} \tag{9.22}$$

$$\frac{p_L - C_L(i)}{p_L} = \frac{k}{\varepsilon_{L,i}}. \tag{9.23}$$

In addition to the known result that surcharges are distributed in such a way that the highest ones are levied on the producers with the lowest factor substitution elasticities, the surcharge increases with the distance to the next producer, and the remote area surcharge increases with the amount transported.

9.6 Summary and Conclusions

In this chapter CO_2 transportation infrastructure was analyzed from an economic point of view. It demonstrated that the structure of the system cost function is determined by the utilization and connection effects which depend on the distribution of sources and sinks. If the utilization effect is strong, then a U-shaped average cost function depending on the capacity is expected, as shown in Example 2 (Middleton and Bielicki 2009a); if it is small, then the average costs are constant or increase. Although the infrastructure costs represent only a small share in CCS system costs, they can play a relevant role for the CCS system, because under decreasing average costs, the infrastructure – organized by a profit-maximizing planner – cannot be operated profitably without market power. In other words, this type of cost function, profitable production by a non-subsidized or unregulated private business is only possible above a lower threshold of avoidance costs (allowance price) and above the corresponding minimum capacity. If the allowance prices are lower, then the supply must be regulated. As Example 2 shows, this is to be expected in particular if long pipelines are necessary for offshore storage. Example 1 (Middleton and Bielicki 2009b) illustrates that the U-shaped average cost functions are less relevant if sources and sinks are evenly distributed. In such a case, the utilization effect is low, and the average costs are constant or increase. It is therefore necessary to examine each individual case in order to deploy CCS technology at minimal costs.

References

Bakken BH, von Streng Velken I (2008) Linear models for optimization of infrastructure for CO_2 capture and storage. IEEE Trans Energy Convers 23:824–833

Baufumé S, Hake JF, Linssen J, Markewitz P (2011) Carbon capture and storage: a possible bridge to a future hydrogen infrastructure for Germany? Int J Hydrogen Energy 36:8809–8821

Benson H, Odgen J (2002) Mathematical programming techniques for designing minimum cost pipeline networks for CO_2 sequestration. Sixth Green House Gas Technologies Conference, Kyoto, Japan

Bumb P, Desideri U, Quattrocchi F, Arcioni L (2009) Cost optimized CO_2 pipeline transportation grid: a case study from Italian industries. World Acad Sci Eng Technol 58:138–145

Dooley JJ, Dahowski RT, Davidson CL, Bachu S, Gupta N, Gale JA (2005) CO_2-Storage supply curve for north America and its implications for the deployment of carbon dioxide capture and storage systems. In: Proceedings of the 7th international conference on greenhouse gas control technologies, September 2004 2005 Vancouver Canada. Elsevier Ltd., pp 593–601

Erich K (1980) Industrieökonomik. Vahlen, München

Fischedick M, Esken A, Pastowski A, Schüwer D, Supersberger N, Nitsch J, Viebahn P, Bandi A, Zuberbühler U, Edenhofer O (2006) RECCS, Strukturell-ökonomisch-ökologischer Vergleich regenerativer Energietechnologien (RE) mit Carbon Capture and Storage (CCS). Abschlussbericht [Online]. Available: http://elib.dlr.de/69910/1/RECCS_Langfassung.pdf

Fritsch M (2014) Marktversagen und Wirtschaftspolitik - Mikroökonomische Grundlagen staatlichen Handelns, 9. Aufl. Vahlen, München. ISBN-10: 3800647710, ISBN-13: 978–3800647712

Geocapacity (2009) Assessing European capacity for geological storage of carbon dioxide, final report. Available: http://www.geology.cz/geocapacity/publications/D42%20GeoCapacity%20Final%20Report-red.pdf

IEA (2008) Energy technology perspectives 2008 – scenarios and strategies to 2050. Available: http://www.iea.org/textbase/nppdf/free/2008/etp2008.pdf

IPCC (2005) Special report on carbon dioxide capture and storage. Available: http://www.ipcc-wg3.de/publications/special-reports/.files-images/SRCCS-WholeReport.pdf

Jezowski J (2010) Review of water network design methods with literature annotations. Ind Eng Chem Res 49:4475–4516

Knieps G (2002) Netzsektoren zwischen Regulierung und Wettbewerb. In: Berg H (ed) Deregulierung und Privatisierung: Gewolltes – Erreichtes – Versäumtes. Schriften des Vereins für Socialpolitik, Gesellschaft für Wirtschafts- und Sozialwissenschaften. Duncker und Humblot, Berlin

Knieps G (2008) Wettbewerbsökonomie: Regulierungstheorie, Industrieökonomie, Wettbewerbspolitik, 3. Aufl. Springer, Berlin/Heidelberg. ISBN-10: 3540783482, ISBN-13: 978–3540783480

Mccoy ST, Rubin ES (2008) An engineering-economic model of pipeline transport of CO_2 with application to carbon capture and storage. Int J Greenh Gas Control 2:219–229

McKinsey (2008) Carbon capture & storage: "Assessing the economics". Available: http://europeangashub.com/upload/extra_files/force_download.php?file=attach_24.pdf

Mendelevitch R, Oei PY, Tissen A, Herold J (2010) CO_2-highways – modeling aspects of a future CO_2 transport infrastructure. Available: http://aei.pitt.edu/15200/01/WD_340_CO2_Highways.pdf

Middleton RS, Bielicki JM (2009a) A comprehensive carbon capture and storage infrastructure model. Proceedings of the 9th international conference on greenhouse gas control technologies (GHGT-9). Energy Procedia, pp 1611–1616

Middleton RS, Bielicki JM (2009b) A scalable infrastructure model for carbon capture and storage: Sim-CCS. Energy Policy 37:1052–1060

Odenberger M, Svensson R (2003) Transportation systems for CO_2 – application to carbon sequestration. Department of Energy Conversion Chalmers University of Technology, Göteborg

Oei PY, Mendelevitch R, Herold J, Tissen A, Hirschhausen CV (2010) CO_2-Autobahnen für Europa? Energiewirtschaftliche Tagesfragen 12:46–52

Oei PY, Herold J, Tissen A (2011) Modeling a carbon capture, transport and storage (CCTS) infrastructure for the industrial sector. Enerday [Online]. Available: http://tu-dresden.de/die_tu_dresden/fakultaeten/fakultaet_wirtschaftswissenschaften/bwl/ee2/lehrstuhlseiten/ordner_veranstaltungen/ordner_enerday/Enerday%202011/ed2011download/oei.pdf. Accessed 8 Apr 2011

Parfomak PW, Folger P (2008) Carbon dioxide (CO_2) pipelines for carbon sequestration: emerging policy issues. CRS report for Congress, prepared for members and committees of Congress, United States, April 19, 2007. http://www.policyarchive.org/handle/10207/bitstreams/3227.pdf

Poiencot B, Brown C (2011) An optimal centralized carbon dioxide repository for Florida. Int J Environ Res Public Health 8:955–975

Wildenborg T, Gale J, Hendriks C, Holloway S, Brandsma R, Kreft E, Lokhorst A (2004) Cost curves for CO_2 storage: European sector. In: 7th international conference on greenhouse gas control technologies (GHGT-7), Vancouver, Canada

Chapter 10
The System Value of CCS Technologies in the Context of CO_2 Mitigation Scenarios for Germany

Dag Martinsen, Heidi Heinrichs, Peter Markewitz, and Wilhelm Kuckshinrichs

Abstract This chapter analyses the system value of CCS in Germany within the context of consistent greenhouse gas reduction scenarios with and without the implementation of CCS technologies. The system value of CCS is determined using additional CO_2 avoidance costs that would occur if climate change mitigation targets were to be met without using CCS even though CCS technology was available. The development of important parameters, assumptions and energy- and climate-policy targets are represented in scenarios. The methodological basis for the scenario calculations is the bottom-up energy system model IKARUS. The energy economics results comprise energy and CO_2 balances, capacity development, and the costs of CO_2 reduction strategies. From this, the system value of CCS and the contribution of all sectors to it are derived.

Keywords System value • Energy system model • CO_2 avoidance costs

10.1 Introduction

Binding greenhouse gas reduction targets necessitate a huge range of greenhouse gas reduction measures covering all energy sectors as well as industry, trade, transport and traffic, and households. More than 40 % of global CO_2 emissions are caused by electricity generation in fossil-fired power plants. This is therefore of particular significance in the context of greenhouse gas reduction.

The German Federal Government has set CO_2 reduction targets of 40 % for 2020 and 80 % for 2050 in relation to levels in 1990. In addition to CO_2 reduction, German energy and climate policy comprises further ambitious targets. These include increasing energy efficiency and increasing the use of renewable energy.

D. Martinsen • H. Heinrichs (✉) • P. Markewitz • W. Kuckshinrichs
Institute of Energy and Climate Research – Systems Analysis and Technology Evaluation (IEK-STE), Forschungszentrum Jülich GmbH, D-52425 Jülich, Germany
e-mail: d.martinsen@fz-juelich.de; h.heinrichs@fz-juelich.de; p.markewitz@fz-juelich.de; w.kuckshinrichs@fz-juelich.de

W. Kuckshinrichs, J.-F. Hake (eds.), *Carbon Capture, Storage and Use*,
DOI 10.1007/978-3-319-11943-4_10

These targets themselves, as well as how to achieve them, must be taken into account when projecting the capacities of fossil power plants.

Current scenarios for the reduction of greenhouse gases in Germany show that CCS technologies can play an important role within the context of national greenhouse gas reduction strategies with binding reduction targets. Analyses show that for CO_2 reduction targets of more than 35 % (for 2030) the use of power plants with CCS can represent an attractive reduction measure from an economic perspective. Sensitivity calculations concerning investments, energy carrier prices, etc. show that this is a robust reduction measure (Martinsen et al. 2007).

Other scenarios deal with pathways of energy supply completely based on renewable energy (e.g. Krewitt et al. 2009 or DLR et al. 2012). The future usage of CCS technology is sometimes explicitly excluded. This is explained by the fact that CCS technologies are not commercially available and that they have been implemented on a power-plant scale today in no more than a few demonstration projects at best. In addition, it is often argued that the implementation of CCS technologies on a commercial scale will come up against considerable acceptance problems, and that the construction of CCS infrastructures for the transportation and storage of CO_2 appears unrealistic. Furthermore, it is often denied that there is a need for the implementation of CCS technologies, because, after all, when the political targets for energy efficiency and renewable energy have been met, a sufficient energy supply will be available.

This is where determining the system value of CCS technologies comes into play. The system value is a term that has been borrowed from the area of the economic analysis of environmental resources. It is calculated on basis of the difference between the values that individuals are willing to pay to ensure continued availability of a specific natural resource and the expected one of future usage. It is therefore a value for the system of being able to use the resource in future. In the case considered here, the reverse applies and the value arises for not using a technology. The system value therefore implicitly indicates the willingness to pay which is necessary should CCS technology not be used.

This chapter analyses the system value of CCS in Germany within the context of consistent greenhouse gas reduction scenarios with and without the implementation of CCS technologies. The development of important parameters, assumptions and energy- and climate-policy targets are represented in scenarios. The methodological basis for the scenario calculations is the bottom-up energy system model IKARUS.

The methodological approach and scenario design are explained in Sect. 10.2. The energy economics results are presented in Sect. 10.3. These comprise energy and CO_2 balances, capacity development, and the costs of CO_2 reduction strategies. From this, the system value of CCS can be derived. In Sect. 10.4, conclusions are drawn.

10.2 Methodological Approach and Scenario Design

10.2.1 System Value

The system value of CCS is determined using additional CO_2 avoidance costs that would occur if climate change mitigation targets were to be met without using CCS even though CCS technology was available (Bauer et al. 2009; Pietzcker et al. 2009; Manger et al. 2009). It basically represents a monetary value for refraining from using climate change mitigation technologies, and can be interpreted as a measure of the necessary willingness to pay for refraining from using this technology.

The system value of CCS technologies is not a statically given variable. It depends on numerous parameters and general assumptions, including first and foremost technical parameters such as the costs and potential of competing technologies, as well as targets for the reduction of CO_2 together with those for the use of other technologies such as renewable technologies, energy-efficient technologies, or nuclear energy. Figure 10.1 is a schematic demonstrating this correlation.

In a multi-option scenario (I), all technical options are allowed, and an energy mix is established which leads to CO_2 reduction with minimal costs. The more stringent the reduction targets, the higher the reduction costs. In an alternative scenario (II), the share of competing technical options must be higher, because CCS technologies are not permitted in this scenario. If we follow the assumption of increasing marginal costs of the technologies for CO_2 reduction, then for the given CO_2 reduction targets, the respective reduction costs in scenario (II) are higher. If the technical alternatives for substituting the use of CCS were further regulated, and existing fossil-fired power plants, for example, were only replaced by highly efficient new ones, the costs for a given CO_2 reduction would increase further. The reverse is also true that the costs for a given CO_2 reduction would be lower if

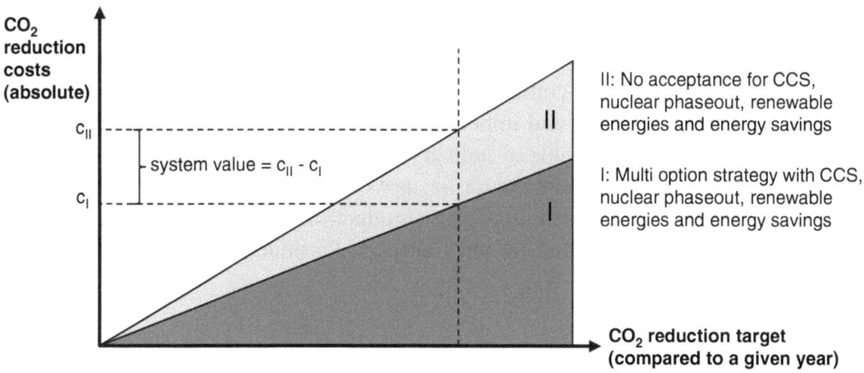

Fig. 10.1 Schematic of the system value of CCS technologies (Linearity used for the purpose of schematic representation)

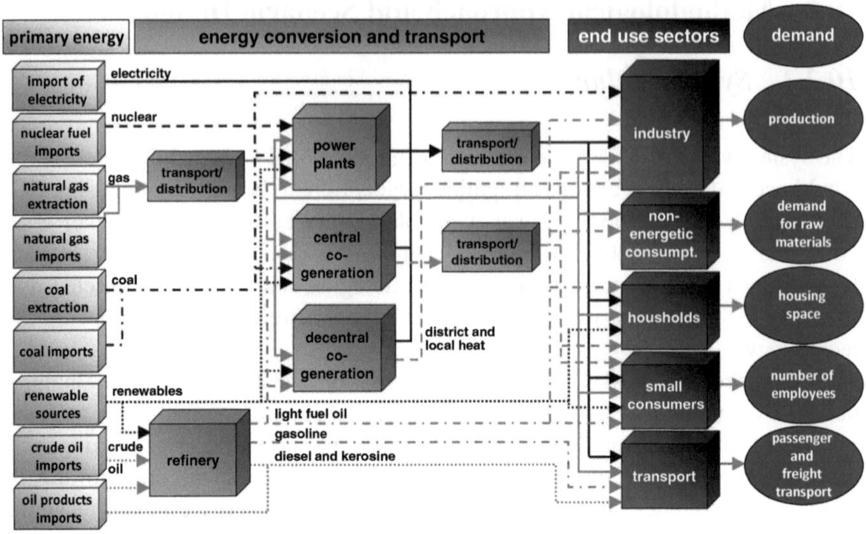

Fig. 10.2 Structure of the IKARUS energy system model (Source: Hake et al. 2009)

restrictions for the technical alternatives were to be relaxed, e.g. by extending the operating time of nuclear power plants.[1]

The respective CO_2 reduction costs of the two cases I and II shown schematically here are indicated by c_I and c_{II}. The associated system value is calculated as $c_{II} - c_I$.

10.2.2 *The IKARUS Energy System Model*

The scenarios were calculated with the energy system model IKARUS. IKARUS is a demand-driven bottom-up energy model, which represents the German energy system and depicts energy technologies in detail (Fig. 10.2).

It depicts the energy flows from the primary energy via the end-use energy to the demand governing energy consumption, and differentiates between the primary energy resources (domestic and import), conversion sector, energy transport, need for end-use energy in the fields of industry, non-energy consumption, households, small consumers, and transport and traffic, as well as demand, which is described by industrial production, energy-intensive production, living space, number of employees, and passenger and freight transport. Particular emphasis is on saving

[1] With regard to the use of nuclear energy, this aspect does not come into play. Both scenarios assume that nuclear energy will be phased out and that no nuclear power stations will produce electricity from 2023 onwards. On the other hand, CCS power plants will effectively only be available from 2020. See Sect. 10.3.

energy using technologies designed to increase energy efficiency by linking the demand and need for end-use energy.

The approach allows the definition of restrictions for the energy system, e.g. regarding the use of certain energy technologies. With respect to climate change mitigation, the emphasis is on defining CO_2 target values respectively upper limits. In principle, the approach would also allow an upper limit to be set for reduction costs.

The model calculates primary and end-use energy consumption, corresponding greenhouse gas emissions, the (necessary) capacity development of technologies, and reveals the total costs. The results for all variables are consistent with the scenario requirements, basic assumptions and technology data. They are reproducible and sensitivity calculations can thus be performed for the main assumptions and parameters.

10.2.3 Scenario Structure, Underlying Data and Basic Assumptions

The following outlines the main assumptions and framework upon which the scenarios are built. The scenarios cover the period 2005–2050.

For the period 2010–2050, it is assumed that the gross domestic product (GDP) increases by 1.4 %/a and that the population decreases by 2050 to around 77 million. Although the GDP is not directly incorporated into the model calculations, it sets the framework for the exogenously determined demand for energy services.

Figure 10.3 shows the demand over time of the most important areas of demand determining energy consumption. The development of energy services depends heavily on the sector being considered. While demand in passenger transport services remains almost constant, it is assumed that freight transport services almost double by 2050. The gross value added of industry also increases considerably (+84 % by 2050) and is characterized by structural changes in favour of less energy-intensive sectors. Here, steel and aluminium production would be particularly affected as would the cement industry. Living space increases moderately by 25 % in the period 2005–2050. The number of employees in the sectors of commerce, trade and services drops by around 17 % due to the underlying demographic development.

For CCS power plants and for power plants based on renewable energy, upper limits up to 2050 have been set in the model because their potential is limited by factors such as usable amounts of biomass and the maximal amounts of CO_2 that can be stored. For CCS, both new plants and the retrofitting of existing plants are considered. Figure 10.4 shows the maximal possible expansion of the capacity of these power plant types as installed net capacity up to 2050.

Costs for investments and net efficiencies for the most important fossil power plant types as well as other technical and ecological data are selected analogously to the data in Chap. 7. The analysis also incorporates fixed and variable operating

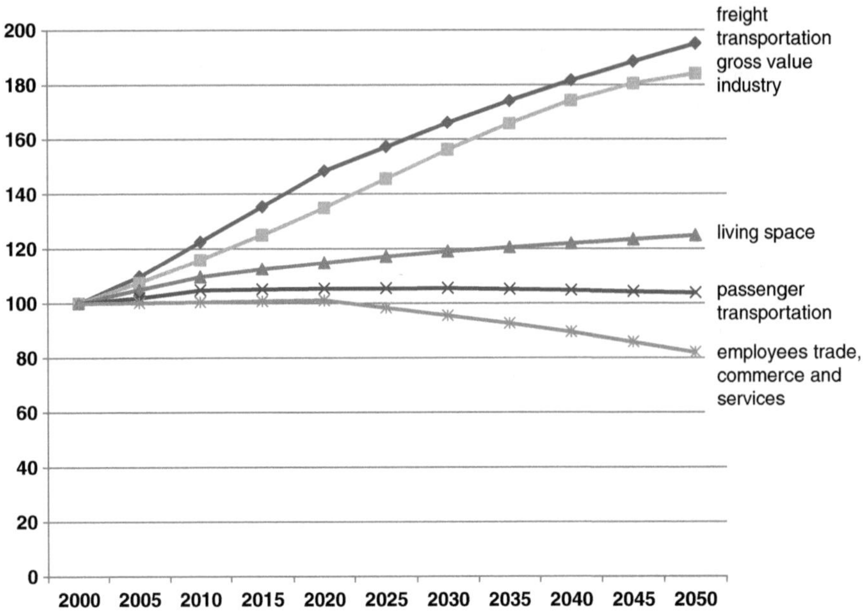

Fig. 10.3 Development of demand according to energy services (Source: Hake et al. 2009)

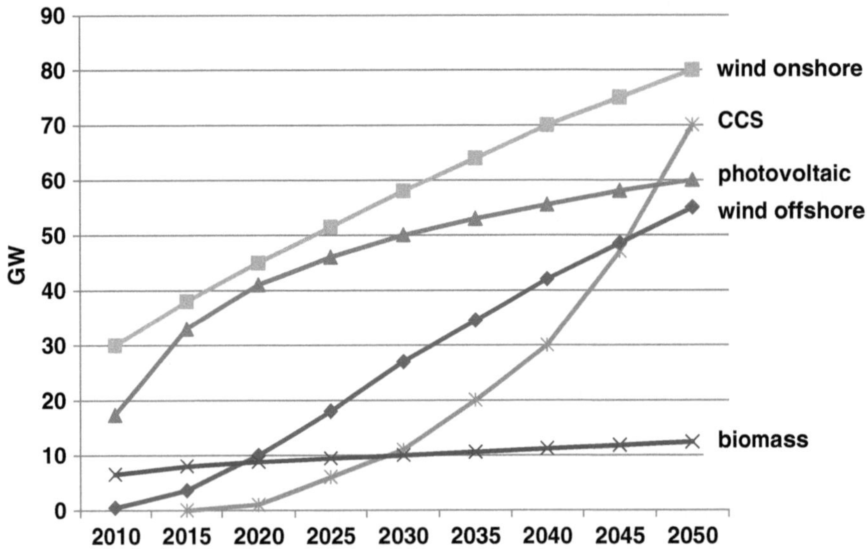

Fig. 10.4 Upper limits for net installed power plant capacity

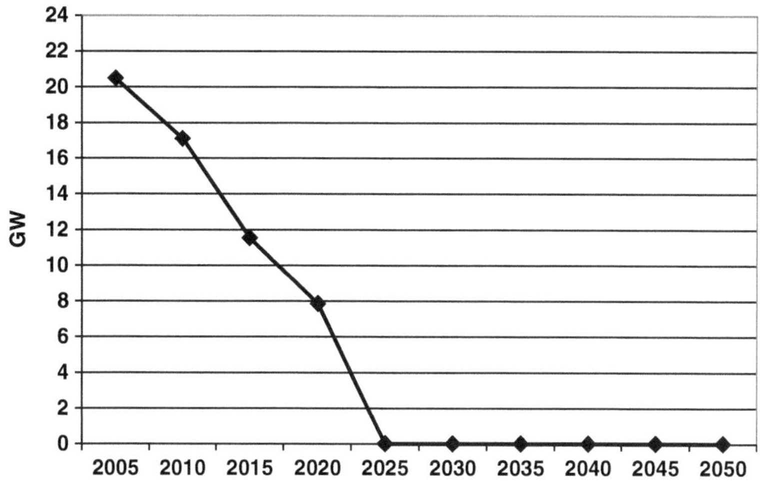

Fig. 10.5 Limitation of installed net nuclear power plant capacity

costs, which also include mean transportation and storage costs for CO_2. Furthermore, the subsequent calculations assume the phasing out of nuclear power. In the model, this means reducing the (net) nuclear power plant capacities in accordance with current legislation, as shown in Fig. 10.5.

The assumed price development of the most important imported energy carriers is shown in Fig. 10.6 in monetary value as of 2010. The real crude oil price in 2050 is equal to US$$_{2010}$ 130/bbl (with US$ 1.3/€ in 2050).

With respect to the import of solar power from North Africa, it was assumed that this will be available in larger quantities from 2030 onwards, and that the price will decrease over time. However, there is a cap on the maximal quantities that can be imported:

- Import price of € 0.19/kWh in 2030 decreasing to € 0.15/kWh in 2050 (see Komendantova et al. 2010; Williges et al. 2010).
- Upper limit for imports in 2050: 70 TWh or approx. 20 % of the total electricity needed.

The following scenarios were generated with the IKARUS model:

- REF: reference scenario without CO_2 reduction targets
- CA: CO_2 reduction targets with a CCS option
- CD: CO_2 reduction targets without a CCS option

In the scenarios with CO_2 reduction targets (CA, CD), energy-related CO_2 emissions are limited after 2010, as shown in Fig. 10.7. By 2050, the energy-related CO_2 emissions may not exceed 23 % of the 1990 level (temperature adjusted). The development over time is mapped based on the mid-term targets of −40 % in 2020 and −55 % in 2030.

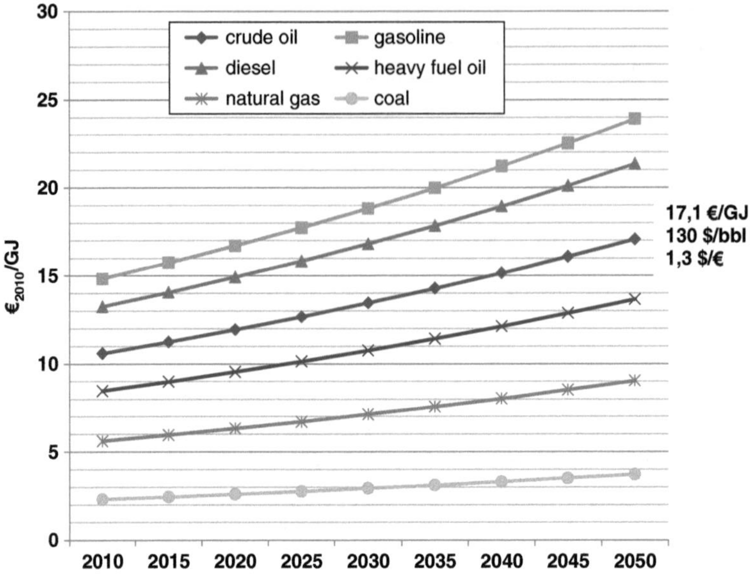

Fig. 10.6 Price development of imported energy carriers

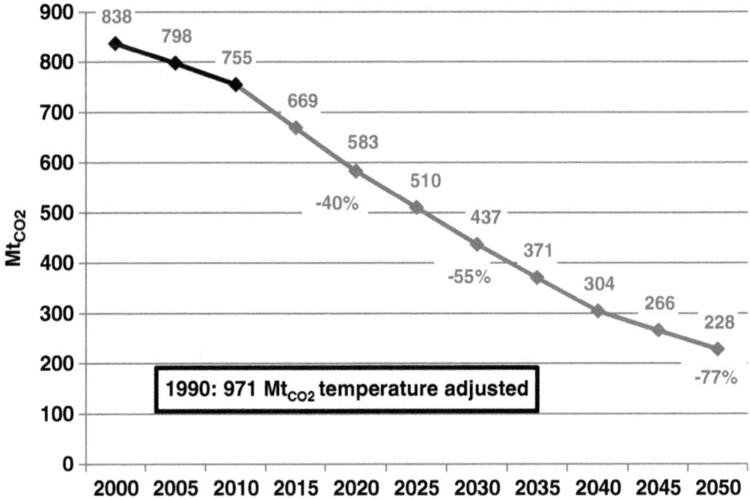

Fig. 10.7 CO_2 restrictions in scenarios CA and CD (Source: Hake et al. 2009)

10.3 Energy Economics Results

10.3.1 Energy and CO_2 Balances

For the scenarios defined in the previous section, the following will be compared for 2005–2050: primary energy balances according to energy carriers, electricity generation and power plant capacities according to kind/type, end-use energy by sector, and CO_2 emissions broken down into sectors.

10.3.1.1 Primary Energy

Even without a CO_2 reduction target, i.e. in the reference scenario (REF), a clear drop in primary energy demand can be seen (−23 % from 2005 to 2050). Renewables account for a relatively constant share, while oil becomes less important in the primary energy mix. The CO_2 scenarios CA and CD show the same development qualitatively, but over time renewable energy from wind and biomass increasingly replaces the energy carrier hard coal, oil, gas and nuclear energy. In addition, the changes in scenario CD in particular are more pronounced (CA: −28 %, CD: −35 %). In the scenario without CCS (CD), the primary energy efficiency is higher than in the scenario with CCS. This is due to greater energy savings and the larger share of renewable energy in this scenario as chosen by the model. Overall, the share of renewables in primary energy in 2050 increases in both scenarios with CO_2 restrictions to 38 % (CA) and 48 % (CD) (Fig. 10.8).

10.3.1.2 End-Use Energy

Overall, the end-use energy demand decreases in the period 2005–2050 by approx. 16 % (REF), 27 % (CA) and 31 % (CD). The changes in end-use energy consumption are very different in the individual sectors. In the transport and traffic sector, the energy demand only drops slightly (−4.5 % in CA and −6 % in CD) or even grows slightly (+5 % in REF) despite a considerable decrease in the mean specific fuel consumption. This can be explained by the strong growth in freight transport services. In all other sectors, the end-use energy demand drops distinctly due to energy savings measures (e.g. thermal insulation) (Fig. 10.9):

- Households: from −18 % (REF) to −41 % (CD)
- Industry: from −29 % (REF) to −41 % (CD)
- Commerce, trade and services: from −28 % (REF) to −54 % (CD)

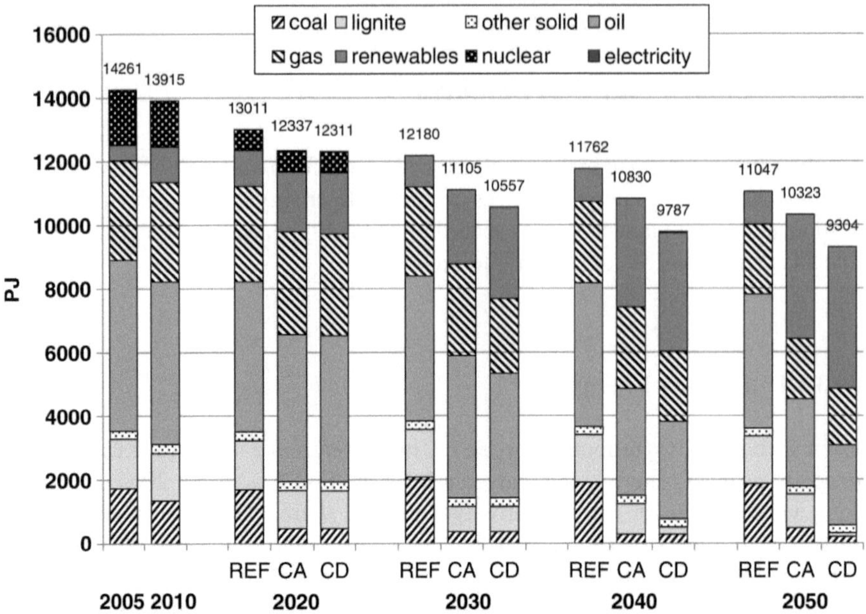

Fig. 10.8 Comparison of primary energy according to energy carriers

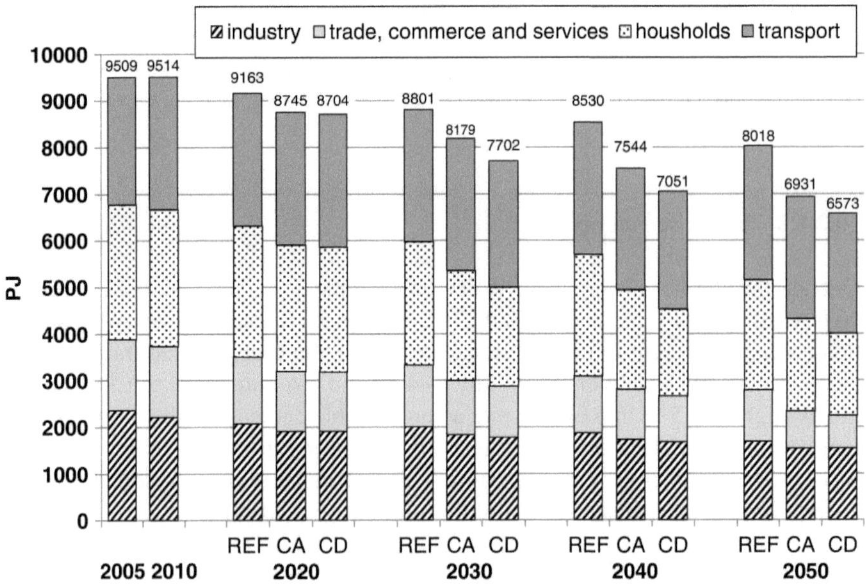

Fig. 10.9 Comparison of end-use energy demand by sector

10.3.1.3 Installed Net Capacity

In the reference scenario, the installed net capacity of the electricity generation plants remains almost constant at slightly more than 150 GW for the period 2010–2050. The following trends can be discerned: The decommissioned nuclear power plant capacity is replaced primarily by building lignite and hard coal power plants. The installed wind power capacity (on-shore and off-shore) remains at 30 GW after 2010 and is not further expanded before 2050. Gas power plants are used as reserve capacity (at very low utilization) for short-term wind fluctuations. In the reduction scenarios, the required power plant capacity is much higher than in the reference scenario, whereby the capacity in the CD scenario increases continuously to more than 300 GW in 2050, while in the CA scenario it initially increases to approx. 260 GW by 2040 and then decreases to almost 220 GW in 2050. The capacity of wind turbines in particular is expanded (max. 87 GW (CA) and 116 GW (CD) in 2040 and 2050, respectively) and PV plants in the scenario without CCS grows slightly to max. 36 GW from 2030 onwards, while biomass power plants (in Fig. 10.10 under 'others') almost constantly account for 10 GW of the installed power plant capacity. In the scenario CA (with CCS), the CCS option for reducing CO_2 is taken from the model, where a total of some 41 GW power plants with CCS are erected by 2050. Of this, lignite CCS accounts for approx. 19 GW, hard coal CCS for approx. 5 GW, and gas CCS for approx. 17 GW. In addition, CCS power plants increase the utilization of the installed power plant fleet, which means that less capacity is required overall in 2050. In scenario CD (without CCS), the existing

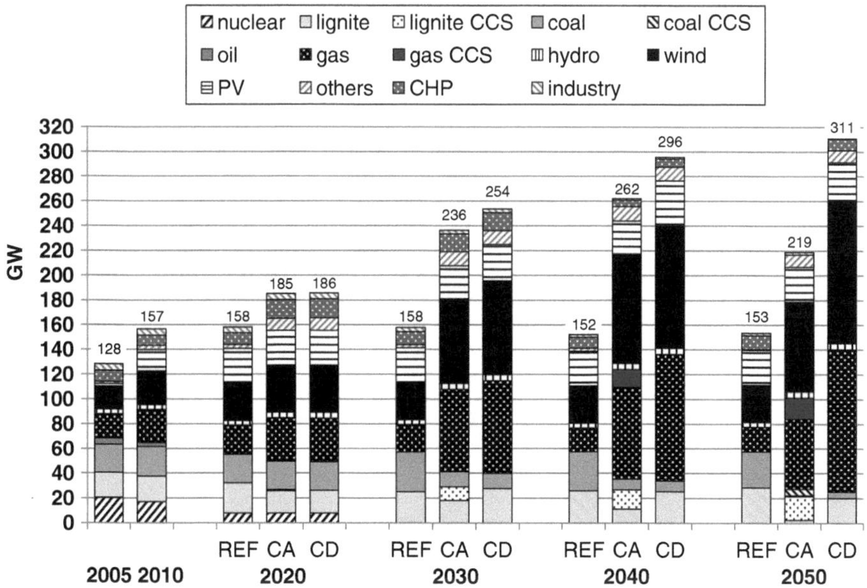

Fig. 10.10 Comparison of installed net capacity by power plant type

lignite and hard coal power plants are hardly used at all in the later periods. However, compared to the scenario with CCS, additional capacities such as PV and wind are incorporated in the model. Overall, the increase in intermittent power plant types leads to a greater need for reserve capacity, which is provided by gas power plants.

10.3.1.4 Net Electricity Generation

In contrast to the strong growth in power plant capacity, electricity generation in the reference scenario experiences a minimal drop by 2050 (−4 %) (Fig. 10.11).

In the reduction scenarios, there is either a transition to CCS power plants (CA) or wide-reaching electricity savings measures are introduced (CD), which in scenario CA results in almost no drop in the net electricity generation (−2 %), but in a clear drop in scenario CD (-8 %). The share of electricity generated from renewables increases considerably in the reduction scenarios from approx. 11 % in 2005 (approx. 18 % in 2010) to nearly 50 % (CA) and even almost 80 % (CD) in 2050. In scenario REF without CO_2 restrictions, the share of electricity from renewables increases slightly to approx. 18 % in 2050. In the CO_2 scenario without CCS, a large share – in 2050 up to approx. 68 % – of the total energy demand is covered by wind and biomass power. In contrast, the CCS power plants in the scenario with CCS (CA) cover approx. 50 % of the demand for electricity. Overall, in the CO_2 reduction scenarios, electricity generation in 2050 is almost CO_2-free.

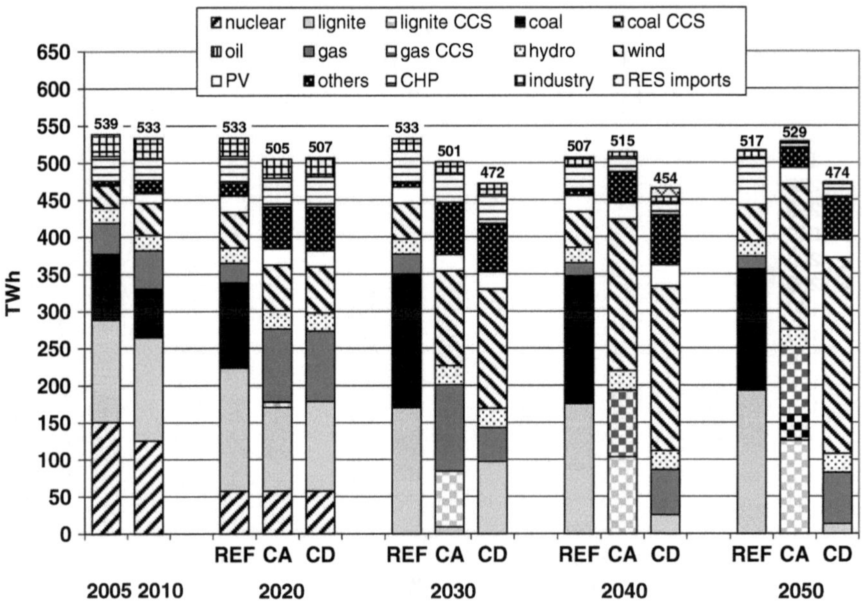

Fig. 10.11 Comparison of net electricity generation by power plant type

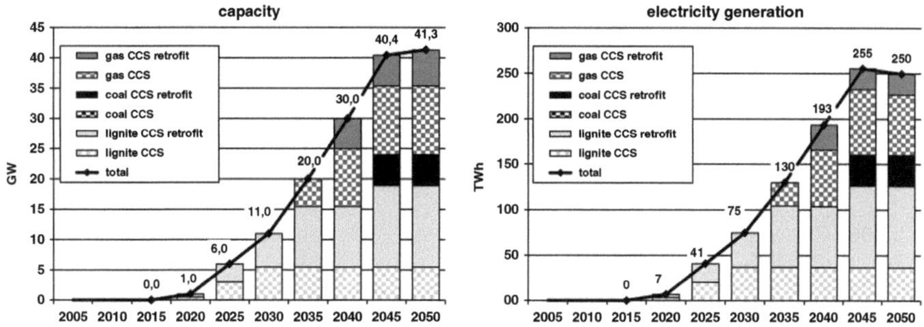

Fig. 10.12 Installed net CCS capacity and net CCS electricity generation according to power plant type in the CCS scenario (CA)

10.3.1.5 Installed Net CCS Capacity and CCS Electricity Generation

Figure 10.12 shows the development of CCS capacity for the reduction scenario CA in more detail. We assume that CCS technology will be available from 2020, and the upper capacity limit therefore expands from this point on. In addition to differentiating according to power plant type, the model also differentiates between new plants and carbon capture retrofits for existing power plants (built after 2005). For lignite power plants – because of inexpensive domestic lignite and the base-load requirements – the option of CCS new plants and that of CCS retrofitting are both selected by 2050. Hard coal power plants are almost only retrofitted because the construction of new hard coal power plants is avoided in the model. The same should hold for gas power plants, but as these have to be available as operating reserve, they are retrofitted and new plants with CCS are constructed.

10.3.1.6 CO_2 Emissions

Figure 10.13 shows the CO_2 emissions broken down into sectors. Even in the reference scenario (REF) without a CO_2 reduction target, there is a decrease in CO_2 emissions by 2050 of approx. −14 % compared to levels in 2005. This can be explained mainly by developments in the sectors industry, commerce, trade and services, and households. The most important drivers are increasing energy prices, which induce energy savings measures and structural changes in industry. In contrast, the CO_2 emissions in the sectors of transport and traffic and electricity generation remain constant. In the electricity sector, the emissions increase temporarily (in 2030) to around 30 $MtCO_2$ due to the phasing out of nuclear energy.

Although there is an emission cap in the CO_2 scenarios with and without CCS (CA and CD), the model selects cost-optimized sector-independent measures, which are essential for compliance with the upper limit, i.e. the sectoral breakdown shown in Fig. 10.14 is a result of the model calculation. In relation to the reference scenario, the emissions are halved in the transport and traffic sector by 2050 and

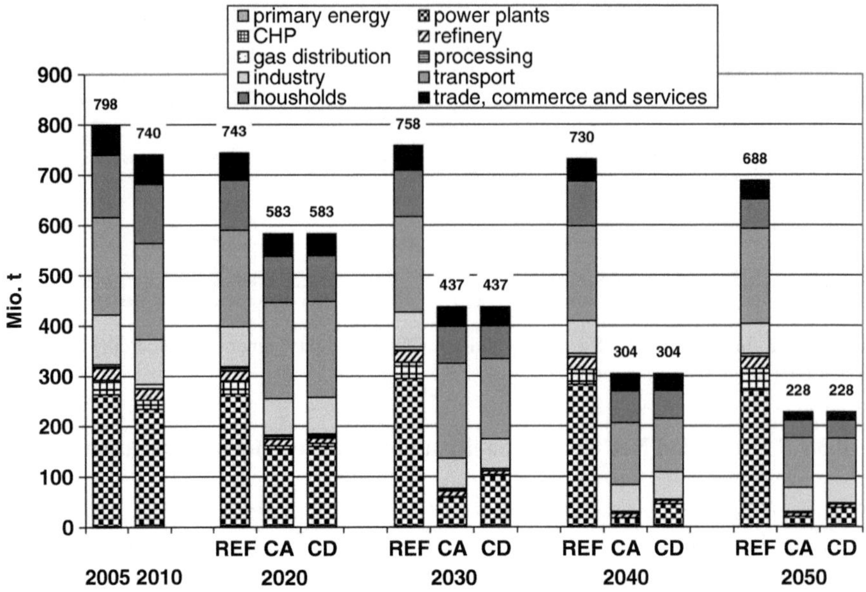

Fig. 10.13 Comparison of CO_2 emissions by sector

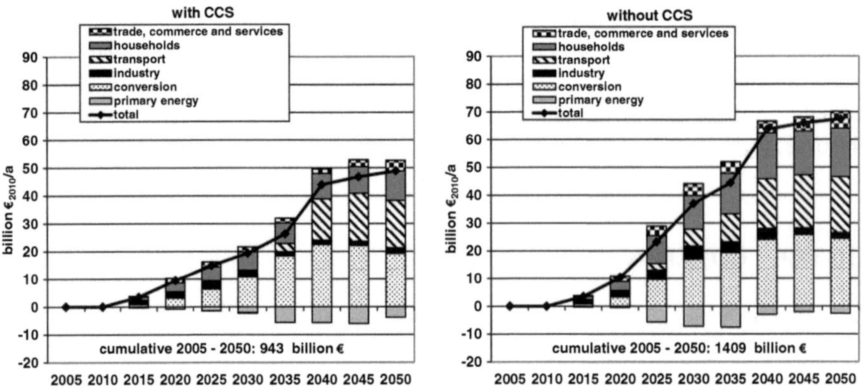

Fig. 10.14 Annual additional costs of the CO_2 scenarios in relation to the reference scenario

there is an even stronger CO_2 reduction in the electricity sector. Compared to 2005, the following reductions are calculated for 2050 in scenarios CA and CD:

- Electricity sector: 87–93 % (REF: +4 %)
- Industry: 51–52 % (REF: 40 %)
- Commerce, trade and services: 72–73 % (REF: 27 %)
- Households: 71 % (REF: 53 %)
- Transport and traffic: 49–59 % (REF: 2 %)
- Total: 71 % (REF: 14 %)

10.3.1.7 Comparison of CO_2 Reduction Scenarios

The comparison of sectoral CO_2 emissions and of the measures for the two CO_2 reduction scenarios with CCS (CA) and without CCS (CD) is particularly interesting here. In summary, the following can be concluded:

By 2050, in the scenario without CCS (CD) the amount of CO_2 in the conversion sector will increase while end-users will emit less CO_2 than in the scenario with CCS (CA). In other words, more measures affecting end-users must be introduced in order to compensate for the additional emissions in the electricity sector and to achieve the overall reduction target. As the CO_2 reduction measures affecting end-use sectors are generally more expensive than measures in the conversion sector (particularly electricity generation), additional costs arise here compared to the scenario with CCS (see section on system value of CO_2, see also Fig. 10.14). The realignment of these measures and the associated additional costs correspond to a displacement of the reduction loads ('displacement solution') in order to achieve the overall reduction target. However, sector-specific changes also occur at times (mainly savings), which have no impact on CO_2 emissions. In practice, this affects energy carriers that do not emit CO_2, such as renewables, local and district heating networks, as well as electricity. This also gives rise to additional costs, which do not lead to CO_2 reduction but should be interpreted as the result of minimizing the total costs without a CCS option ('CO_2 neutral solution').

10.3.2 Cost of Reduction Strategies

10.3.2.1 CO_2 Reduction Costs

Figure 10.14 shows a breakdown of the annual additional costs (monetary value 2010) that arise because of CO_2 reduction targets in relation to the reference scenario according to sector. The additional costs in the scenario without CCS are higher over the whole period than the costs of the scenario with CCS, i.e. the CCS option is used and cuts the costs.

In the scenario with CCS, additional costs arise especially in the sectors of energy conversion (including extra costs for CCS), transport and traffic, and households. Savings measures decrease the demand for primary energy and thus the primary energy costs.

In the scenario without CCS, additional costs arise in the household and transport and traffic sectors, as well as to a smaller extent in the industrial sector. In the conversion sector, the additional costs in the reduction scenarios are very similar. However, in the scenario without CCS (CD), the cost savings for primary energy carriers decrease continuously as a result of the increase in wind power and the associated decrease in fossil power plants (see Fig. 10.11).[2]

[2] For reasons of space, it is not possible to discuss the individual measures and the resulting additional costs or cost reductions here.

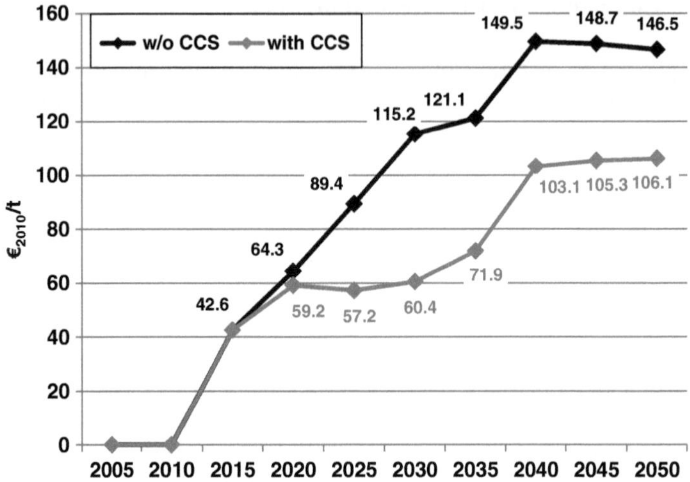

Fig. 10.15 Average specific CO_2 reduction costs

Overall, the integral additional costs for the period 2005–2050 are approx. €$_{2010}$ 940 billion for the scenario with CCS and approx. €$_{2010}$ 1,410 billion for the scenario without CCS in relation to the reference scenario.

Particularly in the long term, the discounting of costs becomes more important.[3] If the additional costs are discounted to 2005 at a constant discount rate of 5 %/a, then we get the following actual cash values in monetary values as of 2010:

- With CCS: €$_{2010}$ 203 billion
- Without CCS: €$_{2010}$ 302 billion

The mean specific CO_2 reduction costs that make up some of the additional costs in Fig. 10.14 are shown in Fig. 10.15. Until 2050, the CO_2 costs increase irregularly with a tendency to level off after 2040 to €$_{2010}$ 106/t in the scenario with CCS and with a tendency to decrease after 2040 to €$_{2010}$ 147/t in the scenario without CCS. The difference in the specific CO_2 reduction costs between the scenarios with and without CCS increases from 2020 to 2030 from approx. €$_{2010}$ 5/t to approx. €$_{2010}$ 55/t. For the period thereafter, this difference becomes smaller and is approx. €$_{2010}$ 40–49/t.

The marginal CO_2 reduction costs are much higher (up to approx. €$_{2010}$ 430/t in the scenario with CCS and approx. €$_{2010}$ 580/t in the scenario without CCS).

[3] For more information on modelling discounting and selecting discount rates, see the extensive discussions in the specialist literature (Cairns 2006; Dasgupta 1982; Hellweg et al. 2003; Kenley and Armsteasd 2004; Newel and Pizer 2004; Rabl 1996), which comprise the economic, engineering, and scientific perspectives.

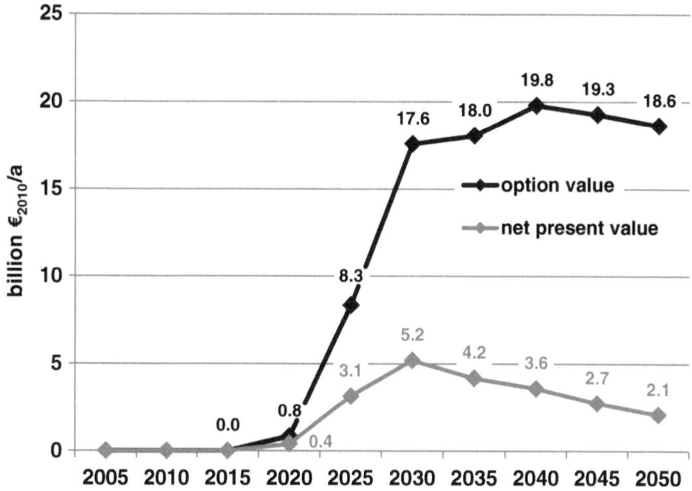

Fig. 10.16 System value of CCS per annum

10.3.2.2 CCS System Value

From the difference between the additional costs with and without CCS in Fig. 10.14, the current system value and actual cash value of CCS is derived and shown in Fig. 10.16 over time. The current system value increases rapidly from $€_{2010}$ 0.8 billion/a in 2020 to $€_{2010}$ 17.6 billion/a in 2030, after which it increases slightly until 2040 and then decreases slightly to € 18.6 billion/a in 2050. The corresponding actual cash value$_{2005}$ increases from $€_{2010}$ 0.4 billion/a in 2020 to $€_{2010}$ 5.2 billion/a in 2030 before subsequently decreasing continuously to $€_{2010}$ 2.1 billion/a in 2050. The cumulative system value for the period 2005–2050 amounts to approx. $€_{2010}$ 466 billion. The corresponding actual cash value$_{2005}$ is approx. $€_{2010}$ 101 billion.

Figure 10.17 shows the contributions of the sectors to the CCS system value. All end-use sectors contribute to the system value in such a way that the use of CCS helps to avoid relatively expensive savings measures. Such a contribution is also made by primary energy, where additional costs for the import of biomass products (e.g. bioethanol) can generally be avoided when CCS is implemented. However, these are offset by additional costs for fossil fuels, which results in a negative sectoral contribution of the primary sector to the CCS system value by 2035. In the conversion sector, additional costs are mainly due to the increased expansion of renewable energy capacity (e.g. wind).

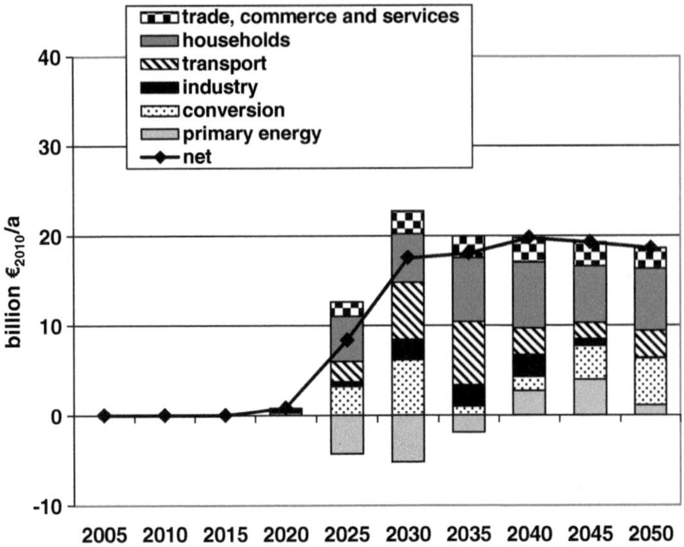

Fig. 10.17 Sectoral contributions to CCS system value

10.4 Summary and Conclusions

This chapter analysed the system value of CCS technologies in Germany within the context of consistent greenhouse gas reduction scenarios. In this context, the system value of a technology is determined by the additional avoidance costs that would occur when climate change mitigation targets are to be achieved without these technologies. The system value is therefore an implicit measure of the level of willingness of society to pay for refraining from the use of CCS technologies.

The methodological basis for calculating the system value of CCS technologies is the IKARUS energy system model, a bottom-up approach with detailed depictions of the technical energy supply structures in Germany for scenario-based analysis of CO_2 reduction strategies. The approach allows the variation of important parameters and general assumptions for which different developments are possible in future.

The system value of CCS technologies was analysed within the framework of a reference scenario without CO_2 reduction targets (REF) and two scenarios with CO_2 reduction targets (CA: without CCS; CD: with CCS). For renewable energy, the framework was extended e.g. via electricity imports from the DESERTEC Initiative, while for nuclear energy the decision as of early 2010 to phase out nuclear energy is implemented. The period considered is from 2005 to 2050.

The cumulative system value (in current values, with no discounting) for CCS technologies is $€_{2010}$ 466 billion for the period 2005–2050. If additional costs are discounted at a constant discount rate of 5 %/a, the result is an actual cash value$_{2005}$ of the cumulative system value of $€_{2010}$ 101 billion. For actual cash value analysis, the development of costs over time and the level of the discount rate are important. The later the costs are incurred over time (burden on future generations) and the higher the discount rate (high preference for the present), the lower the actual cash value.

The system values presented here are calculated by balancing across all model sectors (end-use sectors, conversion sector, primary energy sector incl. imports). All end-use sectors (industry, households, transport and traffic, commerce, trade and services) contribute to the system value in such a way that the implementation of CCS (in the conversion sector) helps to prevent relatively expensive savings measures. In the same way, the primary energy sector including imports also plays a role, where most of the additional costs associated with the import of biomass products (e.g. bioethanol) are avoided when CCS is implemented, but additional costs are incurred for fossil fuels, which predominate until 2035. Despite the additional costs caused by CCS technologies, the conversion sector also contributes to the system value because an additional increase in renewable energy capacity is avoided. To summarize, all sectors contribute to the system value even if to a different extent.

References

Bauer N, Edenhofer O, Leimbach M (2009) Low-stabilisation scenarios and technologies for carbon capture and sequestration. Energy Procedia 1:4031–4038

Cairns J (2006) Developments in discounting: with special reference to future health events. Resour Energy Econ 28:282–297

Dasgupta P (1982) Resource depletion, research and development, and the social rate of discount. In: Lind RC, Arrow KJ, Corey GR, Sen AK, Dasgupta P (eds) Discounting for time and risk in energy policy. The John Hopkins University Press, Baltimore/London, pp 273–324

Deutsches Zentrum für Luft- und Raumfahrt (DLR), Fraunhofer Institut für Windenergie und Energiesystemtechnik (Fraunhofer IWES), Ingenieurbüro für neue Energien (IFNE) (2012) Langfristszenarien und Strategien für den Ausbau der erneuerbaren Energien in Deutschland bei Berücksichtigung der Entwicklung in Europa und global. Studie im Auftrag des BMU

Hake JF, Hansen P, Heckler R, Linßen J, Markewitz P, Martinsen D, Weber K (2009) Projektionsrechnungen bis 2050 für das Energiesystem von Deutschland – im Rahmen des VDI-Projektes "Future Climate Engineering Solutions". Forschungszentrum Jülich, Institute of Energy Research – Systems Analysis and Technology Evaluation, Research Report 5–2009, Jülich

Hellweg S, Hofstetter TB, Hungerbühler K (2003) Discounting and the environment – should current impacts be weighted differently than impacts harming future generations? Int J LCA 8 (1):8–18

Kenley CR, Armsteasd DC (2004) Discounting models for long-term decision making. Syst Eng 7 (1):13–24

Komendantova N, Patt A, Barras L, Battaglini A (2010) Perception of risks in renewable energy projects: the case of concentrated solar power in North Africa. Energy Policy doi:10.1016/j.enpol.2009.12.008 (online first)

Krewitt W, Teske S, Simon S, Pregger T, Graus W, Blomen E, Schmid S, Schäfer O (2009) Energy [R]evolution 2008 – a sustainable world energy perspective. Energy Policy 37(12):5764–5775

Manger S, Pietzcker R, Bauer N, Bruckner T, Luderer G (2009) Option values of concentrating solar power and photovoltaics for reaching a 2 °C climate target. www.gfz-potsdam.de

Martinsen D, Linssen J, Markewitz P, Vögele S (2007) A future CO_2 mitigation option for Germany? – a bottom-up approach. Energy Policy 35(4):2110–2120

Newel RG, Pizer WA (2004) Uncertain discount rates in climate policy analysis. Energy Policy 32:519–529

Pietzcker R, Manger S, Bauer N, Luderer G, Bruckner T (2009) The role of concentrating solar power and photovoltaics for climate protection www.aaee.at

Rabl A (1996) Discounting of long-term costs: what would future generations prefer us to do? Ecol Econ 17:137–145

Williges K, Lilliestam J, Patt A (2010) Making concentrated solar power competitive with coal: the costs of a European feed-in tariff. Energy Policy 38(6):3089–3097

Chapter 11
Public Acceptance

Diana Schumann

Abstract Public acceptance is considered to be crucial for the large-scale demonstration and application of CO_2 capture and storage (CCS). Hence, with the further development of the technologies the number of studies on public acceptance of CCS is continuously increasing. However, the majority of these studies use the term "acceptance" without defining it, although it can represent different contents. Hence, in the first section of this article the research subject "public acceptance of CCS" is defined and delimited. Subsequently, an overview about the methods of CCS acceptance research is given and the key findings are explained. In the second section, the awareness, knowledge, initial attitudes as well as risk and benefit perceptions in Germany for different groups and regions are described. Then it is explained which factors influence the initial attitudes of CCS among the German public. The results are discussed with regard to future communication and future research on CCS.

Keywords CO_2 capture and storage (CCS) • Public acceptance • Initial attitudes • Risk perceptions • Benefit perceptions

11.1 Introduction

Public acceptance is regarded as an important prerequisite for the large-scale demonstration and implementation of CCS technologies in commercial power plant operation (see e.g. Ashworth et al. 2010; De Best-Waldhober et al. 2010; Fischer et al. 2010; Schumann et al. 2010; Tokushige et al. 2007). With the refinement and further development of the technologies, the number of scientific studies concerned with investigating the public acceptance of CCS in different countries has grown continuously.[1]

[1] Overviews of studies on CCS acceptance can be found in e.g. (Ashworth et al. 2010; De Best-Waldhober et al. 2008; Schulz et al. 2010; Sharp et al. 2009; Terwel et al. 2011).

D. Schumann (✉)
Institute of Energy and Climate Research – Systems Analysis and Technology Evaluation (IEK-STE), Forschungszentrum Jülich GmbH, D-52425 Jülich, Germany
e-mail: d.schumann@fz-juelich.de

W. Kuckshinrichs, J.-F. Hake (eds.), *Carbon Capture, Storage and Use*,
DOI 10.1007/978-3-319-11943-4_11

The majority of studies focus on investigating awareness and knowledge of CCS as well as initial attitudes towards it, and on identifying factors and conditions that may be relevant for the acceptance of the technologies. In these studies, the concept of 'acceptance' is usually used without being defined, because it is assumed to be a self-explanatory concept as a result of its use in everyday language. Studies on risk and technology acceptance, however, have made it clear that there are different forms of acceptance depending on the technology at hand (Renn 2005).

The first section of this chapter will therefore define and delimit the 'public acceptance of CCS' as a subject of research. Then, an overview of the methods of CCS acceptance research will be presented, and the key findings that have already been achieved will be discussed.

Against this background, the second section will discuss awareness and knowledge of CCS in Germany, as well as initial attitudes towards it, and risk/benefit perceptions. This analysis will be broken down into different groups of people and regions. Subsequently, the factors that influence initial attitudes towards CCS among the German public will be explored. The findings will finally be discussed within the context of the future communication of CCS and of the need for further research.

11.2 Public Acceptance of CCS as a Subject of Research

The concept of 'acceptance' is not just a key concept in scientific research, but is also a word used frequently in everyday language. In studies on risk and technology acceptance, it is therefore often used as if its meaning was clear and its definition was unambiguous. Even in studies on the public acceptance of CCS, acceptance is rarely defined as a concept, although it can be used to mean different things. In the following, the 'public acceptance of CCS' will therefore first be defined and delimited as a subject of research. Then, the methods most frequently used to investigate CCS acceptance will be outlined together with existing key findings.

11.2.1 Definition and Delimitation of the Subject of Research

In general, acceptance can be understood as passive or active approval (Schubert and Klein 2006). The public acceptance of new technologies, however, can be broken down into three different forms depending on what technology sector is being dealt with (Renn 2005). In the case of product and everyday technology, acceptance is shown by purchasing the respective products. In the case of work technology, acceptance is reflected in the active use of a product by the employees in a company. In the case of large-scale technologies,[2] which include energy

[2] Large-scale technologies are also referred to as 'external technology' ('technology as your neighbour') by Renn and Zwick (1997).

technologies such as CCS technologies, acceptance means that the respective facilities are tolerated by those concerned (Renn 2005). It is not necessary for those concerned to have a positive attitude towards the technology.

Decisions on the development and use of work and large-scale technologies are usually not made by the actors directly or indirectly affected by the use of the technologies (Gloede and Hennen 2005). For this reason, acceptance in the case of work and large-scale technologies also means the 'active or passive approval of decisions or actions of others' (Schubert and Klein 2006). It is expressed in the attitudes and behaviour of individual or complex actors[3] and can be measured at a given point in time (Dierkes and von Thienen 1982).

The *public acceptance of CCS* can therefore be defined as the passive or active approval of the development, the large-scale demonstration or the implementation of CCS technologies, which is reflected in the attitudes and behaviour of individual or complex social actors, and which can be measured at a certain point in time (Schumann and Simon 2009).

A distinction must be made between the concept of 'acceptance' and that of 'acceptability'. The evaluation of the acceptability of a technology looks at the normative judgement of whether and under what conditions the external effects of technological developments (e.g. noise, emissions, risks) would be considered acceptable. The basis for this judgement is either provided by rational criteria or by democratically legitimated decisions (Grunwald 2005). The first case involves acceptability in the sense of rational theory, and the second acceptability in the sense of the theory of democracy.

Both concepts share the fact that they involve the determination of unreasonable demands and what can reasonably be expected when a technology is being implemented. Acceptability in the sense of the theory of democracy, however, is the result of democratically legitimated decision-making processes, in which binding regulations on the social distribution of unreasonable demands and what can reasonably be expected are agreed upon, while acceptability in the sense of rational choice theory is the result of a risk assessment based on scientific and technical risk assessments taking into account intuitive risk perceptions (Renn et al. 2007).

The acceptability of CCS in the sense of rational theory can thus be defined as individual or collective risks of the development, the large-scale demonstration or the implementation of CCS technologies, which would be assessed as acceptable on the basis of normative criteria. Its determination is the subject matter of scientific and technical risk assessments, which predominantly focus on the risks associated with storing CO_2 in geological formations or in the ocean (see e.g. IPCC 2005).

The acceptability of CCS in the theory of democracy can be defined as agreement on unreasonable demands and what can reasonably be expected in relation to the development, the large-scale demonstration or the implementation of CCS

[3] A collection of individuals can be understood as a complex actor, if all of those concerned intend to acquire a shared product, to achieve a shared objective, or realize a shared interest (Scharpf 2000).

technologies, as well as their social distribution, which has been achieved in democratically legitimated decision-making processes.

A systematic analysis of the prerequisites and the conditions required to achieve acceptability for the implementation of CCS technologies in the sense of the theory of democracy has not yet been performed. In addition, passive or active approval of the development, the large-scale demonstration or the implementation of CCS technologies cannot yet be reliably measured because there is still a lack of knowledge of CCS technologies among the general public.

The subject of previous CCS acceptance studies has therefore been the investigation of the awareness and knowledge of CCS among the general public as well as initial attitudes towards it, and the identification of conditions and factors that could have an impact on the future approval or rejection of the technologies. The methods that are used for this are outlined in the next section.

11.2.2 Methods of CCS Acceptance Research

Since the beginning of 2000, around 40 studies have been performed on CCS acceptance (see Schulz et al. 2010). Most of these studies are country case studies and only three of them are comparative studies. Table 11.1 lists the countries for which case studies have already been conducted.[4]

The first comparative study comprised a comparative analysis of surveys, which were conducted in coordination with each other in 2003 and 2004 by four research institutes in the USA (Massachusetts Institute of Technology, MIT), the UK (University of Cambridge, Judge Business School), Sweden (Chalmers University of Technology), and Japan (Mizuho Information & Research Institute) (Reiner et al. 2006a, b). Although each of the four institutes was responsible for the surveys they conducted, particular importance was placed on asking the same questions in all of the countries (Reiner et al. 2006a).

The second comparative study involved a survey of European stakeholders[5] within the framework of the project 'Acceptance of Carbon Dioxide Capture Storage, Economics, Policy and Technology' (ACCSEPT), which involved Det Norske Veritas (DNV), Energy Centre of the Netherlands (ECN), Institute for European Environmental Policy (IEEP), Baker & McKenzie, and Tyndall Centre for Climate Change Research. It was funded within the European Sixth Framework Programme (De Coninck et al. 2009; Shackley et al. 2007, 2009).

---

[4] As research on CCS acceptance continues to develop dynamically, Table 11.1 is not intended to be exhaustive.

[5] 'Stakeholders' are understood as complex actors who have a professional interest in CCS (De Coninck et al. 2006; Shackley et al. 2009; Van Alphen et al. 2007). Stakeholders are therefore mainly organizations that represent industry, non-governmental organizations (NGOs), politics, administration or scientific institutions.

Table 11.1 Studies on CCS acceptance by country

Country	Studies
Australia	Ashworth et al. (2006a, b, 2009), Ashworth and Gardner (2006), and Miller et al. (2007)
France	Ha-Duong et al. (2009b)
Germany	Denkstelle Hamburg (2009), Duetschke (2010), Fischedick et al. (2008), Pietzner et al. (2010), and Schumann et al. (2010)
Japan	Itaoka et al. (2004, 2006, 2009, 2010), Tokushige et al. (2006, 2007), and Uno et al. (2004)
Canada	Sharp (2005) and Sharp et al. (2006, 2009)
Netherlands	Daamen et al. (2006), De Best-Waldhober et al. (2008, 2009a, b, 2010), De Coninck et al. (2006), Huijts et al. (2007), and Van Alphen et al. (2007)
Sweden	AGS (2007)
Switzerland	Wallquist et al. (2009, 2010)
Spain	Solá et al. (2008)
USA	Curry et al. (2004, 2007), Fleishman et al. (2010), and Palmgren et al. (2004a, b)
UK	Curry et al. (2005), Gough and Shackley (2006), Gough et al. (2006), Roberts and Mander (2010), Shackley et al. (2004a, 2007, 2009), and Shackley and McLachlan (2006)

The third comparative study was a project initiated by the Fossil Energy Coalition (FENCO ERA-NET) and conducted in six countries: Germany, Greece, Netherlands, Norway, Romania and the UK (Schumann 2009). Within this project, two methods were applied, and they were implemented in the same way in all of the countries participating: (1) a comparative study of the effectiveness of two methods of CCS communication, and (2) a representative survey of the general public with regard to their knowledge of and attitudes towards CCS on a regional and/or national level (Daamen et al. 2011; Pietzner et al. 2011; Reiner et al. 2010a; Schumann 2009).

Most of the CCS acceptance studies performed to date have focused on investigating awareness and knowledge of CCS among the general public, and initial attitudes towards it, as well as the identification of conditions and factors that could have an impact on the acceptance of technologies. This is due to the fact that acceptance of CCS cannot yet be reliably measured because the general public knows very little about CCS technologies (cf. Curry et al. 2007; Duetschke 2010; Ha-Duong et al. 2009a; Itaoka et al. 2009; Miller et al. 2007; Pietzner et al. 2011; Reiner et al. 2006a).

In order to ascertain awareness and knowledge of CCS, as well as initial attitudes towards it, moderated group discussions (e.g. focus groups, citizens' panels or regional dialogues) e.g. (Ashworth and Gardner 2006; Ashworth et al. 2006b; Roberts and Mander 2010; Shackley et al. 2004a, b), qualitative in-depth interviews e.g. (Wallquist et al. 2009) or standardized surveys e.g. (Ashworth and Gardner 2006; Ashworth et al. 2006b; Ha-Duong et al. 2009a; Huijts 2003; Itaoka et al. 2006; Miller et al. 2007; Pietzner et al. 2011; Reiner et al. 2010a; Tokushige et al. 2007) were predominantly conducted.

The advantages of moderated group discussions in this context are that lay people are informed about a topic with which they were previously unfamiliar, discuss it intensively, and form opinions on it during this discussion. In addition, group discussions also offer an opportunity to ask the experts who spoke about the topic directly for clarification.

The disadvantages of moderated group discussions are that even a good moderator cannot always prevent situations where individual people do not form their own opinion but instead adapt their own views to conform to the assumed group opinion (Janis 1972). Furthermore, group discussions are disadvantageous in that the number of participants is usually very small,[6] which means that the findings cannot be generalized. Group discussions are therefore suitable for exploring the awareness, knowledge and initial attitudes of lay people concerning CCS, but are rather unsuitable for identifying causal relationships between relevant influencing factors and attitudes towards CCS.

Suitable methods for investigating causal relationships between relevant influencing factors and attitudes towards CCS include statistical methods, experimental approaches, structural equation modelling (SEM) or agent-based simulations. The main statistical methods that have been used in CCS acceptance studies to date include regression analysis, analysis of variance (ANOVA), mediation analysis, and path analysis (Huijts et al. 2007; Miller et al. 2007; Sharp et al. 2009; Terwel et al. 2011; Tokushige et al. 2007; Wallquist et al. 2010). The data basis for these statistical methods generally comprised the findings of standardized surveys.

A major advantage of using the findings of standardized surveys for the application of statistical methods is that compared to group discussions a larger number of cases can be incorporated into the analysis. The extent to which the statistical analyses can be generalized depends on whether the surveys were conducted among selected sections of the population or whether the surveys were representative surveys of the population based on random samples.

However, standardized surveys are not (yet) suitable for measuring the acceptance of CCS because opinions of respondents on CCS in conventional opinion polls are generally 'pseudo-opinions' (Bishop et al. 1980; De Best-Waldhober et al. 2008). These are opinions expressed by individuals on a topic despite the fact that they know little or nothing about this topic. Studies have shown that pseudo-opinions on CCS are very unstable and can be easily changed by information or by slight alterations in the general mood (Daamen et al. 2006; De Best-Waldhober et al. 2008).

Due to this, standardized surveys are mainly used today to assess awareness and knowledge of CSS among lay people, as well as their initial attitudes towards

---

[6] For example, a group of six to ten people is recommended for focus groups (cf. e.g. Lamnek 1995).

it. These findings can then be used, for example, in SEM analyses to investigate causal relationships between relevant influencing factors, and changes in initial attitudes towards CCS (Schumann et al. 2011).

Further approaches that can be used, despite the low level of knowledge among the general public, to identify conditions and factors that could have an effect on the acceptance of CCS are experimental approaches and agent-based simulation. Experimental approaches have been applied in CCS acceptance studies mainly to investigate the influence of trust in stakeholders on the acceptance of CCS (Terwel 2009; Terwel et al. 2011), the influence of stakeholder collaboration on the effectiveness of CCS communication (Ter Mors et al. 2010; Ter Mors 2009), and the effectiveness of different methods of communication (Daamen et al. 2011). Agent-based simulation has been applied in CCS acceptance research to date to investigate the influence of communication on knowledge of CCS and attitudes towards it among the general public (Schumann and Simon 2009, 2010).

To summarize, CCS acceptance research therefore uses the whole spectrum of quantitative and qualitative methods of empirical social research as well as experimental approaches and agent-based simulation. The key findings of these studies will be outlined in the following section.

11.2.3 Key Findings of CCS Acceptance Research

An important cross-national finding of previous CCS acceptance studies is that the general public has very little awareness of CCS to date (Curry et al. 2007; Duetschke 2010; Ha-Duong et al. 2009a; Itaoka et al. 2009; Miller et al. 2007; Pietzner et al. 2011; Reiner et al. 2010a). In one of the first representative surveys on CCS, which was conducted in 2003 in the USA, only 4 % of those surveyed had even heard of CCS (Curry 2004). This survey was replicated in 2003/2004 in Sweden, the UK, and Japan. The findings showed that the proportion of respondents who had heard of CCS at some point in time was 5 % in the UK, 15 % in Sweden, and 22 % in Japan (Reiner et al. 2006a). Other representative surveys conducted in the period from 2004 to 2006, also showed that the proportion of respondents who had heard of CCS at some stage was very low: 9–25 % in the Netherlands, 27–29 % in Australia, 11 % in Canada, and 34 % in France (Ashworth et al. 2006b; Daamen et al. 2006; Ha-Duong et al. 2009a; Sharp 2005).[7]

However, over the last few years, awareness of CCS has increased considerably. In the USA, although awareness in 2006 had only increased by one percentage point compared to 2003 from 4 to 5 % (Curry et al. 2007), the findings of representative

[7] Although all studies revealed a low awareness of CCS, only the proportions of the coordinated surveys in the USA, the UK, Sweden, and Japan can be compared directly, because identical questions were asked in these surveys in order to ascertain the awareness of CCS (Reiner et al. 2006a).

Table 11.2 Awareness of CCS among the general public (values in percent)

Country	No, never heard of it	Yes, heard of it but don't know much about it	Yes, heard of it and know quite a bit or a lot about it	Total
Germany (n = 1,017)	62.0	28.3	9.7	100
Greece (n = 1,000)	76.5	18.7	4.8	100
Netherlands (n = 1,109)	50.0	44.5	5.5	100
Norway (n = 1,000)	37.4	45.2	17.4	100
Romania (n = 1,002)	75.7	21.4	2.9	100
UK (n = 1,040)	61.9	31.8	6.3	100
Total	*60.4*	*31.9*	*7.7*	*100*

Source: Based on Pietzner et al. (2011)

surveys conducted at the end of 2009/beginning of 2010 as part of the comparative FENCO ERA-NET study show that CCS was no longer unknown among the general public (cf. Table 11.2). The proportion of respondents in the UK who had heard of CCS at some point in time, for example, increased by 33 percentage points compared to the study in 2003/2004 and was 38.1 % in 2009.[8]

In Norway, where CO_2 that arises during the purification of natural gas from the Sleipner West Reservoir has been captured since 1996 and stored in the Utsira Reservoir,[9] 62.6 % of the respondents answered that they had heard of CCS at some point in time. In Greece and Romania, where no concrete CCS projects are yet planned, awareness of the technologies was also much higher (23.5 % and 24.3 %, respectively) than in the first studies conducted in the USA, the UK, Sweden, and Japan.

The fact that awareness of CCS has increased over time, however, does not mean that knowledge of the technologies has increased among the general public. Even in the first comparative analysis of the four representative surveys in the USA, the UK, Sweden and Japan, it was clear that lay people often have misconceptions of CCS technologies: for example, between around a quarter and over half of the respondents in the four countries answered that CCS could help to reduce smog or acid rain. In addition, only some 25 % in the USA and approx. 50 % in the UK knew that CCS can help to mitigate global warming. In Sweden, around 60 % of respondents knew this, while in Japan the figure was above 75 % (Reiner et al. 2006a).

[8] When comparing the proportions for the UK from the study in 2003/2004 and the study in 2009, it must be noted that although the questions were formulated in a similar manner in both studies, they were not identical (cf. Curry et al. 2005; Reiner et al. 2010b).

[9] For more details, see http://www.statoil.com/en/TechnologyInnovation/NewEnergy/Co2Management/Pages/SleipnerVest.asp

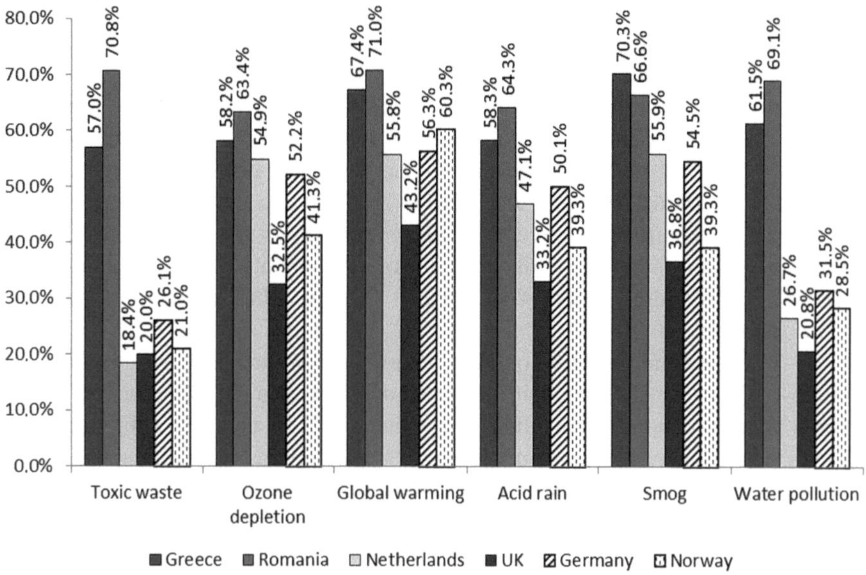

Fig. 11.1 Environmental problems which the general public believe could be reduced by CCS (Share of respondents who answered 'Can be reduced by CCS')

The findings of the comparative FENCO ERA-NET study revealed that misconceptions of CCS were still widespread 6 years later (cf. Fig. 11.1). For instance, in the UK only 43.2 % knew that CCS could contribute to mitigation of global warming, while 71 % were aware of this in Romania, and 67.4 % in Greece. However, in Greece and Romania, the conception that CCS could reduce toxic waste, ozone depletion, acid rain, smog, and water pollution was more common than in the other countries.

In addition, similar proportions of individuals who were surveyed in the Netherlands, the UK, Germany, and Norway believed that CCS could reduce ozone depletion, global warming, acid rain, and smog, which leads us to believe that lay people find it difficult to distinguish between the four environmental problems because they know only very little about them. Overall, it can therefore be assumed that lay people tend to have a vague conception of CCS technologies and the environmental problems they can help to mitigate.

While CCS acceptance studies thus largely come to the same conclusion that the general public knows very little about CCS, the findings of the investigations are ambiguous concerning initial attitudes and the way in which information changes these initial attitudes. For example, (Curry et al. 2007; Shackley et al. 2004b) revealed that citizens who tended to have a negative initial attitude towards CCS developed more positive attitudes after they received information on the technologies. However, (Palmgren et al. 2004b) showed that citizens with a moderately negative initial attitude towards CCS tended to develop even more negative attitude of the technologies after receiving information. De Best-Waldhober et al. (2006)

and Itaoka et al. (2006) also found that citizens developed more negative attitudes towards CCS than they originally had after they were provided with information.

The findings of the comparative FENCO ERA-NET study made clear that changes in initial attitudes depended on the content of the information provided: in four of the six countries, the attitudes of respondents became more positive after they received positive information and more negative after they received negative information (Pietzner et al. 2011). In Romania, however, respondents' attitudes became more positive than beforehand after they received negative information, and in Germany, the attitudes of respondents who received positive information became more negative towards CCS than before they had received the information.

All of these studies thus confirm the aforementioned research findings on pseudo-opinions by showing that initial attitudes towards CCS can be very easily altered by information. Against this background, CCS acceptance studies therefore investigated what type of information and what methods of communication are most effective in helping the general public to form stable, consistent, and well-considered opinions (Daamen et al. 2011; Schumann 2009; Ter Mors et al. 2009, 2010).

The findings of these studies show that citizens perceived the information provided jointly by the stakeholders as being of a higher quality than information provided by individual stakeholders (Ter Mors et al. 2009, 2010). With respect to the development of stable, consistent and well-considered opinions, it was shown that both focus groups and information-choice questionnaires (ICQs)[10] can be used to help lay people develop such highly informed attitudes towards CCS. However, citizens who completed an ICQ tended to have more stable and consistent attitudes towards the technologies and were more convinced of their opinions than citizens who participated in a focus group (Daamen et al. 2011). The dissemination of information on CCS was thus more effective in the form of an ICQ than within the framework of a focus group.

Due to the well-known studies which have shown that acceptance of new technologies (e.g. genetic engineering and nanotechnologies) are influenced to a large extent by trust as well as by the perception of risks and benefits (Siegrist 2000; Siegrist et al. 2007), the significance of these factors was also investigated in CCS acceptance studies (e.g. Huijts et al. 2007; Miller et al. 2007; Sharp et al. 2009; Terwel et al. 2011; Tokushige et al. 2007; Wallquist et al. 2010). For instance, (Tokushige et al. 2007) showed that a high level of trust in the operators of CO_2 storage facilities led to a lower risk perception and a higher perception of the benefits of CCS. In addition, trust had an indirect impact on the acceptance of

[10] An ICQ is a specific, informative, computer-based questionnaire that aims to allow respondents to use the information provided in order to develop their own opinions (De Best-Waldhober et al. 2008). Prior to the study outlined here, ICQs were used in the Netherlands, for example, to ascertain awareness of CCS and the development of informed opinions on different CCS options and on CCS as an alternative to other CO_2 reduction strategies (De Best-Waldhober and Daamen 2006; De Best-Waldhober et al. 2008).

CCS, which was positively influenced by the perception of benefits and negatively influenced by the perception of risks (Tokushige et al. 2007).

Huijts et al. (2007) demonstrated that the trust placed by citizens in CCS stakeholders depends on the expertise and intentions ascribed to the stakeholders, as well as on how well the objectives and views of the trusting citizen conform to those of the stakeholder to be trusted. Furthermore, (Huijts et al. 2007) reported that the general public placed the most trust in environmental organizations, while economic actors were considered the least trustworthy.

While (Huijts et al. 2007; Tokushige et al. 2007) used data from surveys for their investigation, (Terwel et al. 2011) applied an experimental approach in order to explore how acceptance of CCS is influenced by trust in CCS stakeholders. They differentiated between competence-based and integrity-based trust,[11] and found that lay people who placed more competence-based trust in advocates of CCS believed CCS technologies to have lower risks and greater benefits than lay people who showed less competence-based trust. Lay people who placed less integrity-based trust in advocates of CCS had a more negative attitude towards the technologies than lay people who showed more integrity-based trust.

Terwel et al. (2011) also found that the general public placed more trust in environmental organizations than in industrial stakeholders. The reason for this was that lay people assumed stakeholders to have different organizational motives: while industrial stakeholders were viewed as organizations motivated by their own interests, environmental organizations were viewed as actors pursuing social interests in general. As a result, lay people placed more (integrity-based) trust in environmental organizations than in industrial stakeholders (Terwel et al. 2011).

In contrast to (Huijts et al. 2007; Terwel et al. 2011; Tokushige et al. 2007), (Schumann et al. 2011) performed a comparative SEM analysis to determine the influence of knowledge, trust, and risk/benefit perception on the stability of initial attitudes towards CCS in Germany and Norway. The findings of the analysis, which was conducted using data from representative public surveys, showed that the stability of initial attitudes towards CCS is influenced in the same way in both countries by knowledge and the appraisal of risks: the greater the knowledge of CCS or the greater the perceived risk, the more stable the initial attitudes towards the technologies. In Norway, the stability of initial attitudes was also influenced by the appraisal of benefits. Citizens who believed the technologies to have smaller benefits had more stable attitudes than people who believed the technologies to have greater benefits.

The influence of knowledge on the perception of risks and benefits as well as the influence of trust on the stability of initial attitudes was different in the two countries: in Germany, individuals who were better informed believed the benefits

[11] "Competence-based trust refers to trust originating from perceived organizational expertise and experience concerning the issue at hand. Integrity-based trust refers to trust based on beliefs about whether or not the organization is open, honest, and truly concerned with public interests (cf. Metlay 1999)."

of CCS to be smaller, while in Norway, individuals who were better informed believed the risks of CCS to be smaller. A high level of trust in environmental organizations among the German citizens led to a higher risk perception and among the Norwegian citizens to a higher perception of benefits. Trust in political decision makers and political parties gave rise to a higher perception of the benefits of CCS among the respondents in Germany, while this had no influence on the perception of risks and benefits in Norway.

11.3 Public Acceptance of CCS in Germany

The first scientific study conducted in Germany on the public acceptance of CCS was a joint project on socioeconomic research on the public acceptance of CO_2 capture and storage (CCS) on a national and international level. This project was funded by the Federal Ministry of Economics and Technology (BMWi), and was coordinated by the Wuppertal Institute for Climate, Environment and Energy (WI). It was jointly implemented between 2006 and 2007 by the Fraunhofer Institute for Systems and Innovation Research (ISI), BSR Sustainability GmbH, and IEK-STE at Forschungszentrum Jülich (Cremer et al. 2008; Fischedick et al. 2009). The methods used in the project were literature and document analysis, media analysis, and a stakeholder survey. A representative survey of the general public on the acceptance of CCS was not conducted as part of this project.

The first studies on ascertaining the awareness and attitudes of the German population towards CCS were conducted in 2009. On behalf of the information centre for climate-smart coal-fired power plants 'IZ Klima', the market research company 'Denkstelle Hamburg' conducted a representative study on the level of knowledge of the general public on the topics of climate protection, CO_2, CCS, and associations with the topic of electricity. The findings of the study revealed that only 1 % of respondents were able to describe CCS technologies without any assistance (Denkstelle Hamburg 2009). With respect to the evaluation of the individual process steps of CCS technologies, 80 % of the respondents saw no difficulties with capturing the greenhouse gas in the power plant process. The transport of the captured CO_2, on the other hand, was considered problematic by more than 70 % – and storage by more than 75 % – of the study participants (Denkstelle Hamburg 2009).

At the beginning of 2009, the above-mentioned FENCO ERA-NET project was implemented in Germany under the title of 'CCS Communication' by IEK-STE in cooperation with WI. As part of this project, three representative public surveys were performed in Germany at the end of 2009 on the topics of the environment, energy sources, and CCS. For the first time, a nationwide survey was conducted parallel to two regional surveys. An area along the Rhine was chosen to represent a region where a CCS demonstration power plant was planned, and the northern region of Schleswig-Holstein was chosen to represent a region with potential storage reservoirs for CO_2.

The three representative surveys concentrated on assessing awareness and knowledge of CCS among the general public, their initial attitudes towards CCS in general and towards the official approval of a CCS demonstration power plant, as well as changes in these attitudes after receiving brief information on the three process steps of capture, transport, and storage. The formation and change of attitudes towards CCS were placed in context by investigating how important the topic of 'environment' was for the general public, and by identifying what attitudes they had towards the use of different energy sources.[12]

The data from these representative surveys were used for this study in order to investigate awareness and knowledge of CCS, initial attitudes towards it, as well as the perception of risks and benefits in Germany among different groups of people and in different regions. The factors that influence initial attitudes towards CCS among the German population were also explored.

11.3.1 Awareness and Knowledge of CCS

The German population is no longer unaware of CCS. 38 % of the German population said that they had heard of CCS at some point in time (cf. Fig. 11.2). However, awareness of CCS differed depending on gender, educational background, and region. Figure 11.2 shows that around half of the men questioned but only around a quarter of the women questioned said that they had heard of CCS at some point in time. In addition, 15.2 % of men said that they knew 'quite a bit or a lot' about CCS, while only 4.3 % of women gave the same answer. Statistical tests show that these gender-specific differences are significant.[13]

With respect to educational background, Fig. 11.3 shows that awareness of CCS among individuals who do not (yet) have a professional qualification is much lower than among individuals who have a professional qualification or an academic degree. 80.8 % of the respondents with no professional qualification said that they had never heard of CCS, while only 42.9 % of respondents with a university degree said the same. These findings, which are also statistically significant, agree with findings from innovation research, which demonstrated that individuals who are among the first to be aware of new technologies have a higher level of education than individuals who become aware of the technologies later (Rogers 2003).

Differentiated by region, Fig. 11.4 shows that more than half of the respondents in Schleswig-Holstein said that they had heard of CCS at some point in time. This indicates that the protests against CCS in Schleswig-Holstein (Uken 2009), for

[12] A full English version of the questionnaire used can be found in Reiner et al. (2010b).

[13] Statistically significant means that there is a relationship between two or more variables with a certain probability that random choice cannot explain the result. It can therefore be assumed that statistically significant relationships do not just apply to the respective survey participants, but that they also apply to groups of individuals represented by the survey respondents.

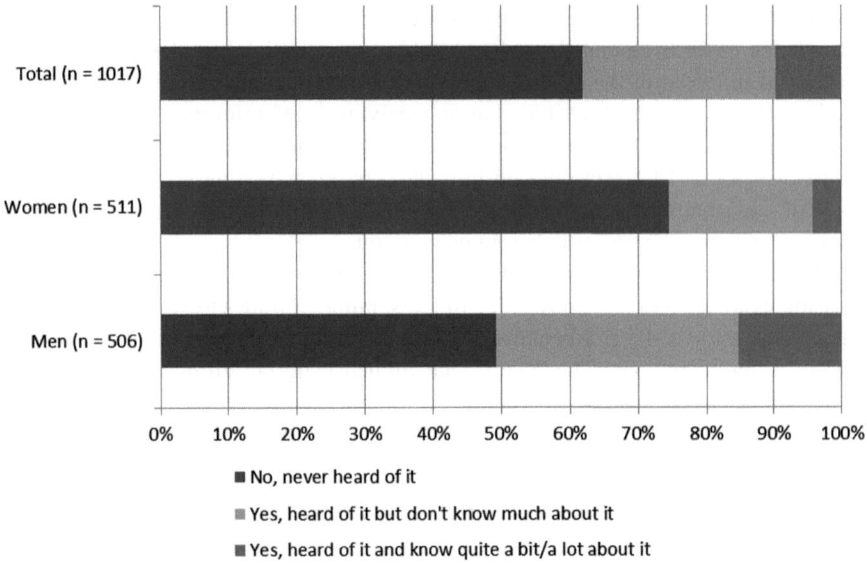

Fig. 11.2 Awareness of CCS broken down according to gender (Source: WI/IEK-STE survey 2009)

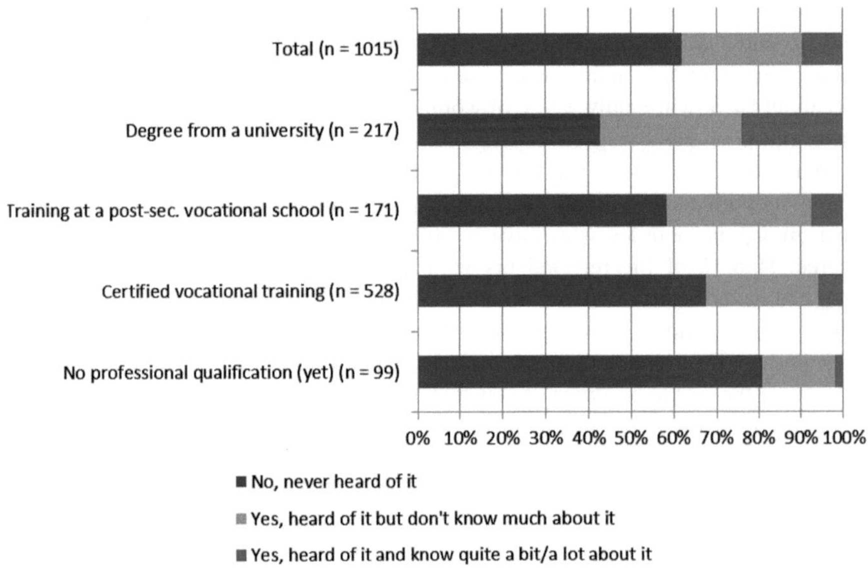

Fig. 11.3 Awareness of CCS depending on professional qualifications (Source: WI/IEK-STE survey 2009)

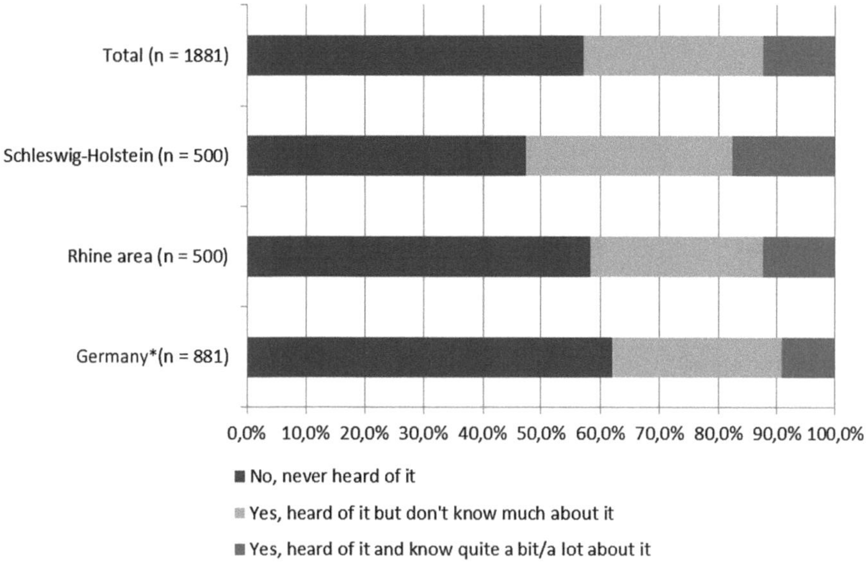

Fig. 11.4 Awareness of CCS in different regions (Source: WI/IEK-STE survey 2009)

example, meant that the awareness of at least the concept of CO_2 capture and storage there was much higher than in the Rhine area and in Germany.[14] There, some 38 % and 42 % of respondents, respectively, said that they had heard of CCS at some point in time. The regional differences in knowledge of CCS were also statistically significant.

As already explained, when citizens say that they have heard of CO_2 capture and storage at some point in time, this does not necessarily mean that they also know what it actually involves. Table 11.3 shows that of the respondents who said that they had heard of CCS at some point in time, only slightly more than half of them knew that CCS could help to mitigate global warming. Statistical tests reveal that there are no significant differences between knowledge of CCS among men and women or between individuals with different levels of professional qualifications. The differences between the regions, on the other hand, are statistically significant, and indicate that more respondents in the areas along the Rhine and in northern Schleswig-Holstein knew that CCS can help to mitigate global warming than respondents in the region Germany.

[14] As citizens of the regions along the Rhine and in northern Schleswig-Holstein also participated in the nationwide survey, they were taken out of the nationwide data set for the regional comparison. The region termed 'Germany*' is thus the region 'Germany without the regions along the Rhine and in northern Schleswig-Holstein'. It comprises 881 cases, while the full nationwide data set with the two regions comprises 1,017 cases.

Table 11.3 Knowledge of CCS

Global warming...	Can be reduced by CCS		Cannot be reduced by CCS		Don't know		Total
Gender	**Number**	**%**	**Number**	**%**	**Number**	**%**	**Number**
Men	151	58.8	75	29.2	31	12.1	257
Women	67	51.5	50	38.5	13	10.0	130
Total	*218*	*56.3*	*125*	*32.3*	*44*	*11.4*	*387*
Professional qualification	**Number**	**%**	**Number**	**%**	**Number**	**%**	**Number**
No professional qualification (yet)	10	52.6	7	36.8	2	10.5	19
Certified vocational training	97	56.4	52	30.2	23	13.4	172
Training at a post-secondary vocational school	38	53.5	26	36.6	7	9.9	71
Degree from a university	73	58.9	39	31.5	12	9.7	124
Total	*218*	*56.5*	*124*	*32.1*	*44*	*11.4*	*386*
Region	**Number**	**%**	**Number**	**%**	**Number**	**%**	**Number**
Rhine area	141	67.5	43	20.6	25	12.0	209
Schleswig-Holstein	175	66.5	67	25.5	21	8.0	263
Germany*	191	56.8	107	31.8	38	11.3	336
Total	*507*	*62.7*	*217*	*26.9*	*84*	*10.4*	*808*

Source: WI/IEK-STE survey 2009
Details of respondents who answered 'Yes, have heard of CCS'

11.3.2 Initial Attitudes Towards CCS

Although the majority of the general public had little or no knowledge of CCS, they still had initial attitudes towards the technologies. In order to identify these initial attitudes, the survey participants were asked to rate whether they would use CCS to mitigate global warming on a scale of 1 (=definitely would not use it) to 7 (=definitely would use it).[15] To determine how the initial attitudes changed after information was provided, this question was asked twice: once before and once after the respondents received information on CCS.

Table 11.4 details the initial attitudes towards CCS before and after information was received, as well as changes in these attitudes depending on gender, professional qualification, and region. In general, the mean values show that the respondents tended to have a neutral view of CCS technologies before information was received. Women had a more negative initial opinion of CCS than men. This difference is statistically significant.

[15] In the surveys, initial attitudes towards official approval of a CCS demonstration power plant were also investigated although this cannot be discussed here. The findings can be found in Schumann et al. (2010).

Table 11.4 Mean values and standard deviations (SD) of initial attitudes towards CCS before and after information was received

	Before information was received		After information was received		Change
Gender	**Mean**	**SD**	**Mean**	**SD**	**Mean**
Men (n = 506)	4.25	1.94	4.01	1.94	−0.24
Women (n = 511)	3.72	1.88	3.71	1.86	−0.01
Total (n = 1,017)	*3.99*	*1.93*	*3.86*	*1.91*	*−0.13*
Professional qualification	**Mean**	**SD**	**Mean**	**SD**	**Mean**
No professional qualification (yet) (n = 99)	4.01	1.94	4.14	1.97	0.13
Certified vocational training (n = 528)	3.89	1.91	3.74	1.89	−0.15
Training at a post-secondary vocational school (n = 171)	4.13	1.87	4.18	1.94	0.05
Degree from a university (n = 217)	4.10	2.01	3.78	1.96	−0.32
Total (n = 1,015)	*3.99*	*1.93*	*3.86*	*1.90*	*−0.13*
Region	**Mean**	**SD**	**Mean**	**SD**	**Mean**
Rhine area (n = 500)	3.84	1.83	3.65	1.81	−0.19
Schleswig-Holstein (n = 500)	3.16	1.93	3.09	1.89	−0.07
Germany* (n = 881)	4.04	1.94	3.89	1.91	−0.15
Total (n = 1,881)	*3.75*	*1.94*	*3.61*	*1.90*	*−0.14*

Source: WI/IEK-STE survey 2009
Scale of 1 (=definitely would not use it) to 7 (=definitely would use it)

After receiving information, both men and women had more negative opinions of CCS than beforehand. These negative changes in attitude, which are more pronounced among the men than the women, are only statistically significant in the group of male respondents.

Differentiated by professional qualifications, no statistically significant differences could be determined for initial attitudes after information had been received. After receiving information, respondents with no professional qualifications and respondents with training at a post-secondary vocational school had a slightly more positive view of the technologies than beforehand. However, this change in attitude is not statistically significant. Respondents with certified vocational training or an academic qualification displayed more negative attitudes after receiving information than beforehand. This negative change in attitude, however, is only statistically significant for the group of individuals who had a degree from a university.

Differentiated by region, the mean values in Table 11.4 show that respondents in Schleswig-Holstein had a much more negative opinion of CCS before receiving information than respondents in the Rhine area or in Germany. This difference is statistically significant. After receiving information, the respondents in all regions had more negative attitudes towards the technologies than beforehand. This change in attitude, however, is only statistically significant for the areas along the Rhine and Germany.

Overall, initial attitudes towards CCS thus tend to be neutral, and women have a more negative opinion of the technologies than men. In addition, citizens of the Schleswig-Holstein region have a more negative initial opinion than citizens of the area along the Rhine and Germany. Professional qualifications have no impact on initial attitudes towards CCS.

Statistically significant changes in attitude after the receipt of information were found among the male respondents, individuals with degrees from a university, and in the areas along the Rhine and in Germany. The changes in attitude among women, individuals with no professional qualification, individuals with certified vocational training or training at a post-secondary vocational school, and citizens of the region Schleswig-Holstein were not significantly significant, thus allowing us to assume that these groups only marginally changed their opinions after receiving information.

11.3.3 Perception of the Risks and Benefits of CCS

Previous studies on the acceptance of risks and technologies verified that the acceptance of technologies by the general public is greatly influenced by the intuitive perception of risks, as well as by the perception of benefits and trust (Renn 2005; Renn and Zwick 1997; Siegrist 2000; Siegrist et al. 2007). In order to investigate how the risks and benefits of CCS are perceived by the German public, participants in the representative survey were first given brief information on the three steps of the CCS process chain: capture, transport, and storage. The respondents were then asked to rate the personal risk, the risk for society, personal benefits, and benefits for society on a scale of 1 (=very low) to 7 (=very high) for each of the three process steps.

Table 11.5 shows the perception of personal risks associated with CO_2 capture, transport, and storage depending on gender, professional qualification, and region. In general, the mean values illustrate that the personal risks associated with the three process steps were assessed quite neutrally, but that the personal risk associated with storage was considered slightly higher than that of capture or transport.

In addition, women had a more negative overall opinion of CCS than men, and they considered the risks to be higher for all three process steps. These gender-specific differences in the perception of risks are statistically significant.

With respect to professional qualification, the mean values show that individuals with a degree from a university considered the personal risk associated with capture, transport, and storage to be smaller than individuals with no professional qualifications, those with certified vocational training or those with training in a post-secondary vocational school. However, these qualification-specific differences are only statistically significant with respect to the risk perception of transport and storage.

Clear regional differences are visible in the perception of personal risks associated with CO_2 storage. These were assessed as significantly higher in the region in

Table 11.5 Mean values and standard deviations (SD) of the perception of personal risks associated with the three CCS process steps

	Capture		Transport		Storage	
Gender	**Mean**	**SD**	**Mean**	**SD**	**Mean**	**SD**
Men (n = 506)	3.50	1.67	3.48	1.89	3.68	1.92
Women (n = 511)	3.81	1.73	3.76	1.95	4.00	1.90
Total (n = 1,017)	*3.66*	*1.71*	*3.62*	*1.93*	*3.84*	*1.92*
Professional qualification	**Mean**	**SD**	**Mean**	**SD**	**Mean**	**SD**
No professional qualification (yet) (n = 99)	3.68	1.50	3.75	1.90	3.92	1.71
Certified vocational training (n = 528)	3.70	1.69	3.72	1.92	3.95	1.95
Training at a post-secondary vocational school (n = 171)	3.83	1.65	3.70	1.86	3.92	1.90
Degree from a university (n = 217)	3.44	1.86	3.29	1.97	3.47	1.92
Total (n = 1,015)	*3.66*	*1.71*	*3.63*	*1.93*	*3.84*	*1.92*
Region	**Mean**	**SD**	**Mean**	**SD**	**Mean**	**SD**
Rhine area (n = 500)	3.61	1.75	3.91	1.85	3.55	1.88
Schleswig-Holstein (n = 500)	3.60	1.91	3.60	1.98	4.48	1.96
Germany* (n = 881)	3.61	1.72	3.53	1.91	3.79	1.92
Total (n = 1,881)	*3.61*	*1.78*	*3.65*	*1.92*	*3.91*	*1.95*

Source: WI/IEK-STE survey 2009
Scale of 1 (=very low) to 7 (=very high)

Schleswig-Holstein than in the areas along the Rhine or in Germany. The personal risk associated with transport, on the other hand, was considered greater in the area along the Rhine than in Schleswig-Holstein or Germany. No significant regional differences could be ascertained in the perception of risks associated with capture.

With respect to personal benefits associated with the three process steps, the mean values in Table 11.6 show that the personal benefits of CO_2 storage and transport are generally considered lower than the personal benefits of capture. The perceptions of men and women were not statistically different.

Statistically significant differences in the appraisal of personal benefits associated with the three process steps depending on professional qualification are only visible for opinions on capture. The benefits of this process step were considered much smaller by individuals with training at a post-secondary vocational school and individuals with a degree from a university than by individuals with no professional qualification or those with certified vocational training. Qualification-specific differences in appraisals of transport and storage are not statistically significant.

Regional differences were visible in the evaluation of personal benefits associated with CO_2 storage. These were smallest in Schleswig-Holstein. Appraisals of CO_2 capture and transport in the regions did not differ statistically from each other.

To summarize, the personal risk of CO_2 storage is considered higher than the personal risk of capture and transport. Accordingly, the personal benefits associated with storage are considered smaller than the personal benefits associated with the

Table 11.6 Mean values and standard deviations (SD) of the perception of personal benefits associated with the three CCS process steps

	Capture		Transport		Storage	
Gender	**Mean**	**SD**	**Mean**	**SD**	**Mean**	**SD**
Men (n = 506)	3.45	1.70	3.18	1.77	3.18	1.78
Women (n = 511)	3.60	1.76	3.35	1.72	3.29	1.72
Total (n = 1,017)	*3.53*	*1.73*	*3.26*	*1.75*	*3.23*	*1.72*
Professional qualification	**Mean**	**SD**	**Mean**	**SD**	**Mean**	**SD**
No professional qualification (yet) (n = 99)	3.69	1.52	3.65	1.65	3.45	1.67
Certified vocational training (n = 528)	3.66	1.78	3.25	1.77	3.25	1.81
Training at a post-secondary vocational school (n = 171)	3.26	1.55	3.22	1.72	3.12	1.60
Degree from a university (n = 217)	3.34	1.80	3.18	1.74	3.17	1.74
Total (n = 1,015)	*3.53*	*1.73*	*3.27*	*1.75*	*3.23*	*1.75*
Region	**Mean**	**SD**	**Mean**	**SD**	**Mean**	**SD**
Rhine area (n = 500)	3.41	1.70	3.19	1.78	3.10	1.76
Schleswig-Holstein (n = 500)	3.43	1.83	3.17	1.75	3.02	1.78
Germany* (n = 881)	3.54	1.73	3.29	1.75	3.28	1.76
Total (n = 1,881)	*3.48*	*1.75*	*3.23*	*1.76*	*3.16*	*1.77*

Source: WI/IEK-STE survey 2009
Scale of 1 (=very low) to 7 (=very high)

other two process steps. Gender-specific differences are only visible in the perception of risks but not in the perception of benefits. For all three process steps, women considered the personal risk to be higher than men. Qualification-specific differences exist in the perception of risks associated with transport and storage and in the perception of benefits associated with capture. In contrast, regional differences can be seen in the perception of risks associated with storage and transport as well as in the perception of benefits associated with storage.

11.3.4 Factors Influencing Initial Attitudes Towards CCS

In the previous sections, the results of bivariate statistical analyses were used to show that the awareness of CCS in Germany as well as the initial attitudes and perception of risks and benefits differed in the general public depending on gender, professional qualification, and region. However, as bivariate analyses can only investigate the relationship between two variables (e.g. between gender and risk perception or region and the perception of benefits), they cannot be used as a basis to answer the question of whether initial attitudes towards CCS are more heavily influenced by gender, professional qualification, regional affiliation or other factors. This question can only be answered by investigating the simultaneous influence of several factors on initial attitudes towards CCS. This necessitates multivariate statistical analyses, which also comprise regression analysis.

In order to answer the question of which factors considerably influence initial attitudes towards CCS, a multiple linear regression was therefore performed. The dependent variable in this regression model is thus the 'initial attitude towards CCS', which is an index comprising the three variables 'attitude towards capture in general', 'attitude towards transport in general', and 'attitude towards storage in general'.[16] These attitudes were surveyed by providing study participants with information on each of the process steps, and asking them to assess the personal and social risks and benefits.

As influencing factors (=independent variables), the regression model incorporated sociodemographic characteristics (gender, age, professional qualification), general knowledge of the environment and science, of activities that increase CO_2 emissions, and of CCS, attitudes towards the use of renewable energy, coal and nuclear energy, trust in information provided by political decision-makers/parties, companies, scientists and NGOs, perception of the personal and social risks associated with CCS, as well as the perception of the personal and social benefits of CCS (see Table 11.7).[17] In order to investigate whether the impact of these factors is different depending on whether the respondents live in the 'storage region' of Schleswig-Holstein, in the 'CCS demonstration plant region' along the Rhine or in the 'rest' of Germany, the regression analysis was performed for these three regions separately.

Table 11.7 shows the standardized beta coefficients and significance of the linear regression as well as R^2 for the whole model by region.[18] The R^2 values reveal a high quality of the regression model, as between 56 % and 61 % of the variance in initial attitudes towards CCS can be explained by the factors incorporated.[19]

[16] The prerequisite for combining several items to form one index is their internal consistency. Internal consistency is measured with Cronbach's α, which can have theoretical values ranging from 0 (=no internal consistency) to 1 (=full internal consistency). To evaluate the three process steps, a reliability analysis showed a Cronbach's α value of 0.785 and thus high internal consistency, which allows the three items to be combined in one index.

[17] Most independent variables are indices which combine multiple items. Reliability analyses were also performed for these in advance, and it was found that the individual items could be combined based on their internal consistency.

[18] The beta coefficients are regression coefficients that would result when both the dependent and explanatory variables are transformed into z values before the implementation, and thus standardized (Brosius 1998). In contrast to the non-standardized regression coefficients, which are documented with the complete results of the regression analysis in Schumann (2011), beta coefficients can be compared directly.

[19] R^2 indicates how high the proportion of variance of a dependent variable is, which is explained by independent variables integrated in the linear regression model. R^2 can have values ranging from 0 to 1, where the variance of dependent variables with a value of 1 is fully explained by the independent variables integrated in the regression model.

Table 11.7 Standardized regression coefficients and significance of the linear regression depending on the region

Influencing factors	Germany*		Rhine area		Schleswig-Holstein	
	Standardized beta coefficients	Significance	Standardized beta coefficients	Significance	Standardized beta coefficients	Significance
Gender	−0.028	0.224	−0.049	0.131	−0.019	0.564
Age	0.043	0.047*	0.072	0.019*	0.016	0.613
Professional qualification	−0.010	0.652	0.010	0.742	−0.027	0.413
General knowledge of the environment and science	0.027	0.269	−0.019	0.561	0.017	0.622
Knowledge of activities that increase CO_2 emissions	−0.019	0.411	−0.024	0.467	0.034	0.313
Knowledge of CCS	−0.044	0.073	0.038	0.257	−0.040	0.251
Attitude towards the use of renewable energy	0.058	0.014*	0.009	0.791	0.034	0.300
Attitude towards the use of coal	0.033	0.171	0.076	0.023*	0.058	0.071
Attitude towards the use of nuclear energy	0.039	0.105	0.094	0.005*	−0.031	0.362
Trust in information from political decision-makers and parties	0.035	0.202	0.032	0.383	−0.013	0.707
Trust in information from companies	0.034	0.192	0.069	0.057	0.128	0.000**

Trust in information from scientists	−0.007	0.770	0.071	0.042*	0.022	0.539
Trust in information from NGOs	−0.006	0.831	−0.001	0.974	−0.040	0.272
Perception of personal risks associated with CCS	−0.089	0.036*	−0.026	0.625	−0.224	0.000**
Perception of social risks associated with CCS	−0.181	0.000**	−0.214	0.000**	−0.050	0.441
Perception of personal benefits associated with CCS	0.185	0.000**	0.220	0.000**	0.227	0.000**
Perception of social benefits associated with CCS	0.453	0.000**	0.404	0.000**	0.413	0.000**
R^2 for model as a whole	0.610		0.580		0.562	

Source: WI/IEK-STE survey 2009

Dependent variable: initial attitude towards CCS, ** $p < 0.001$, * $p < 0.05$

The beta coefficients show that cross-regional initial attitudes towards CCS are most heavily influenced by the perception of the social benefits of the technologies. This influence is positive, which means that initial attitudes become more positive towards the three process steps the greater the respondents consider the benefits of CCS for society to be.

A positive perception of the personal benefits of CCS also has a positive effect on the overall evaluation of the three process steps. This correlation was slightly stronger in the areas along the Rhine and in Schleswig-Holstein than in Germany*.

The importance of the perception of personal and social risks for initial attitudes towards CCS varies from region to region. In Germany*, the perception of social risks has a stronger, negative impact on attitudes towards CCS than the perception of personal risks. This means that attitudes become more negative towards the three process steps the greater the respondents consider the social risks of CCS to be. This relationship is also true for the perception of personal risks, but the impact is much weaker than that of the perception of social risks.

In the area along the Rhine, the perception of social risks is important for opinions on the CCS process steps, while in the region in Schleswig-Holstein, the perception of personal risks is more important. These negative correlations between risk perception and attitudes towards CCS illustrate that respondents in the area along the Rhine have a more negative opinion of CCS the higher they consider the social risks of the technologies to be. In contrast, respondents in the region in Schleswig-Holstein have a more negative opinion of the technologies the higher they consider the personal risks of CCS to be. These findings indicate that the above-mentioned discussion about the storage of CO_2 led to a situation where the citizens of Schleswig-Holstein believe themselves to more directly affected by CCS than citizens of other regions and thus assess the personal risks of CCS as being greater.

In addition to the perception of benefits and personal risks, only the trust in information from companies is statistically relevant for initial attitudes towards CCS in the region in Schleswig-Holstein. Individuals who trust information from companies have a more positive opinion of CCS than individuals who place little or no trust in information from companies.

In the area along the Rhine, trust in information from scientists and the opinions of respondents on nuclear energy and the use of coal had a weak, positive influence on the evaluation of CCS technologies: the more trust respondents had in information from scientists and the more positive their attitudes towards nuclear energy or the use of coal, the more positive their opinions of CCS were. In contrast, attitudes towards CCS in Germany* were more positive when the respondents had more positive opinions of the use of renewable energy. However, this is only a weak correlation.

Of the sociodemographic characteristics in the regression model, only age had an impact on initial attitudes towards CCS. However, this correlation, which revealed that older respondents had a more positive opinion of CCS than younger respondents, was also weak. The characteristics gender and professional qualification, on the other hand, had no statistical influence in the regression model on attitudes towards the CCS process steps. With respect to the findings outlined above, this finding illustrates that the attitudes of men and women and of individuals with different levels of professional qualification towards CCS can be very different. The factors that decisively shape initial attitudes towards CCS, however, are the perception of social and personal benefits of the technologies, followed by the perception of social and personal risks associated with CCS.

11.4 Summary and Conclusions

In this chapter, the 'public acceptance of CCS' was defined and delimited as a subject of research, and it was shown that the acceptance of technologies cannot yet be measured reliably because the general public does not know enough about CCS technologies. CCS acceptance research studies therefore focus on investigating awareness and knowledge of CCS and initial attitudes towards it among the general public. Such studies also concentrate on identifying factors that have an impact on initial attitudes towards the technologies as well as on analysing the impact of information and methods of communication on changes in and the stability of initial attitudes.

With respect to how well known CCS is among the general public, the findings of international and national studies confirm that at least awareness of the concept of 'CO_2 capture and storage' has increased considerably over the course of time. In Germany, for example, 38 % of respondents said that they had heard of CCS at some point in time. Men and individuals with a higher professional qualification were more likely to have heard of the technologies at some point in time than women or individuals with a lower professional qualification. These findings correlate with findings from research on the diffusion of innovations, which also showed that more highly qualified individuals were aware of new technologies at an earlier stage than individuals with a lower level of professional qualification.

The increasing awareness of the concept of 'CO_2 capture and storage', however, is not accompanied by an increase in knowledge of the technologies. As the findings of international and national studies show, misconceptions about CCS (still) abound among the general public. This can be explained by the fact that lay people find it difficult to distinguish environmental problems, such as ozone depletion, global warming, acid rain or smog, from each other. Strategies aiming to inform the general public about CCS should therefore consider the fact that citizens are often not only unfamiliar with CCS but also with the environmental problems that the technologies could help to mitigate. The communication of CCS as a potential option for climate change mitigation therefore often necessitates not only providing citizens with basic information on the technologies but also with basic information on important environmental problems such as global warming, acid rain or ozone depletion.

In addition, it is important for information on CCS and the communication of CCS that citizens have initial attitudes towards the technologies even though they know little or nothing about CCS. In Germany, these initial attitudes towards CCS are on average (still) mainly neutral, although women are more sceptical of the technologies than men.

The regional differences in initial attitudes before and after the receipt of information demonstrate that citizens of Schleswig-Holstein do not only have more negative attitudes towards CCS than citizens of the area along the Rhine or of the 'rest' of Germany, but that the debate about CO_2 storage in Schleswig-Holstein (cf. Chap. 12 by Fischer) has already led to negative attitudes towards CCS being formed in this region, which are not necessarily initial attitudes any more but rather stable opinions. As the present findings also suggest, these negative attitudes in Schleswig-Holstein are mainly related to the fact that citizens here consider the personal risks associated with CO_2 storage to be much greater than citizens of the other regions.

This assumption is supported both by bivariate analyses as well as by the results of multiple regression. The regression results also illustrate that initial attitudes towards CCS in all regions are most heavily influenced by the perception of the social benefits of the technologies, and that this influence is positive: the greater the social benefits of CCS are considered to be, the more positive the initial attitudes towards the technologies.

How stable these attitudes are regarding the benefits of CCS or how easily they can be changed by new information cannot be analysed using the present findings as a basis. The influence of information on the perception of the benefits of CCS, as well as the influence of the perception of benefits on the stability of attitudes towards CCS, must therefore be systematically investigated in future studies in order to assess the importance of the perception of benefits as an indicator for evaluating the future public acceptance of CCS in Germany.

References

AGS (2007) Public and stakeholder attitudes towards energy, environment and CCS. Pathways to Sustainable European Energy Systems; AGS, The Alliance for Global Sustainability, Göteborg

Ashworth P, Gardner J (2006) Understanding and incorporating stakeholder perspectives to low emission technologies in New South Wales. Centre for Low Emission Technology, Kenmore

Ashworth P, Littleboy A, Pisarski A, Beath A, Thambimuthu K (2006a) Understanding and incorporating stakeholder perspectives to low emissions technologies in Australia. In: GHGT 8, Trondheim, Norway

Ashworth P, Pisarski A, Littleboy A (2006b) Understanding and incorporating stakeholder perspectives to low emission technologies in Queensland. Centre for Low Emission Technology, Kenmore

Ashworth P, Carr-Cornish S, Boughen N, Thambimuthu K (2009) Engaging the public on carbon dioxide capture and storage: does a large group process work? Energy Procedia 1:4765–4773

Ashworth P, Boughen N, Mayhew M, Millar F (2010) From research to action: now we have to move on CCS communication. Int J Greenh Gas Control 4:426–433

Bishop GF, Oldendick RW, Tuchfarber AJ, Bennett SE (1980) Pseudo-opinions on public affairs. Public Opin Q 44:198–209

Brosius F (1998) SPSS 8. Professionelle Statistik mit Windows. MITP-Verlag GmbH, Bonn

Cremer C, Fischedick M, Esken A (2008) Sozioökonomische Begleitforschung zur gesellschaftlichen Akzeptanz von Carbon Capture and Storage (CCS) auf nationaler und internationaler Ebene. Available www.cooretec.de/index.php/index.php?index=17&file=43

Curry TE (2004) Public awareness of carbon capture and storage: a survey of attitudes toward climate change mitigation. Master of Science in Technology and Policy, Massachusetts Institute of Technology, Cambridge, MA

Curry T, Reiner D, Ansolabehere S, Herzog HJ (2004) How aware is the public of carbon capture and storage?. In: GHGT 7, Vancouver, Canada

Curry T, Reiner D, Figueoredo MD, Herzog HJ (2005) A survey of public attitudes towards energy & environment in Great Britain. Massachusetts Institute of Technology, Laboratory for Energy and the Environment, Cambridge, MA

Curry T, Ansolabehere S, Herzog HJ (2007) A survey of public attitudes towards climate change and climate change mitigation technologies in the United States: analyses of 2006 results. Massachusetts Institute of Technology, Laboratory for Energy and the Environment, Cambridge, MA

Daamen D, de Best-Waldhober M, Damen K, Faaij A (2006) Pseudo-opinions on CCS technologies. In: GHGT 8, Trondheim, Norway

Daamen DDL, Terwel B, ter Mors E, Reiner SD, Næss R et al (2011) Scrutinizing the impact of CCS communication on opinion quality: focus group discussions versus Information-Choice Questionnaires: results from experimental research in six countries. Energy Procedia 4:6182–6187

De Best-Waldhober M, Daamen D (2006) Public perceptions and preferences regarding large scale implementation of six CO_2 capture and storage technologies. Centre for Energy and Environmental Studies, Faculty of Social Sciences, Leiden University, Leiden

De Best-Waldhober M, Daamen D, Faaij A (2006) Public perceptions and preferences regarding large-scale implementation of six CO_2 capture and storage technologies: well-informed and well-considered opinions versus uninformed pseudo-opinions of the Dutch public. Leiden University, Leiden

De Best-Waldhober M, Daamen D, Hendriks C, de Visser E, Ramírez A, Faaij A (2008) How the Dutch evaluate CCS options in comparison with other CO_2 mitigation options. Results of a nationwide Information-Choice Questionnaire survey. Leiden University, Leiden

De Best-Waldhober M, Daamen D, Faaij A (2009a) Informed and uninformed public opinions on CO_2 capture and storage technologies in the Netherlands. Int J Greenh Gas Control 3:322–332

De Best-Waldhober M, Daamen D, Ramirez A, Faaij A, Hendriks C, De Visser E (2009b) Informed public opinions on CCS in comparison to other mitigation options. Energy Procedia 1:4795–4802

De Best-Waldhober M, Paukovic M, Brunsting S, Daamen D (2010) Awareness, knowledge, beliefs, and opinions regarding CCS of the Dutch general public before and after information. In: GHGT-10, Amsterdam, The Netherlands

De Coninck H, Anderson J, Curnow P (2006) Acceptability of CO_2 capture and storage. Energy research Centre of the Netherlands (ECN). Available http://www.ecn.nl/docs/library/report/2006/c06026.pdf

De Coninck H, Flach T, Curnow P, Richardson P, Anderson J, Shackley S, Sigurthorsson G, Reiner D (2009) The acceptability of CO_2 capture and storage (CCS) in Europe: an assessment of the key determining factors: part 1. scientific, technical and economic dimensions. Int J Greenh Gas Control 3:333–343

Denkstelle Hamburg (2009) CCS-Akzeptanzstudie für das IZ Klima Berlin 2008/2009 [Online]. Available http://www.presseportal.de/pm/67576/1410708/iz_klima

Dierkes M, von Thienen V (1982) Akzeptanz und Akzeptabilität der Informationstechnologien. Wissenschaftszentrum, Berlin

Duetschke E (2010) What drives local public acceptance – comparing two cases from Germany. In: GHGT-10, Amsterdam, The Netherlands

Fischedick M, Pietzner K, Kuckshinrichs W, Schumann D (2008) Gesellschaftliche Akzeptanz von CO_2-Abscheidung und -Speicherung in Deutschland. Energiewirtschaftliche Tagesfragen 58:20–23

Fischedick M, Pietzner K, Supersberger N, Esken A, Kuckshinrichs W, Zapp P, Linßen J, Schumann D, Radgen P, Cremer C, Gruber E, Schnepf N, Roser A, Idrissova F (2009) Stakeholder acceptance of carbon capture and storage in Germany. Energy Procedia 1:4783–4787

Fischer W, Hake J-F, Kuckshinrichs W, Schenk O, Schumann D (2010) Carbon capture and storage – Politische und gesellschaftliche Positionen in Deutschland. Technikfolgenabschätzung – Theorie und Praxis 19:38–46

Fleishman LA, de Bruin WB, Morgan MG (2010) Informed public preferences for electricity portfolios with CCS and other low-carbon technologies. Risk Anal 30:1399–1410

Gloede F, Hennen L (2005) Einführung in den Schwerpunkt. Technikfolgenabschätzung – Theorie und Praxis 14:4–12

Gough C, Shackley S (2006) Towards a multi-criteria methodology for assessment of geological carbon storage options. Clim Chang 74:141–174

Gough C, Shackley S, Holloway S, Bentham M, Bulatov I, Mclachlan C, Klemes J, Purdy R, Cockerill T (2006) An integrated assessment of carbon dioxide capture and storage in the UK. In: GHGT 8, Trondheim, Norway

Grunwald A (2005) Zur Rolle von Akzeptanz und Akzeptabilität von Technik bei der Bewältigung von Technikkonflikten. Technikfolgenabschätzung – Theorie und Praxis 14:54–60

Ha-Duong M, Nadaï A, Campos AS (2009a) A survey on the public perception of CCS in France. Int J Greenh Gas Control 3:633–640

Ha-Duong M, Nadaï A, Sofia Campos A (2009b) A survey on the public perception of CCS in France. Energy Procedia 1:4757–4764

Huijts N (2003) Public perception of carbon dioxide storage: the role of trust and affect in attitude formation. Eindhoven University of Technology, Eindhoven

Huijts NMA, Midden CJH, Meijnders AL (2007) Social acceptance of carbon dioxide storage. Energy Policy 35:2780–2789

IPCC (2005) In: Metz B, Davidson O, De Coninck H, Loos M, Meyer L (eds) Special report on carbon dioxide capture and storage. Cambridge University Press, New York

Itaoka K, Saito A, Akai M (2004) Public acceptance of CO_2 capture and storage technology: a survey of public opinion to explore influential factors. In: GHGT 7, Vancouver, Canada

Itaoka K, Saito A, Akai M (2006) A path analysis for public survey data on social acceptance of CO_2 capture and storage technology. In: GHGT 8, Trondheim, Norway

Itaoka K, Okuda Y, Saito A, Akai M (2009) Influential information and factors for social acceptance of CCS: the 2nd round survey of public opinion in Japan. Energy Procedia 1:4803–4810

Itaoka K, Saito A, Akai M (2010) A study on roles of public survey and focus groups to assess public opinions for CCS implementation. In: GHGT-10, Amsterdam, The Netherlands

Janis I (1972) Victims of groupthink: a psychological study of foreign-policy decisions and fiascoes. Houghton Mifflin, Boston

Lamnek S (1995) Qualitative Sozialforschung Band 2. Methoden und Techniken. Beltz/Psychologie Verlags Union, Weinheim

Metlay D (1999) Institutional trust and confidence: a journey into a conceptual quagmire. In: Cvetkovich GT, Löfstedt RE (eds) Social trust and the management of risk. Earthscan, London

Miller E, Bell L, Buys L (2007) Public understanding of carbon sequestration in Australia: socio-demographic predictors of knowledge, engagement and trust. Aust J Emerg Technol Soc 5:15–33

Palmgren CR, Bruine De Bruin W, Keith DW, Morgan MG (2004a) Public perceptions of oceanic and geological CO_2 disposal. In: GHGT 7, Vancouver, Canada

Palmgren CR, Morgan MG, Bruine De Bruin W, Keith DW (2004b) Initial public perceptions of deep geological and oceanic disposal of carbon dioxide. Environ Sci Technol 38:6441–6450
Pietzner K, Schumann D, Esken A (2010) CO_2-Abscheidung und -Speicherung aus gesellschaftlicher Sicht. ÖkologischesWirtschaften 4:39–42
Pietzner K, Schumann D, Tvedt SD, Torvatn HY, Næss R, Reiner DM, Anghel S, Cismaru D, Constantin C, Daamen DDL, Dudu A, Esken A, Gemeni V, Ivan L, Koukouzas N, Kristiansen G, Markos A, ter Mors E, Nihfidov OC, Papadimitriou J, Samoila IR, Sava CS, Stephenson MH, Terwel BW, Tomescu CE, Ziogou F (2011) Public awareness and perceptions of carbon dioxide capture and storage (CCS): insights from surveys administered to representative samples in six European countries. Energy Procedia 4:6300–6306
Reiner D, Curry T, Figueoredo MD, Herzog HJ, Anasolabehere S, Itaoka K, Akai M, Johnsson F, Odenberger M (2006a) An international comparison of public attitudes towards carbon capture and storage technologies. In: GHGT 8, Trondheim, Norway
Reiner DM, Curry TE, Defigueiredo MA, Herzog HJ, Ansolabehere SD, Itaoka K, Johnsson F, Odenberger M (2006b) American exceptionalism? Similarities and differences in national attitudes toward energy policy and global warming. Environ Sci Technol 40:2093–2098
Reiner DM, Pietzner K, Schumann D, Tvedt T, Næss R et al. (2010a) Measuring regional attitudes towards proposed CCS plants: a four country comparison. In: GHGT-10, Amsterdam, The Netherlands
Reiner DM, Stephenson M, West J, Firth R, Boulouta I (2010b) Scrutinising the impact of CCS communication on the general and local public: results of a comparative study of CCS communication methods in the United Kingdom. Research report
Renn O (2005) Technikakzeptanz: Lehren und Rückschlüsse der Akzeptanzforschung für die Bewältigung des technischen Wandels. Technikfolgenabschätzung – Theorie und Praxis 14:29–38
Renn O, Zwick MM (1997) Risiko- und Technikakzeptanz. Springer, Berlin [u.a.]
Renn O, Schweizer P-J, Dreyer M, Klinke A (2007) Risiko. Über den gesellschaftlichen Umgang mit Unsicherheit. Oekom Verlag, München
Roberts T, Mander S (2010) Assessing public perceptions of CCS: benefits, challenges and methods. In: GHGT-10, Amsterdam, The Netherlands
Rogers EM (2003) Diffusion of innovations. Free Press, New York/London/Toronto/Sydney
Scharpf FW (2000) Interaktionsformen. Akteurzentrierter Institutionalismus in der Politikforschung. Leske + Budrich, Opladen
Schubert K, Klein M (2006) Akzeptanz. In: Schubert K, Klein M (eds) Das Politiklexikon, 4., aktual. Aufl. Dietz, Bonn
Schulz M, Scheer D, Wassermann S (2010) Neue Technik, alte Pfade? Zur Akzeptanz der CO_2-Speicherung in Deutschland. GAIA 19:287–2986
Schumann D (2009) Public acceptance of carbon dioxide capture and storage: research approaches for investigating the impact of communication. GAIA 18(3):261–263
Schumann D (2011) Gesellschaftliche Akzeptanz von CO_2-Abscheidung und -Speicherung (CCS). STE Preprint 25/2011
Schumann D, Simon A (2009) Communication of CO_2 capture and storage (CCS): simulating the impact on knowledge and public acceptance. In: Edmonds B, Gilbert N (eds) The sixth conference of the European Social Simulation Association (ESSA 2009), University of Surrey, Guildford, UK, 14–18 September 2009
Schumann D, Simon A (2010) The impact of persuasive communication on public attitudes regarding CO_2 capture and storage (CCS): challenges for the empirical parameterization of an agent-based simulation model. In: 8th European workshop on Multi-Agent Systems (EUMAS 2010), Paris. http://lipade.math-info.univ-paris5.fr/eumas10/accepted-papers.html
Schumann D, Pietzner K, Esken A (2010) Umwelt, Energiequellen und CCS: Regionale Unterschiede und Veränderungen von Einstellungen der deutschen Bevölkerung. Energiewirtschaftliche Tagesfragen 60:52–56

Schumann D, Tvedt SD, Torvatn HY (2011) How do knowledge, trust and perceptions of risks and benefits influence the stability of initial perceptions of CO_2 Capture and Storage (CCS)? A comparative structural equation modelling analysis of data from representative surveys in Germany and Norway. STE Research Report 05/2011

Shackley S, Mclachlan C (2006) Trade-offs in assessing different energy futures: a regional multi-criteria assessment of the role of carbon dioxide capture and storage. Environ Sci Pol 9:376–391

Shackley S, Gough C, Mclachlan C (2004a) The public perceptions of carbon dioxide capture and storage in the UK. In: GHGT 7, Vancouver, Canada

Shackley S, Mclachlan C, Gough C (2004b) The public perceptions of carbon capture and storage. In: Research TCFCC (ed) Working paper, Tyndall Centre for Climate Change Research, Manchester

Shackley S, Waterman H, Godfroij P (2007) Stakeholder perceptions of CO_2 capture and storage in Europe: results from a survey. Energy Policy 35:5091–5108

Shackley S, Reiner D, Upham P, Coninck HD, Sigurthorsson G, Anderson J (2009) The acceptability of CO_2 capture and storage (CCS) in Europe: an assessment of the key determining factors part 2. The social acceptability of CCS and the wider impacts and repercussions of its implementation. Int J Greenh Gas Control 3:344–356

Sharp J (2005) Public attitudes toward geological disposal of carbon dioxide in Canada. Simon Fraser University, Burnaby

Sharp J, Jaccard M, Keith D (2006) Public attitudes toward geological disposal of carbon dioxide in Canada. In: GHGT 8, Trondheim, Norway

Sharp JD, Jaccard MK, Keith DW (2009) Anticipating public attitudes toward underground CO_2 storage. Int J Greenh Gas Control 3:641–651

Siegrist M (2000) The influence of trust and perceptions of risks and benefits on the acceptance of gene technology. Risk Anal 20:195–203

Siegrist M, Cousin M-E, Kastenholz H, Wiek A (2007) Public acceptance of nanotechnology foods and food packaging: the influence of affect and trust. Appetite 49:459–466

Solá R, Sala R, Oltra C, Gamero N (2008) A proposal to address the governance of carbon capture and storage technologies in Spain: trust, communication, and public involvement. Editorial CIEMAT, Madrid

Ter Mors E (2009) Dealing with information about complex issues: the role of source perceptions. Dissertatiereeks Kurt Lewin Institut 2009–13

Ter Mors E, Weenig MWH, Ellemers N, Daamen DDL, De Best-Waldhober M (2009) Public information: on why and when multiple information sources are more effective than single information sources in communication about CCS. Energy Procedia 1:4715–4718

Ter Mors E, Weenig MWH, Ellemers N, Daamen DDL (2010) Effective communication about complex environmental issues: perceived quality of information about carbon dioxide capture and storage (CCS) depends on stakeholder collaboration. J Environ Psychol 30:347–357

Terwel BW (2009) Origins and consequences of public trust: towards an understanding of public acceptance of carbon dioxide capture and storage. Dissertatiereeks Kurt Lewin Institut 2009–12

Terwel BW, Harinck F, Ellemers N, Daamen DDL (2011) Going beyond the properties of CO2 capture and storage (CCS) technology: how trust in stakeholders affects public acceptance of CCS. Int J Greenh Gas Control 5:181–188

Tokushige K, Akimoto K, Tomoda T (2006) Public perceptions on the acceptance of CO_2 geological storage and the valuable information for the acceptance. In: GHGT 8, Trondheim, Norway

Tokushige K, Akimoto K, Tomoda T (2007) Public perceptions on the acceptance of geological storage of carbon dioxide and information influencing the acceptance. Int J Greenh Gas Control 1:101–112

Uken M (2009) Der neue Endlager-Streit. Zeit-Online. Available http://pdf.zeit.de/online/2009/26/ccs-protest.pdf

Uno M, Mori Y, Tokushige K, Furukawa A (2004) Exploration of public acceptance regarding CO_2 underground sequestration technologies. In: GHGT 7, Vancouver, Canada

Van Alphen K, Voorst QVT, Hekkert MP, Smits REHM (2007) Societal acceptance of carbon capture and storage technologies. Energy Policy 35:4368–4380

Wallquist L, Visschers VHM, Siegrist M (2009) Lay concepts on CCS deployment in Switzerland based on qualitative interviews. Int J Greenh Gas Control 3:652–657

Wallquist L, Visschers VHM, Siegrist M (2010) Impact of knowledge and misconceptions on benefit and risk perception of CCS. Environ Sci Technol 44:6557–6562

Part III
Framework for Energy and Climate Policy

Chapter 12
No CCS in Germany Despite the CCS Act?

Wolfgang Fischer

Abstract The implementation of the EU Directive on CCS into German law was a controversial political process with different and unusual lines of conflict – between, but also within parties both at the federal and state level, between federal States (Länder) and the federal government, between NGOs and scientific advisory boards. Because the German federal political system opens up the opportunity for States to influence the policy making process strongly, some States succeeded in implementing provisions into the Federal CO_2-storage law which provides States the principal opportunity to bar storage projects. Therefore, it is almost certain that in the next years not a single CCS-demonstration project will be realized in Germany.

Keywords Carbon capture and storage • CCS • EU-directive • Policy • Regulation • Germany • Parties • NGOs • Government

12.1 Introduction

Carbon capture and storage (CCS) has been a topic in science since the 1980s (Marchetti 1989, p. 7), but it was only with the introduction of policies aiming to reduce carbon dioxide (CO_2) emissions during the late 1990s that it began to become important for research and then politics. CCS technologies are an important component in the European Union's energy and environmental policy. The EU CCS Directive[1] provides member states with a legal framework that they must flesh out with national legislation. After a long and conflict-ridden process, the German legislature passed a CCS act at the end of June 2012. This is a compromise between advocates and opponents of CCS, allowing the research and development of CCS technologies but practically prohibiting the demonstration of their commercial usage. This legislative process is an example of the problems associated with

[1] Access to EU regulations can be gained at http://eur-lex.europa.eu

W. Fischer (✉)
Institute of Energy and Climate Research – Systems Analysis and Technology Evaluation (IEK-STE), Forschungszentrum Jülich GmbH, D-52425 Jülich, Germany
e-mail: wo.fischer@fz-juelich.de

W. Kuckshinrichs, J.-F. Hake (eds.), *Carbon Capture, Storage and Use*,
DOI 10.1007/978-3-319-11943-4_12

implementing policies in a federal political system in view of divergent energy policy concepts within society and politics, opposing interests of federal states and local acceptance problems. Election campaigns and party political tactics exacerbate the divergences further. Section 12.2 describes the legal and political framework conditions of German CCS policy, while Sects. 12.3, 12.4 and 12.5 look at contradictory policies and conflicts between and within parties on a national and state level as well as between social organizations. Section 12.6 outlines the legislative process, the most important landmarks of which are shown in Table 12.1, while Sect. 12.7 looks at whether CCS is doomed to failure in Germany from a political perspective.

12.2 The EU Sets the Framework and the Deadlines

With the adoption in 2009 of the CCS Directive on the geological storage of CO_2 (European Commission 2009), which is part of the EU's climate action package (Reichert 2009), the EU created the first legal framework in the world for permanent storage and for uniform standards governing the approval, environmental protection and safety of CO_2 storage facilities. The Directive also specifies the framework conditions, criteria, selection methods, approval, operation, closure, and monitoring of a CO_2 storage site, determines the procedure in the case of leakage, and defines the process for closure of a storage site and the transfer of responsibility to a national authority. The central importance of CCS for the integrated energy, climate and industrial policies of the EU is reflected in the fact that the Directive passed through the complex EU legislative process in a short period between January 2008 and April 2009 (Chiavari 2010). Two deadlines are fundamental in the CCS-Directive: 25 June 2011 was the deadline for transposing the Directive into national legislation, and 31 December 2015 is the deadline for CCS demonstration projects to become operational in order to benefit from funding within the European Energy Programme for Recovery (EEPR)[2] and receive subsidies within the EU Emissions Trading System (New Entrants Reserve, NER300)[3] via the European Investment Bank. Six CCS projects have been selected in 2009 by the EU-Commission for funding within the EEPR: Bełchatów (Poland), Compostilla (Spain), Don Valley, formerly Hatfield (United Kingdom), Jänschwalde (Germany), Maasvlakte (Netherlands), each to the tune of € 180 million, and Porto Tolle (Italy) with subsidies totalling € 100 million.

In Germany, responsibility for CCS is shared between the Federal Ministries of Economics and Technology (BMWi), Education and Research (BMBF) and the

[2] http://ec.europa.eu/energy/eepr/

[3] http://www.ner300.com/

Table 12.1 Important landmarks for German CCS policy

01.01.2006	5th Energy Research Programme with a focus on CCS (BMWi 2005)
12.09.2006	European Commissioner for the Environment announces proposal on establishing a legal basis for CCS (Dimas 2006)
10.01.2007	European Commission calls for legal framework and demonstration plants (European Commission 2007)
23.08.2007	Meseberg energy and climate package adopted by grand coalition (Bundeskabinett 2007): demands legal framework created for CCS, testing of 2–3 CCS-power plants in Germany
19.09.2007	Report on the status of development and prospects of CCS technologies (BMWi BMU BMBF 2007)
23.01.2008	Draft CCS Directive of the European Commission (Europäische Kommission 2008).
19.11.2008	Law office drafts a bill for CO_2 plants on behalf of Alstom, Babcock Borsig, EnBW, E.ON, RWE, Vattenfall (Skrylnikow 2010)
12.12.2008	Political agreement in the European Council on the Directive
17.12.2008	European Parliament passes CCS Directive (co-decision procedure)
06.04.2009	Reading of the Directive in the Council of the European Union
01.04.2009	German federal cabinet passes the bill regulating the capture, transportation, and permanent storage of CO_2 (CO2ATSG-E) (Deutscher Bundestag 2009b)
23.04.2009	Publication of CCS Directive 2009/31/EC (European Union 2009)
06.05.2009	First reading of the CCS draft bill in the Bundestag (Deutscher Bundestag 2009c)
15.05.2009	Position of the Bundesrat (2009): criticism and call for changes in the CCS draft bill
25.05.2009	Hearing in the Bundestag Committee on the Environment: call for a law on CCS-demonstration plants only (Grefe 2009; Skrylnikow 2010)
12.06.2009	Response of federal government to a member's question about the bill (Deutscher Bundestag 2009a)
14.06.2009	State parliament decision: bill rejected in the Bundesrat (Schleswig-Holsteinischer Landtag 2009)
24.06.2009	CCS bill taken off the Bundestag agenda (IZ Klima 2009)
25.06.2009	EU CCS Directive enters into force
17.10.2009	Coalition agreement in federal state Schleswig-Holstein: rejection of CO_2 storage (CDU Schleswig-Holstein/FDP Schleswig Holstein 2009)
24.10.2009	CDU/CSU and FDP agree in the federal coalition agreement (Koalitionsvertrag 2009a): speedy implementation of the CCS Directive
05.11.2009	(Koalitionsvertrag 2009b): Call for the testing and demonstration of CCS technology
19.11.2009	State parliament and state governments of (Schleswig-Holsteinischer Landtag 2009) call for a state clause
26.07.2010	Federal government proposes CCS storage bill (KSpG) (Referentenentwurf 2010)
26–27.08.2010	CCS hearing in BMWi: call for state "opt-out" clause (IZ Klima 2010; Liebing 2010)
27.09.2010	Federal cabinet takes the CC bill off the Parliament agenda, opposition expected in the Bundesrat (Baethge 2010)
28.09.2010	Federal government energy concept emphasizes importance of CCS for process-related emissions (Bundesregierung 2010)
14.02.2011	Federal government submits an amended CCS bill (Bundesregierung 2011b)

(continued)

Table 12.1 (continued)

23.02.2011	Bundestag question time session: disagreement, particularly on the state "opt-out" clause (Deutscher Bundestag 2011d)
23.02.2011	State parliament in Schleswig-Holstein unanimously calls for binding state clause (Schleswig-Holsteinischer Landtag 2011)
13.04.2011	Federal cabinet approves CCS Bill 17/5750 (Deutscher Bundestag 2011b)
12.05.2011	1st discussion of Bill 17/5750 in the Bundestag (Deutscher Bundestag 2011e)
25.05.2011	CCS hearing in the Bundestag Committee on the Environment (IZ Klima 2011)
27.05.2011	Bundesrat rejects proposal submitted by Saxony and Brandenburg to abandon the state clause, but does request certain amendments to the bill (Bundesrat 2011; IZ Klima 2011)
06.06.2011	Public CCS hearing in the Bundestag Committee on the Environment (Deutscher Bundestag 2011a)
25.06.2011	CCS Directive must be transposed into national legislation by EU member states in accordance with Art. 39 CCS Directive
30.06.2011	Submission of a report to the European Commission on the implementation of the CCS Directive in accordance with Art. 27 (Europäische Kommission 2011), submitted by the federal government in August 2011
07.07.2011	2nd and 3rd discussion of the bill in the Bundestag (Deutscher Bundestag 2011b)
	(Deutscher Bundestag 2011c), majority in the Bundestag votes in favour of CCS bill
13.09.2011	Majority of Bundesrat Committee on the Environment rejects the bill, mainly because of the state clause and the regulation of post-closure liability for closed storage sites; appeal to the Mediation Committee recommended
23.09.2011	No majority vote for CCS bill in the Bundesrat, no appeal to the Mediation Committee by the Bundesrat
26.10.2011	Federal government calls on the Mediation Committee
08.11.2011	13th session of the Mediation Committee: first adjournment of the topic of CCS
22.11.2011	1st continuation of the 13th session of the Mediation Committee: adjournment of the topic of CCS
05.12.2011	Vattenfall: plans for Jänschwalde dropped and exploratory activities on storage in the region ended
14.12.2011	2nd continuation of the 13th session: adjournment of the topic of CCS
End 2011	A CCS bill should be adopted and in force (original objective of the federal government)
January 2012	Informal consultations between federal government, opposition and state governments on the CCS bill
08.02.2012	3rd continuation of the 13th session: adjournment
13.06.2012	4th continuation of the 13th session: adjournment
27.06.2012	5th continuation of the 13th session: agreement on a compromise with a narrow majority
28.06.2012	Bundestag passes the amended CCS act with a coalition majority
29.06.2012	Bundesrat passes the CCS act with a majority

Source: Adaptation of Léger (2011, p. 35f)

See also Léger (2011, S. 35f)

The report provides a basis for the Meseberg decisions (main points of the German integrated energy and climate program); the decisions emphasize the need for a national legal framework for CCS and for the generation of public acceptance for CCS

For details on the parliamentary process, see the Parliamentary Material Information System for the Bundestag and the Bundesrat http://dipbt.bundestag.de/extrakt/ba/WP17/351/35126.html

Environment, Nature Conservation and Nuclear Safety (BMU). Within research and development, BMWi is responsible for power plant engineering, carbon capture, carbon transportation, and the CO_2-reduction technologies initiative.[4] BMBF funds the research and development (R&D) programme GEOTECHNOLOGIEN,[5] which focuses on technologies for storing CO_2 and includes the research project CO_2 Storage by Injection into a Natural Saline Aquifer at Ketzin (CO2SINK), and the follow-up project CO_2 Reservoir Management (CO2MAN),[6] which is being run by the German Research Centre for Geosciences in Potsdam and other institutions close to the town of Ketzin in Brandenburg. Furthermore, the research and development project CLEAN (CO_2 Large-Scale Enhanced Gas Recovery in the Altmark Natural Gas Field)[7] investigates whether conventional non-extractable natural gas could be recovered by injecting compressed CO_2. While approval of the pilot storage site in Ketzin was granted in accordance with mining law, the legal basis for the construction and operation of a demonstration CO_2 storage site is approval on the basis of a national CCS act (GFZ 2011). It is politically significant that BMWi and BMU both play a pivotal role in formulating the key points of the CCS Act. As BMU favours renewable energy, and CCS in the power plant sector is met in the ministry and its Federal Environmental Agency[8] with a certain degree of scepticism and indeed opposition, friction with BMWi, which is more open towards using fossil energy carriers for electricity generation, is not surprising in relation to the formulation of certain provisions of the CCS act.[9]

12.3 Political Parties Attempt a Balancing Act

The parties have been establishing their positions on CCS at a national level since 2005 (cf. Fischer et al. 2010). In the party programmes for the Bundestag (Germany's federal parliament) election in 2009, the CDU/CSU (common parliamentary group of Christian Democratic Union and Christian Social Union), the FDP (Free Democratic Party), and the SPD (Social Democratic Party) declared their general support for the development and implementation of CCS, largely in

[4] http://www.cooretec.de/

[5] See http://www.geotechnologien.de

[6] http://www.co2ketzin.de/

[7] http://www.clean-altmark.org

[8] http://www.umweltbundesamt.de

[9] Inner-administrative conflicts are not discussed here. In 2009, when the first draft of the bill was being compiled, the differences between the two ministries were quite small (Götze 2009). However, the subsequent scepticism in BMU towards CCS, combined with increasing opposition towards the use of coal, appears to have intensified.

line with many of the state governments at that time. The Greens (Bündnis90/DieGrünen) and particularly the Left Party (Die Linke), both small parties, are vehemently opposed to CCS. With the exception of the Greens, who also oppose CCS in their regional party branches and in state governments where the party held seats, the positions of the federal parties do not always conform to those adopted by their parties in the federal states or locally in areas with a potential for CCS projects. This often forces parties to perform a balancing act between federal, state, and local politics. Since 2009, the positions of some parties have altered slightly at the federal level, as demonstrated in the debates on CCS and CCS legislation in the Bundestag and Bundesrat.[10]

The Greens continue to oppose CCS at the federal and state levels as being too hazardous and too expensive. Although they believe that there is no place for coal-fired power plants equipped with CCS in a future based on renewable energy, they have become more open towards CCS research during the course of the CCS debate, provided that it is limited to process-related CO_2 emissions from industry. The stance of the federal SPD in relation to CCS in general has been unclear since 2009 despite rejection of the CCS bill in 2011 and abstention from voting on the legislative compromise on CCS in the Bundestag in 2012: the party has yet to clarify its position on the future role of coal-fired power plants in the long term. When he was minister for the environment in the grand CDU/CSU-SPD coalition (2005–2009), the later SPD party chairman Gabriel advocated a speedy implementation of the EU Directive with reference to the objectives of the grand coalition government. In the context of its integrated energy and climate programme (Bundesregierung 2007), CCS was an option for lignite and hard coal power plants, and a package of measures was drawn up to fund R&D on CCS technology lines, the testing of two or three CCS power plants, and the drafting of a legal framework in order to bring CCS to market maturity by 2020 (BMWI/BMU 2007). It would appear that Gabriel follows such a pro-CCS policy. In January 2011, during an official visit to the CCS pilot power plant at the 'Schwarze Pumpe' site in Brandenburg, he appealed together with the trade union for the swift adoption of a CCS act that would permit the construction of a Vattenfall demonstration power plant in Brandenburg (Wendler 2011). However, since 2009, the voices have become louder of those in the SPD who believe that the fossil era will be long gone by the time CCS is ready for implementation (Sattler 2011), who are critical of the CCS-technologies, and who oppose them.[11] Despite this, the SPD has neither

[10] The representative body of Germany's federal states, participating in legislation and administration of the federation; http://www.bundesrat.de

[11] In the SPD's newsletter 'Aktuell' of 17 May 2011, the lack of acceptance of CCS was explained by the fact that the technologies 'have not yet been tested on a large scale and are not yet commercially exploitable. In addition, the national storage potential is limited and a conflict of usage sometimes exists with possible energy storage technologies, and there is currently a lack of acceptance in Germany for CO_2 storage and pipelines.' (SPD-Parteivorstand 2011, p. 14).

officially nor explicitly altered its position – in no small part due to the pro-CCS state government in Brandenburg and the trade unions in the lignite mining districts – of using public support to implement CCS projects in the power plant sector in order to test and verify the technology and thus pave the way for the option of exporting it to newly industrialized countries in future. However, the conceivable options of eliminating process-related emissions as well as the use of CCS in biomass power plants are becoming more important for the SPD at the federal level and for the Greens. At the same time, state governments in which the SPD hold seats are opposed to CO_2 demonstration storage sites in their federal state – but not necessarily in Germany in general – (e.g. Mecklenburg-Western Pomerania, and more hesitantly Saxony-Anhalt), while other governments with SPD involvement (Brandenburg, coalition with the Left Party) or with the SPD alone (Hamburg) are protagonists or sympathizers of CCS demonstration projects, and even allow the exploration of potential storage sites. Nevertheless, none of the state governments with SPD involvement voted in favour of the CCS bill in the Bundesrat in autumn 2011. Only when a compromise was reached on the CCS bill in June 2012 did it get the vote of SPD-governed states in the Bundesrat, but this approval did not induce the SPD parliamentary group to vote in favour of the bill in the Bundestag.

In the new federal government (2009–2013), the three ruling parties CDU/CSU and FDP support the development and demonstration of CCS in principle from the perspective of implementing it in coal-fired power generation, securing technological leadership in power plant construction, and using it to help mitigate climate change as part of an international strategy. However, their policy has evolved further in response to the changed framework conditions of the energy economy and hefty criticism of CCS. Although the parties have not explicitly dissociated themselves from the objectives of CCS development and demonstration, the energy concept published by the CDU/CSU-FDP government in September 2010 can be understood as a shift of emphasis in response to the unclear role of future coal-fired power generation with respect to decision to extend the service lives of nuclear power plants and to increase renewable energy capacity: it states that CCS is first and foremost an option for energy-intensive industries with high process-related CO_2 emissions, and only secondarily an option for fossil power plants (Bundesregierung 2010). In the energy concept of 6 June 2011, centring on phasing out nuclear energy after Fukushima, there is no mention of CCS (Bundesregierung 2011a). The Ethics Commission (Ethik-Kommission 2011), which legitimizes this political anti-nuclear transition, refers briefly to CCS as a conceivable option for reducing global CO_2 emissions, but has a rather negative attitude towards CCS in general. Despite this, from the 2011 federal government's point of view, the development and testing of CCS remains an option for (inter)national energy and climate policy, as demonstrated by the 6th Energy Research Programme (BMWi 2011). But the markets for CCS and other power plant technologies are seen abroad rather than at home. The government aims to implement in Germany at least one of the demonstration projects planned within the EU, which in turn requires the implementation of the CCS Directive (BMU 2011). It also wants to make resources from the Energy and Climate Fund[12] available for CCS-ready power plants for a set period of time and under certain conditions.[13] This generally positive viewpoint of the federal government, however, does not prevent party colleagues, particularly those in Schleswig-Holstein and Lower Saxony, who are in CDU-FDP coalitions (until May 2012 in Schleswig-Holstein), from embarking on a collision course with the federal government and of championing a 'state clause' that would allow federal states to prohibit any storage of CO_2 on their territories (including territorial waters extending 12 nautical miles from the baseline). Only with the legislative compromise, which also weakened the state clause, do the coalition parties succeed in voting almost unanimously at the federal and state levels in favour of the Act.

The Left Party is also divided. At the federal level, it proposes a bill that would prohibit CCS pursuant to Article 4 (1) of the Directive (Bulling-Schröter et al. 2011). At the same time, the Left Party actively pursues a pro-CCS policy in Brandenburg with the SPD in a coalition government – albeit increasingly unwillingly -, which is supported in the Bundesrat by the CDU-FPD state government of Saxony and the SPD government ("Senate") in Hamburg, aiming to overturn the 2011 CCS bill not least because of the state clause.

12.4 The Federal States Have Conflicting Interests

A CCS act requires the approval of the Bundesrat because the federal states enforce the act and implement the approval procedures for CO_2 storage sites. This allows the states to influence federal legislation in advance. However, at times, they pursue contrary objectives and policies, which are influenced by factors such as those related to the energy economy and geological features, i.e. the suitability of a state as a storage region. If we take these two factors as a basis,[14] the states can be divided into groups:

1. States with large CO_2 storage potential and low-CO_2 power generation (Schleswig-Holstein, Lower Saxony, Mecklenburg-Western Pomerania);

[12] The fund finances climate protection programmes, R&D, technologies etc.; its source of funding is predominantly the auctioning of CO2 emission certificates (EU emission trading system, EU ETS).

[13] http://www.bundestag.de/presse/hib/2012_03/2012_106/03.html. When suitable storage reservoirs are available for example, subsidies can be granted to cover max. 10 % of the subsidizable costs.

[14] The rough categorization of potential is based on (Knopf et al. 2010). The CO_2 intensity of power generation does not take into account the shutting down of nuclear power plants since Fukushima (categorization according to http://www.statistikportal.de/Statistik-Portal/linksuebersicht.asp). For more on process-related emissions, which at least generally reflect the CO_2 intensity of power generation: http://www.ugrdl.de/pdf/ugrdl_analyse_2007.pdf and (VIK 2010, 97).

2. States with moderate to average storage potential and above-average CO_2-intensive power generation (Saxony, Brandenburg, Saxony-Anhalt);
3. States with minor or no storage sites, which can be split into two groups: (a) those with CO_2-intensive power generation (North Rhine-Westphalia, Saarland, Hamburg, Bremen, Berlin), and (b) those with low-CO_2 generation (Bavaria, Thuringia, Baden-Württemberg, Hesse, Rhineland-Palatinate).

If this 'objective' structure of interests is taken as the basis, the CO_2-intensive states are generally expected to advocate CCS, and those among them with no storage sites to press for liberal and easy-to-implement storage legislation without a state clause (3a). The low carbon states in the group 1, in contrast, will oppose CCS because they will not benefit from the CCS technologies, and they have no reason to offer their territories as CO_2 disposal areas in view of the local resistance. CO_2-intensive states with some storage potential (2) will probably campaign for a widespread application of CCS and the opening of the entire country for storage projects. Group 3b) is only indirectly affected (neither CO_2 intensive nor storage capabilities), and that gives these countries a great deal of flexibility in CCS policy.

In reality, some of the states do follow a CCS policy reflecting these interests (in particular states from group 1 and 2, partly from 3). However, this 'objective' structure of interests is sometimes modified by long-term energy policy objectives of the parties (such as 100 % electricity from renewables), by tactical party and election considerations, and most importantly by the balance of political power and local resistance to (potential) pipelines and storage sites – the intensity of which surprised the energy industry and politicians (Markusson et al. 2012).

Most state governments, similar to politics at the federal level, were initially predominantly open to CCS technologies in view of many planned coal power projects.[15] But the stance of some governments began to change at the latest from 2008 onwards when the German power supply company RWE announced its plans to lay a CO_2 pipeline from a planned CCS power plant in Hürth (North Rhine-Westphalia) to Schleswig-Holstein, where CO_2 storage was to be explored. In parts of the north of Germany, resistance to the exploration project and CO_2 storage in general began to spread. Along the potential pipeline routes and in particular in Schleswig-Holstein, and then in Lower Saxony, which has already

[15] The SPD-led government in Mecklenburg-Western Pomerania, for example, thus considered turning the hard-coal power plant in Lubmin, which was in the planning phase, into a CCS model project. After Dong Energy decided not to go ahead with the construction of the plant due to increasing resistance to fossil power generation and CO_2 storage, the state government began to alter its position to a more anti-CCS stance. In May 2012, the state parliament prohibited CO_2 storage. The state government of NRW headed by Minister-President Rüttgers (CDU) supported the (failed) plans for a CCS power plant in Hürth and a CO_2 pipeline to Schleswig-Holstein.

dealt with conflicts on the nuclear repository Asse, Konrad and in particular the Gorleben-project,[16] this opposition in the form of citizens' initiatives supported by non-governmental organizations (NGOs) has continued to shape the policies of the state governments and all parties. Despite party political differences on single formulations of a federal act that precludes the storage of CO_2, all political powers in these states are of the same opinion: more and more electricity is being produced from renewables in both states, and they are not willing to store CO_2 produced in coal-fired power plants in other states. CDU Minister-President of Lower Saxony McAllister, who is "not convinced" of CCS, summed up the position of the northern states, saying that it's unacceptable that half the state "is in a state of turmoil" because of storage projects (Zabka 2011). As Mecklenburg-Western Pomerania and Saxony-Anhalt[17] make it clear that they too are against storage in their states because of the (expected) opposition, the large land-based storage sites in Germany are politically closed because of a lack of acceptance. The threat made by Schleswig-Holstein and Lower Saxony to block a federal act in the Bundesrat with no state clause providing each state with a legally secure option of closing such sites proves successful – the clause is introduced in 2011 into the federal draft law. Their annoyance is heightened when in the run-up to the vote in the Bundesrat in 2011, the attempt proves unsuccessful to prevent the CDU-FDP government in Saxony, which is seeking storage options in other federal states because of its lignite power plants, from joining forces with the SPD-Left Party state government in Brandenburg, who successfully endeavours to find partners to block the state clause (Land Brandenburg 2011).

The government in Brandenburg, basically in harmony with the CDU and FDP in the state, is the only federal state to pursue a comprehensive CCS policy from the power plant to CO_2 storage. According to the coalition agreement (Koalitionsvertrag 2009a, p. 25), 'CCS technology is an important option. The coalition will support the testing and demonstration of this technology in Brandenburg.' Their reasons include the fact that lignite mining and power plants are extremely important for the regional labour market, decisions must be made on the future role of lignite, and an appropriate CCS act would enable the construction of the Vattenfall demonstration power plant with investments of € 1.5 billion. But fractions of the Left Party in the state parliament have distanced themselves from the CCS and lignite policies of their state government (Fröhlich 2011a, b). On top of this, there is local opposition to storage (Schramm 2010). Therefore, the state government faces a problem: As no other state indicates that they intend to permit

[16] The 'frozen' and heatedly disputed final repository project for high level radioactive waste.

[17] The state government meets with some opposition in the potential storage region of Altmark, and sees "no pressing need in the state for the application" of CCS technology (Landtag Sachsen-Anhalt 2011, p. 187). Members of the state parliament bring about an amendment to the state development plan (Treblin 2010): in contrast to earlier drafts, the approved version (Land Sachsen-Anhalt 2010) no longer contains a reference to 'securing' the almost depleted natural gas reserve in the state for CO_2 storage.

CO_2 storage, a pro-CCS course unique to Brandenburg would be impossible to implement domestically in light of the local resistance and criticism from CCS-opponents within the Left. The only way to overcome resistance is to open entire Germany for CO_2 storage, at least in principle. Therefore, in June 2011, Brandenburg and Saxony try to overturn the state clause in the Bundesrat (2011), but they are not supported by the majority of federal states.[18] This defeat leads to the impression among observers that the CCS bill with its state clause could be passed by the Bundesrat. However, Brandenburg does not give up and before the final vote in the Bundesrat on the bill in 2011 mobilizes also the Hamburg Senate, which agreed with Vattenfall that the Moorburg coal-fired power plant being constructed at the time would be retrofitted with CCS (post-combustion) when this is technically and economically possible and the legal prerequisites are in place.[19] Hamburg would be reliant on storage sites outside of the city state. The opponents of the federal bill are then joined by other states where coalitions involving the Greens are in power – but for exactly the opposite reasons. As the Greens oppose CCS in the power plant sector in general, the state governments in which they also hold seats therefore oppose the CCS bill in the Bundesrat, which pleases their SPD coalition partners who for tactical reason do not want to vote against their SPD party colleagues in the Bundestag – i.e. with "no".[20] This also applies to the new SPD-Green state government in NRW, whose 2010 coalition agreement states that a 'national CCS bill' has 'failed', and CCS technology will be of no practical relevance in NRW for reducing CO_2 emissions over the coming years (SPD/Grüne 2010, p. 34). However, controversial energy policy concepts loom behind this agreement: SPD Minister-President Kraft voices the interests of the state as the largest fossil energy producer for Germany, and emphasizes the need to use domestic lignite in the long term, thus 'irritating the Green coalition partner' (Hebel 2011), who envisages an energy supply based purely on renewables. In the new NRW coalition agreement (SPD/Grüne 2012), the problem is avoided by simply not directly mentioning lignite or CCS.[21]

A negative majority therefore emerges in September 2011 against the CCS bill in the Bundesrat for different reasons: it comprises CCS advocates who believe the bill to be too restrictive; fundamental opponents of CCS and coal; critics of individual passages of the bill and the state clause; tacticians who do not (yet) want to decide on the role that CCS should play in the future; and those who want to

[18] http://www.iz-klima.de/aktuelles/archiv/news-2011/mai/news-30052011/

[19] This is the first (provisional) retrofitting obligation for a power plant in Germany.

[20] A vote in the Bundesrat plenary only seeks approval of a law (only yes-votes are counted). If there is no yes majority, the law is rejected. Therefore, there is neither an explicit no-vote nor an explicit abstention. The voting behaviour of the states is not documented.

[21] While general sections of the agreement on transforming the energy sector call for a speedy transition to an end-to-end supply using renewables in the electricity sector, the concrete description of the power plant sector (planned plants and those being constructed) stipulates that 'coal-fired power plants (must) contribute to electricity and heat supply for a long time to come.' (SPD/Grüne 2012, p. 57). This implicitly keeps the future of lignite (and of CCS?) open.

cause difficulties for the federal government both at home and within the EU by rejecting the bill. An appeal to the Mediation Committee[22] is rejected by this negative majority – they do not pursue a shared objective.

12.5 Social Actors Fail to Find Agreement

Non-governmental organizations with their conflicting expectations are intensively involved in the social and political debate on CCS,[23] and some NGOs are locally active, working with citizens' initiatives[24] to block storage projects and to (successfully) influence communal[25] and state politics. Due to the fact that coal is associated with the concept of 'climate killer' by the general public and to a certain extent by politicians because it releases CO_2, only the demonstration of the technological and economic potential of CCS can politically legitimize the use of coal in the power plant sector in the long term in Germany. Opponents and advocates of the use of coal are both aware of this. This is the reason why they also struggle for a conceptual basis and metaphors in the CCS discourse, which has attracted growing attention from the national media since 2009 because of the CCS legislation (Lakoff and Wehling 2008): while advocates talk about "CO_2 storage" to form an association with oil and gas storage sites, with which citizens are familiar, opponents speak of "CO_2 repositories" to forge links with nuclear repositories, which are viewed with fear and rejection, in an attempt to weaken acceptance for CO_2 injection even further (Schumann et al. 2010, 2012; Fischedick et al. 2008).

This lack of agreement within society reflects and forces discrepancies, which are revealed in policies on CCS and the use of coal: most of the (energy) economic and industrial associations support CCS and the implementation of the CCS Directive in an act that would allow the demonstration of these technologies and their future commercial application. The Voice of German Industry (BDI) sees potential for climate-smart power plants, and campaigns for an act that will "implement the EU Directive 1:1" because tightening the Directive would impede investments in these technologies (BDI 2010, p. 1). In contrast, associations set up around renewable energy, are fundamentally opposed to CCS (Bundesverband Windenergie 2010) or they at least demand that geothermal energy and energy storage

[22] A joint 50:50 Bundesrat/Bundestag committee, which takes decisions on a majority basis. It only makes proposals to resolve differences, about which both "chambers" have to vote finally.

[23] See the description of groups of actors in Léger (2011, p. 29ff).

[24] Most can be discerned via the website http://ccs-protest.de/, a project established by the citizens' initiative aiming to stop CO_2 storage (CO_2-Endlager stoppen e. V.) in the town of Beeskow in Brandenburg.

[25] A "considerable fraction" of the German association of cities and municipalities is in favour of "making use of the option (in the EU Directive) to prohibit" CO_2 storage sites in the form of a federal act in the "full sovereign territory or in parts" (Deutscher Städte- und Gemeindebund 2010, p. 2).

(e.g. compressed air storage) be given absolute priority over CCS in using underground formations (German Renewable Energy Federation (BEE 2009)). The national association for the energy and water economy BDEW, which comprises some 1,800 companies of different sizes and with different energy or water philosophies, is critically supportive of CCS: despite objections regarding the some provisions of the successive drafts of the CCS bill, BDEW welcomes the fact that a political commitment to CCS technology is enshrined in law (BDEW 2010).

The trade unions, despite several critical queries and reservations, are relatively open to CCS because the technologies could safeguard jobs in the energy sector (German Trade Union Federation (DGB 2010)). Churches, which often own real-estate above potential CO_2 storage sites, have adopted no unified stance.[26] This also applies to scientific advisory boards. The more global their viewpoint, the more CCS comes into play as an option for the energy sector, particularly in newly industrialized countries. Their lowest common denominator is the potential application of CCS in Germany for process-related emissions. For the German Advisory Council on Global Change (WBGU), which has a global perspective and monitors the global trend towards an increased use of coal, CCS could be an option for reducing CO_2 in newly industrialized and developing countries, who will continue to use coal in the long term. In addition, it would be conceivable to use CCS to remove CO_2 from the atmosphere (WBGU 2011, p. 7f, 122ff). Due to a 'risk of acidification' of the ocean, WBGU takes a very critical stance towards the injection of CO_2 into deep ocean sediments (WBGU 2006). In a paper on climate and energy policy, the German Council for Sustainable Development (RNE), which also has a strong international orientation, believes that CCS could help to achieve a future with fewer CO_2 emissions. It believes that Germany should help to research and develop these technologies, and that after 2015 no coal-fired power plants without CCS should be approved in Germany. If CCS is not implemented, RNE believes there is a risk that the prices for emissions allowances will increase drastically. Above all, 'should the CCS option fail, the consequences for global climate policy (...) would be incalculable', because there is no way around the use of coal in many newly industrialized countries (RNE 2008, S. 9). The German Advisory Council of the Environment (SRU), which focuses on Germany, argues in its critical statement on the first draft of the CCS bill (SRU 2009) that initially only the testing of CCS should be allowed by a research bill because there are too many unanswered questions, e.g. the competition for using underground formations, ecological risks, cost efficiency, and acceptance. From a subsequent study by SRU on power supply, it can be inferred that it considers CCS a conceivable option for process-related emissions, while it appears 'appropriate to refrain from using CCS coal for reasons of sustainability' (Sachverständigenrat für Umweltfragen 2011, p. 39).

[26] While some regional churches oppose CCS, the overarching Evangelical Church in Germany (EKD) has adopted a critical and sceptical albeit not fully oppositional stance towards CCS (EKD 2009). The Catholic Church (German Bishops' Conference (Deutsche Bischofskonferenz 2007)) considers CCS as a temporary solution at best.

Contrary to nuclear energy, environmental NGOs also have no unified position with respect to CCS. The more strict opponents among the NGOs, like many other opponents, use three fundamental arguments against CCS: coal-fired power generation is superfluous and hinders the breakthrough of renewables; CO_2 storage endangers people; it is a hazard for groundwater when storage reservoirs leak.[27] Among these CCS opponents, combined with local anti-CCS initiatives, are the Federal Association of Environmental Action Groups (BBU 2010) and BUND (Friends of the Earth Germany), who vehemently oppose the construction of coal-fired power plants[28]: CCS is a 'high-risk' technology, and CO_2 repositories should be 'generally prohibited' in Germany (BUND 2011). BUND believes that the CCS bill was rejected in the Bundesrat because of its 2011 campaign, during which 170,000 protest emails were sent to the state governments. The opponents also include the German League for Nature, Animal Protection and Environment (DNR 2010) and Greenpeace (2010). They believe that CCS is an attempt to 'make coal green' – this is why a CCS bill should be replaced by a 'law phasing out coal' (Totz 2011). Their reaction to the compromise that was reached with the CCS act in 2012 was just as harsh: they announced that they would continue to fight against CCS (BUND 2012). Differentiated positions have been adopted by other environmental NGOs, particularly by Germanwatch ((Treber 2011), German Environmental Relief (DUH 2010, 2011)), the Nature and Biodiversity Conservation Union (NABU 2011) – at least on a federal level – and the World Wide Fund For Nature (WWF 2010, 2011). Despite multifaceted criticism of the provisions of the CCS bill and calls for improvements (such as preventing competition for the use of storage formations, not socializing liability in the case of CCS technology failures), they are all in favour of researching and perhaps even developing CCS for process-related emissions (potentially for Germany too) with a view to the power plant sector in newly industrialized countries. CCS could become essential as a 'bridging technology' on the way towards the 'solar age'. Germanwatch 'therefore welcomes *in principle* (emphasis in original, W. F.) (...) the efforts of the German federal government to swiftly create a legal framework for testing this technology in Germany.' (Germanwatch 2009, p. 2). However, CCS should not be used to increase acceptance for new coal-fired power plants in Germany – and CCS should therefore not be automatically linked to coal as has been the case to date.[29] One reason why the NGOs adopt different positions is – similar to the case with the scientific advisory boards – that some of their policies have a stronger global

[27] When the controversy began, the main objection was the risk associated with a ground-level concentration of escaping CO_2 (danger of suffocation). Only later did the groundwater issue gain attention, and helped CCS critics to win the support of the water industry (cf. Wasserzeitung 2009).

[28] BUND publishes a list of coal-fired power plant projects that 'BUND and its allies' would allegedly have stopped (cf. http://www.bund.net/themen_und_projekte/klima_und_energie/).

[29] For the discussion as to whether acceptance would increase if CCS is associated more strongly with process emissions, biomass power plants or the relatively 'popular' natural gas power plants (cf. Schumann and Pietzner 2012).

context and see CCS as an option for the expanding CO_2-intensive electricity sector in newly industrialized countries. Other NGOs, which are closely linked to citizens' initiatives against CCS, are influenced to a greater degree by these local interests.

12.6 The Legislative Process Is Tedious and Contentious

Due to the divergent interests and objectives, it is not surprising that politicians required several attempts from 2009 onwards before a CCS act was finally implemented.[30]

In April 2009, the grand coalition presented the first draft of a bill implementing the EU Directive (cf. in the following (Viebahn et al. 2012)). It is based on the preparatory work of BMWi (Minister Klos, and from February 2009 zu Guttenberg, both CSU), which 'takes up' and 'refines' the draft of a law office (cf. Table 12.1), and the SPD-led BMU (Gabriel) and is 'based on nuclear and waste-related regulations' (Skrylnikow 2010, p. 547). This draft bill closely follows the EU Directive. And it already contains key provisions that reappear in later versions of the bill. However, the draft does not specify limits for the amounts of CO_2 that may be stored nor is there any mention of a state clause. In fact, it envisages the demonstration of the entire CCS with a view to commercial application. Initially, in the Bundestag it is supported by the CDU/CSU parliamentary group as well as by the SPD, while the oppositional FDP, with a very positive attitude towards CCS, raises objections to regulations in the bill that are too 'restrictive' and 'complicated', and proposes its own draft for the 'swift, safe and unbureaucratic implementation of CCS projects' (FDP 2009). Companies in the electricity sector, supported by sections of industry, press for a law that will provide planning certainty and minister-presidents of states with high coal usage (Saxony, North Rhine-Westphalia, Brandenburg) back the CCS bill. It is met with sharp, fundamental criticism among the Greens, the Left Party, some environmental organizations, renewable energy associations, and among the state governments who are under political pressure from the citizens' initiatives and NGOs in Schleswig-Holstein (grand coalition CDU-SPD until July 2009, followed by CDU-FDP until 2012), Lower Saxony (CDU-FDP), and in Bavaria (CSU-FDP coalition), where influential farmers' associations mobilize in protest against the bill because the long-term impacts of storage on the rights of land owners, the value of property and the marketing of agricultural products cannot be foreseen.[31] The criticism prevents a general consensus within the CDU/CSU, while opposition within the SPD in the

[30] For more information on the legislative process, see the Parliamentary Material Information System for the Bundestag and the Bundesrat, http://dipbt.bundestag.de/

[31] Bavaria has some geological CO_2-storage potential in the Vorlandmolasse north of the Alps, which, according to the state government, should be used for geothermal and natural gas-storage projects only.

federal grand coalition is much smaller because it is not represented in the relevant critical state governments, Brandenburg takes a pro-CCS stance, and coal-friendly trades unions back the technologies. In general, criticism of CCS and of the first bill from the ranks of the Greens, the Left Party, and environmental organizations focuses on the fact that CCS would mean that coal would continue to be exploited causing environmental damage and the capacity of renewables would not be expanded, that CO_2 storage is a hazard for humans and the environment, that there would be hidden subsidies for CCS, that the risks would be unequally distributed by concentrating storage in the more suitable northern part of Germany, and that the liability risk would be shifted to the public sector after closure of the storage sites – points of criticism that are also brought to bear against subsequent amendments to the CCS bill. Critics also assert that the bill basically decides in favour of CCS technologies at an early stage, thus establishing planning certainty for future large-scale industrial application without waiting for answers to the many open questions, and federal states could only at best (or so say the critics in the states) attempt to prevent such (storage) projects within the framework of their regional planning but with very little chance of success. Instead, a law on research power plants, or as other critics put it, a restrictive law on demonstration power plants should be sought rather than a comprehensive regulation at best. Due to internal opposition, the CDU/CSU were unable to come to a unanimous decision on the bill at the federal and state levels. Minister for the Environment Gabriel purposely hits a raw nerve of the coalition partner when he emphasizes that the SPD is ready to pass the CCS bill before the end of the legislative period, which is obviously impossible for CDU/CSU in light of opposition in particular from CDU/CSU led governments. On 24 June 2009, the CDU/CSU parliamentary group therefore remove the bill, the first reading of which had already taken place in the Bundestag, from the agenda and postpone it to the next legislative period. The first attempt at a law on CCS fails.

On 14 July 2010, the new CDU/CSU-FDP coalition[32] proposes a bill for carbon dioxide storage. In principle, it is pro-CCS, and aims to implement the Directive 'as soon as possible' according to the coalition agreement (Koalitionsvertrag 2009a, p. 28). The new draft takes account of criticism from politics and society, and comprises significant changes compared to the preceding draft (cf. in the following (Hellriegel 2010)). With the exception of the state clause and related regulations, which first appear in the next draft, the structure and contents are similar and indeed they are also kept in the subsequent two revisions of the bill. The bill concentrates on the testing and demonstration of CO_2 storage, and limits the annual permissible storage amounts to max. 3 million tonnes of CO_2 per storage site,[33] and nationwide

[32] The SPD lost 11.2 % points in the 2009 election due the labour market and social reforms during the SPD-Green government under Chancellor Schröder; CCS as a mainly regional conflictual issue had no discernible impact on election results.

[33] This is equivalent to the annual CO_2 emissions of a lignite-fired unit producing some 430 megawatts of electricity in base-load mode (8,250 hours/year).

to max. 8 million tonnes of CO_2. The decision on the commercial application of CCS is postponed until 2017, when it will be made on the basis of a progress report. This is the response to the demand that the bill should not automatically permit a transition to the demonstration of commercial application (Skrylnikow 2010). Only this draft contains an explicit provision that is intended to increase the very low public acceptance of storage: in accordance with § 42 (communities' right to compensation), the operators of a storage must pay the communities located above the storage site compensation to the value of 2 % of the average price of the CO_2 emissions allowances (EU emission trading system) which would be required to cover the amount of CO_2 stored in the year of operation.[34] Further on, the bill prescribes strict liability of the storage facility operator, and stipulates prior approval for the closure of a storage site, as well an obligation to perform an environmental impact assessment. Only after the long-term stability of the storage site has been determined and the operator has paid a contribution margin or contribution towards post-closure costs, at the earliest 30 years after the site has been decommissioned, can the operator apply to transfer liability to the respective federal state (cf. Eckardt et al. 2011). Overall, the draft provides for more rights for land owners, whose agreement must be sought for exploratory work,[35] takes better account of alternative uses of underground formations (such as geothermal energy, compressed air storage), and appeases critical states by improving their options for impeding or even prohibiting storage projects via state regional planning procedures. In mid-July 2010, when they announce agreement between their respective ministries on the new bill, ministers Röttgen (BMU, CDU) and Brüderle (BMWi, FDP) declare that there will be no testing of CCS against the will of the federal states concerned (Seidler 2010). Despite this, this draft also fails to silence the multifaceted criticism. It is aimed mainly at the amounts of CO_2 that can be stored, which are considered 'too big', the regulation of the transfer of liability for the decommissioned storage sites to the states, as well as the contribution margin or contribution towards post-closure costs to be paid by the operator. This aspect is summarized by Matthias Miersch, SPD parliamentary spokesperson for environmental policy, during the reading of the bill in 2011 in the Bundestag as follows: the regulation of liability is inadequate, and 30 years after the closure of a storage site, the liability for 'long-term risks' is passed on to the public sector; regulations governing cover for liability and contributions to post-closure costs by the company who stores the CO_2 should also be rejected because they are not based on the full extent of possible damage (Deutscher Bundestag 2011c, p. 13960ff). These regulations in the CCS bill therefore play a role in the SPD rejecting it and the lack of a majority in the Bundesrat. In the context of the regulation of liability for CO_2 storage, it became clear that civil liability problems would emerge, as outlined in the final report of the climate protection dialogue between industry and politics,

[34] Subsequent drafts leave this arrangement up to the federal states.

[35] If this is refused and if public interest prevails, for areas outside of enclosed private premises (buildings, gardens), the owner's agreement may be replaced by that of the authorities.

which also involved insurance companies. According to this report, the capture, transportation, and injection of CO_2 could be 'covered by insurance'. However, it is 'currently impossible to predict what level of liability insurance could be offered for the private sector' in the case of CO_2 release (Dietrich 2009). This reticence is due to the fact that there is little experience regarding the risks associated with storage, the insurance periods are long, and that at best very few storage project are to be expected, which is why the law of large numbers ('risk sharing') would not apply. The federal government's bill continues to meet more sharp criticism among the ranks of some state coalition parties CDU and FDP: Schleswig-Holstein and Lower Saxony object to the lack of an 'opt-out' clause that would allow the states to legally refuse storage projects, and they thus remained opposed to it.

On 14 February 2011, the federal government proposes an amended bill which now includes a state clause: § 2 (5) stipulates that federal states can use state laws to determine regions in which the testing and demonstration of storage is allowed, but also explicitly demands an objective consideration on the part of a state making a decision against storage. Thomas Bareiß, energy policy coordinator of the CDU/CSU parliamentary group, states that although a compromise has been reached in favour of the states, it is "unacceptable that a state can completely ignore a federal act" (Sattler 2011), simply by prohibiting storage throughout the state. Although the amendment represents a move towards appeasing the demands of the opposing states, the resistance in some states, in particular in Schleswig-Holstein, remains unyielding: an opt-out option is effectively exactly what these state governments want, but they reject the state clause as not being precise enough, a position which has the backing of the entire state parliament. At the same time, unrest grows in particular in the governments of Brandenburg and Saxony both of whom see their pro-CCS policies in danger, as both appear to be even more isolated by the state clause.

On 14 April 2011, the cabinet complies with the call of those states demanding a more precise state clause and slightly alters the formulation of the clause: 'the states can determine through state legislation whether to allow the testing and demonstration of permanent storage in certain areas only or whether to prohibit this in certain areas.' (§ 2 (5)). The obligation to consider the pros and cons is no longer part of the paragraph. How restrictive the state clause actually is remains controversial from a legal perspective. Green member of the Bundestag Krischer believes that CO_2 storage could still be implemented despite opposition in states, and refers to the answer given by the federal government to a question he posed and to a report by the Research Services of the Bundestag (Robbe 2011). In his answer to the question, BMWi State Secretary Hohman explains that "when the areas are being designated, the energy-related and industrial options for the use of a potential storage reservoir, the geological features of the area, and other public interests (e.g. environment and tourism) must be weighed up. An area must be chosen on the basis of technical and administrative criteria."[36] Other experts (Robbe 2011, p. 4)

[36] Cited from the answer, as published in (Krischer 2011).

emphasize the need for consideration because they believe that purely "negative planning", which "only aims to prevent an undesired use" – here CO_2 storage – is unlawful. The new state clause demands well-balanced reasons for a decision against storage. Despite this, the wording of the bill politically improves the position of the states who oppose storage, even though the 'full scope of exclusion options continue to be connected with uncertainties.' (Dietrich 2011, p. 8). The option of prohibiting storage using state law[37] therefore appears to be reinforced further with the new § 2 (5). The state governments of both Schleswig-Holstein and Lower Saxony are also of this opinion, and are confident that the law protects their interests in not having to designate parts of their territory not even as testing areas for CO_2 storage (Geisslinger and Havlicek 2011).

But the positions of many actors remain inflexible: those who were against CCS up to now continue to reject the bill, and criticism continues and indeed grows, as the issue is met with more and more public interest. A detailed examination of the state clause apparently reveals that it only covers the territories of the states, including territorial waters extending 12 nautical miles from the baseline, but it does not cover the adjacent exclusive economic zone for which the federal government and the Federal Maritime and Hydrographic Agency (BSH) as the executive authority are responsible. There is possibly huge potential for the storage of CO_2 in ocean sediments, and the federal government could approve such storage without involving the states (Nestle 2011), particularly as the amendments to the London Protocol (on the dumping of wastes at sea) and the OSPAR Convention (protecting and conserving the North-East Atlantic and its resources) safeguard this option under international law (Stoll and Lehmann 2008). The SPD Bundestag parliamentary group, whose general stance concerning CCS is unclear but appears to be gradually moving towards the demand for a 'research bill' with a focus on process-related emissions from energy-intensive industries (Deutscher Bundestag 2011c, p. 13961), continues to challenge the bill and ridicules the state clause as the product of dissent in the CDU/CSU and FDP, while concealing the fact that 2 years ago it pleaded for a much more extensive bill. Supporters of CCS are also dissatisfied, believing the bill to be a de facto rejection of storage projects and the state clause to be an infringement of uniform legislation for the Federal Republic of Germany because the states 'can opt out' of joint national responsibility. Some members of the FDP in the Bundestag therefore reject the bill, referring to the federal party conference resolution in May 2011, which calls for 'the state clause (...) to be amended so that CCS demonstration plants cannot be ideologically excluded from the entire state territory but rather only from parts of the state on the basis of objective criteria' in an effort not to 'completely prevent' CCS. In addition, no 'precedence' should be set 'for other energy projects of national

[37] In the debate on § 2 (5), the federal government states that it is providing every state with the possibility of prohibiting or allowing storage. It thus considers § 2 (5) to be neutral. However, it was also discernible that by deciding to go its own way and allow storage, a state would have to carry a heavy political burden and face massive resistance.

significance such as grid expansion or the search for sites for nuclear repositories' (FDP 2011, p. 8). This resolution is met with vehement opposition in those states where FDP-ministers and parliamentarians have campaigned long and hard for the state clause. To avoid conflict in the run-up to the vote on the CCS bill in the Bundestag, the majority of the FDP Bundestag parliamentary group bows to this pressure and supports the bill (FAZ 2011).

A journalist summarizes the result of the political process as follows: 'The black-yellow *(CDU/CSU-FDP)* state governments do not want to aggravate their voters and are against CCS. The federal government supports their party colleagues at the state level and introduce a state clause. This in turn upsets the red-red *(SPD-Left Party)* coalition in Brandenburg', who fear the dissatisfaction of their voters and party members if Brandenburg was to implement CCS alone. 'Nobody is interested in climate change mitigation or the safety of CO_2 repositories.' (Werdermann 2011, p. 2f).

After a controversial debate, the Bundestag passes the bill on 8 July 2011 in a roll-call vote proposed by the Greens: 306 CDU/CSU members of the Bundestag vote yes and 4 no; 262 SPD, Greens and Left Party members of the Bundestag vote no, and one SPD member abstains. Forty-seven members of the Bundestag do not vote, including the large number of 12 Left Party members (two-thirds of their total members of the Bundestag). The dissidents within the coalition parties represent the poles of the attitudes towards the bill. Von der Marwitz, CDU member of the Bundestag from Märkisch-Oderland-Barnim (Brandenburg), a centre for CCS opponents, believes CCS to be a "false course" in favour of coal-fired power generation, and sees "geological and physical risks" (von der Marwitz 2010), while Josef Rief, CDU (Biberach, Baden-Württemberg), shares the fears of the Farmers' Association (Rief 2011). Both vote against the bill because it would make CCS possible in principle. In contrast, FDP member of the Bundestag Meierhofer (Bavaria), CCS parliament rapporteur, votes against the bill because it impedes CCS and creates a negative precedence (Meierhofer 2011). Member of the Bundestag Skudelny (2011) (FDP, Baden-Württemberg), for whom "our climate policy would be almost impossible in the medium term without CCS", also votes against the bill. Member of the Bundestag Kauch, environmental spokesperson of the FDP parliamentary group, sums up the difficult position for the federal government and the members of the Bundestag during the subsequent reading of the CCS bill: "The introduction of CCS via the state clause is not the path we would have chosen, but it is the only way of getting this law through the Bundesrat." (Deutscher Bundestag 2011c, p. 13966).

However, the Bundesrat blocks this path: on 23 September 2011, only the CDU/CSU-FDP state governments of Bavaria, Hesse, Lower Saxony, and Schleswig-Holstein vote for the bill as passed by the Bundestag – the necessary quorum is not reached.[38] This result was to be expected because in summer 2011 a Bundesrat

[38] The minutes of the Mediation Committee session can be found at http://www.wir-im-niederbarnim.de/?page_id=690

Committee rejected the bill by 10–6 votes mainly because of the 30-year period to be observed for transferring liability for storage sites and the state clause (Matern and Fröhlich 2011). As the Bundesrat does not call upon the Mediation Committee, the federal government does so on 26 October 2011 and the Committee puts the CCS law on the agenda of the 13th session. However, in view of the different positions the official, formal debate about the topic is postponed for some 8 months by the Mediation Committee. In each of the following sessions the CCS law is only mentioned in the agenda as the continuation of the 13th session – but without debating it at all. This 'procedural device' opens up the possibility of using informal avenues to determine how agreement on a bill could be achieved, without directly running the risk of causing the conciliation procedure and the legislative process to fail by immediately and explicitly dealing with the bill.[39] Is an informal agreement reached, the Committee can officially take up the issue. However, the duration (and, of course the result) of this informal process cannot be determined.

This long-drawn-out affair regarding the bill is instrumental in the decision made by Vattenfall at the beginning of December 2011, much to the delight of CCS opponents in Brandenburg throughout Germany, to abandon its plans for the only German CCS demonstration project (Jänschwalde) and the exploration of potential storage sites, as implied in July 2011 (Vattenfall 2011b).[40] The company will continue test operation in the CCS pilot plant at the 'Schwarze Pumpe' site, and will participate in European CCS projects, where it sees a critical need for the expansion of a European infrastructure for transportation and storage. It also intends to construct a new CCS power plant in the 2020s at Jänschwalde (Vattenfall 2011a), which is welcomed by the state government of Brandenburg as it wants to continue research in Ketzin (Landesregierung Brandenburg 2011) – a hope that soon scattered (see below).

Despite the political impasse, the deadline for implementing the EU Directive is 25 June 2011, and since Germany failed to adhere to it, the EU Commission initiated an infringement procedure against Germany in June 2011.[41] For this reason, after the legislative standstill some conceivable legislative developments were discussed in an effort to comply with EU law (Fischer and Hake 2012): a prohibition of CO_2 storage (except for "small-scale" storage projects) with a revision clause in line with the Austrian model[42]; the limitation of research to

[39] § 12 of the rules of procedure governing the Mediation Committee: (1) If a conciliatory proposal is not passed in the second session convened on the same matter, then any member of the Committee may propose termination of the procedure. (2) The procedure is terminated in the subsequent session if there is no majority in favour of the conciliatory proposal. http://www.bundesrat.de/

[40] The cost and consolidation pressure after the shutting down and final decommissioning of the Brunsbüttel and Krümmel nuclear power plants may have influenced Vattenfall's decision.

[41] EU infringement procedures are not rare. In midyear 2012, with the CCS procedure, 81 infringement procedures against Germany are pending, of which 26 concern directives that have not been implemented. To date, no penalty payments have been imposed on Germany; cf. (TAZ 2012).

[42] An evaluation report must be submitted to the Austrian Parliament before 31 December 2018.

process-related emissions and biomass power plants[43]; an act explicitly limited to research (for example, lower amounts that can be stored); an act modified in places to allow federal states the option of "still only" being able to prohibit or allow CO_2 storage within the scope of regional planning.

In early summer 2012 – surprisingly for the general public – a compromise was reached in the Mediation Committee by a narrow majority. It comprises elements of the two latter options. It was preceded by informal consultations between those concerned in the federal and state governments, particularly between Brandenburg, Lower Saxony, Saxony, and NRW. Probably everyone involved in the legislative process wanted to get rid of the problem and to break the political deadlock. The failed CCS bill was amended in four points, which were presented in the Bundesrat by Mediation Committee spokesman, Head of the State Chancellery of Saxony, Johannes Beermann (Bundesrat 2012, p. 292f):

- The purpose of the act is defined anew in § 1, which now contains a "neutral formulation": neither climate change mitigation nor energy production is mentioned, thus "ending discussions on the origin of CO_2 – the energy sector or industry." (Citation Beermann). Instead, it now reads: This act governs the permanent storage of CO_2 in underground geological formations with the purpose of protecting people and the environment today and on behalf of future generations. It is initially limited in scope to researching, testing and demonstrating technologies for the permanent storage of CO_2 in underground geological formations (Deutscher Bundestag 2012a).
- In § 2 (2), the amount of CO_2 which can be stored is capped. Storage sites are only permitted if they can store no more than 1.3 million tonnes of CO_2 per year,[44] and the combined yearly storage capacity in the territory covered by the law may not exceed a total of 4 million tonnes of CO_2.
- The 'state clause' in § 2 (5) is amended: The states may determine whether to allow the testing and demonstration of permanent storage in certain areas only or whether to prohibit this in certain areas. When deciding whether or not to allow CO_2 storage, the states must weigh their decision against alternative uses of a potential storage site, the geological characteristics of the area, and other public interests. The proviso 'via state legislation' is dropped, and an explicit obligation to consider the pros and cons is re-introduced into the paragraph itself, as was the case in the bill up to April 2011. Beermann summarizes this aspect as follows: "Storage can also be allowed or prohibited on the basis of a legal regulation. State legislation is no longer necessary. The actual – geological – features must be considered. This will prevent storage being allowed or forbidden without examining the geological suitability." (Bundesrat 2012, p. 293). Whether the new regulation represents a substantial change or not is open to debate. According to the prevailing opinion of legal experts, the state clause in the failed

[43] The proposal was (unsuccessfully) made as a "compromise" by the federal governments of Baden-Württemberg (Green-SPD), NRW and Rhineland-Palatinate after the bill failed to pass.

[44] This is equivalent to the annual CO_2 emissions of a lignite-fired unit producing some 186 megawatts of electricity in base-load mode (8,250 hours/year).

bill also contained an obligation to consider the pros and cons, which must now be implemented within the scope of regional planning procedures to allow/prohibit CO_2 storage.

- § 31 (1): Liability may be transferred from the operator to the competent authority at the earliest 40 years after the closure of a CO_2 storage site, and not, as was the case in the old bill, after 30 years.

The agreement was underpinned by explanations, in which the federal and state governments affirmed the "expansion of a transportation infrastructure for Europe" (K. Reiche, Parliamentary State Secretary, BMU). This could give rise to the option of "storing CO_2 under the ocean" (R. Christoffers, Minister of Economics Brandenburg). Several federal states expressed their willingness to allow the construction of such pipelines (citation: MAZ 2012a) – declarations, which have a more vocal than substantial relevance in the absence of promising CCS projects, a lack of investors in pipelines projects and probable local resistance to any CCS project.

In a vote that was not a roll-call vote, the Bundestag accepted the recommended of the Mediation Committee on 28 June 2012 with the votes of the coalition parliamentary group. The Left Party and the Greens voted against it. The SPD abstained (Deutscher Bundestag 2012b, p. 22359), a sign of the unclear policy of the federal party with respect to coal use and CCS. On 29 June 2012, a majority of the states in the Bundesrat voted in favour of the new law (Bundesrat 2012, p. 292ff). The states with Greens in the state government did not vote in favour of the bill, as did not the SPD-CDU governed states of Berlin and of Mecklenburg-Western Pomerania, because the federal act invalidates their state "CO_2 exclusion law" (Landtag Mecklenburg-Vorpommern 2012) as federal law takes precedence over state law. Schleswig-Holstein immediately announced that it will find legal ways not allow storage on its territory, other states will follow suit.

As expected, criticism of the act is sharp among CCS opponents both in politics and in society (BUND 2012). As an example for this criticism von der Marwitz, the rather isolated CDU CCS opponent in the federal parliament, is cited: He sees the "shabby compromise" as an effort to provide new lignite-fired/coal-fired power plants with "government subsidies" (von der Marwitz 2012). For Vattenfall, the act comes too late: the (only German) demonstration project in Lausitz is "no longer economically viable", says head of Vattenfall Europe Tuomo J. Hatakka (MAZ 2012b). But the CCS law has at least one positive consequence: the EU infringement procedure against Germany is terminated.

12.7 A Future for CCS?

Germany has a CCS Act, but in no way does this imply a "coalition of the willing",[45] i.e. a coalition of CCS advocates, has ousted the negative coalition. The compromise leaves CCS in a state of principal uncertainty in a long term

[45] According to Schleswig-Holstein Minister for Energy Robert Habeck (Greens) (OZ 2012).

perspective, but renders CCS projects virtually impossible in the short and medium term. The legislative compromise shows that there is neither consensus on a national prohibition nor on the demonstration of the commercial application of CCS. The answer to the question of whether CCS is an option for Germany for reducing CO_2 emissions has been pushed into an uncertain future by the act for CO_2 emitting power sectors and industries that have not been specified in detail. Whether this future will be positive for CCS "appears doubtful" from today's perspective (according to Chancellor Merkel in July 2102, cited in (E-World 2012)) – an assessment which seems to be accurate due to economic and social reasons (von Hirschhausen et al. 2012).

Compared to the first bill in 2009, the adopted CCS act has shrunk to a research law with a theoretical potential for smaller demonstration projects which will not be exploited in Germany in the years to come. If the potential of CCS should be demonstrated for large power plants, the act would have to be amended with respect to the storage amounts. In 2017, the CCS act will be evaluated, and the discussion on CCS could become more intensive again. There are still explicit, and political, advocates of CCS who have their hopes pinned on the future and appropriate framework conditions. They include the state government of Brandenburg, who are increasingly concentrating on renewable energy, but also continuing to endorse lignite (Land Brandenburg 2012, p. 12ff, 36ff, 43ff), as well as the state government in Saxony. Its 10-point plan on energy policy from June 2011, which remains the guiding principle of state policy, declares that the exploitation of lignite 'cannot be disregarded in the medium to long term', and that CCS technologies are 'important'. If there should be no CO_2 storage sites in Germany 'in the foreseeable future', then 'extra-territorial reservoirs' would be an option.[46] Both of these states together with Saxony-Anhalt, where lignite is used to generate electricity, declared themselves committed to CCS at a lignite summit in March 2012 (Rempe 2012). NRW with its SPD – Green government, where around 54 % of Germany's lignite is mined, did not participate in the summit – but the last word has yet to be spoken on lignite policy in this state. Possibly, these commitments to lignite are only "whistling in the dark" in light of the precarious economic situation even of lignite in the electricity sector, a consequence of the German Renewable Energy Law (EEG).

Up to now, there is no concept how a "lignite coalition" could convince a majority of German politicians and the society of CCS as a sensible option. In order to have a chance at all, several energy economy framework conditions have to be taken into account: A clearer picture will emerge over the next few years as to whether the targets for expanding the capacity of renewables will be achieved and how smooth and expensive the transformation of the energy system will be (Mikulcak 2012), what role coal in general and lignite-fired base-load power plants could play in a reliable electricity supply system, and how emissions trading and its CO_2 prices (EU ETS) will develop – important factors that seem to reduce the

[46] 10-point plan on current energy policy, http://www.sachsen-cdu-politik.de/

business options for lignite base-load power. Should it emerge that the EU and its member states are not in a position to implement their extremely ambitious action plan for CO_2 mitigation politically or economically, and if the international climate change mitigation policy does not achieve the high targets set (cf. Geden 2012), the prospects for CCS could become more clouded, because the pressure for a carbon free electricity sector could dwindle. A permanently low price for CO_2 allowances would hinder investments in CCS. EU CCS policy could also have an impact. Energy Commissioner Oettinger is considering whether to introduce "additional CCS regulations" to make CCS "obligatory for new and also old power plants" at some point in the future (Wiese 2012). CCS plays still a key role in the EU plan for a low-carbon economy in 2050 (European Commission 2011b) and the more specific 2050 Energy Roadmap (European Commission 2011a), even though commercial application is not expected until after 2030 and CCS projects are not progressing well in the EU member states. Another obstacle for CCS is the lack of acceptance for the solution of the 'back-end' CCS problem, namely storage.[47] The northern federal states will pull out all the stops to block even potential storage projects. This problem might be less virulent if CO_2 were to be stored in deep ocean sediments, particularly within the context of enhanced oil/gas recovery. Research work is being conducted on storage in deep ocean sediments, the safety of such sites, and on the consequences of leaks for the marine environment, which also applies to regions off the German coast.[48] It remains to be seen whether a 'loop hole' (CO_2 Handel 2011) will emerge for federal storage projects below the seabed. That CCS opponents take this option seriously is reflected in the coalition agreement of the new state government in Schleswig-Holstein: it wants to 'preclude' CCS 'for the whole of Germany – particularly in the exclusive economic zone.' (SPD/Grüne/SSW 2012, p. 37). For this reason, should a European CO_2 transportation infrastructure be created, CO_2 could be exported to onshore storage sites in other countries or injected into their deep ocean sediments. The statements issued in response to the legislative compromise indicate an interest in this option in sections of the political and industrial arenas. Assuming that there is in Germany at least a principal interest in the years to come to capture carbon and that the necessary storage exists (offshore), then the acceptance of CO_2 pipelines through the federal states would also have to be ensured – considering the massive opposition to the planned Hürth pipeline in 2008, this represents a very huge challenge for politics and society. From today's point of view it is doubtful, whether that challenge could be mastered.

Even though there is currently no final legal answer to the question of whether CCS has a future in Germany or not, the development of the energy system, the economic climate combined with the political and social balance of power imply that CCS is doomed to failure.

---

[47] The suggestion to "utilize carbon instead of storing it" (Eckardt et al. 2011) appears illusionary for the foreseeable future because of the low demand for CO_2 (cf. Kuckshinrichs et al. 2010)

[48] http://www.eco2-project.eu/

References

Baethge H (2010) CCS-Gesetz verschoben - Hoffnung für Länderklausel. SHZ, 28. September 2010

BBU (2010) Stellungnahme zum Entwurf eines Gesetzes zur Demonstration und Anwendung von Technologien zur Abscheidung, zum Transport und zur dauerhaften Speicherung von Kohlendioxid. Berlin, Bundesverband Bürgerinitiativen Umweltschutz, 28.8.2010

BDEW (2010) Umsetzung der CCS-Richtlinie in Deutschland. Berlin, 17.6.2010, http://www.iz-klima.de/w/files/positionen/bdew_positionspapier_ccs.pdf

BDI (2010) Stellungnahme zum Referentenentwurf Berlin, 26.8.2010, Dokumenten Nr. D 0271.

BEE (2009) Wege in die moderne Energiewirtschaft. Berlin, http://www.bee-ev.de/Energiepolitik/Bund/BEE-Energiekonzept.php

BMU (2011) Kabinett verabschiedet CCS-Gesetz Berlin, BMU, PresseNr. 053/11 vom 13.4.2011

BMWI (2005) Innovation und neue Energietechnologien. Bundesministerium für Wirtschaft und Arbeit, Berlin

BMWI (2011) Forschung für eine umweltschonende, zuverlässige und bezahlbare Energieversorgung. Bundesministerium für Wirtschaft und Technologie, Berlin

BMWI BMU BMBF (2007) Entwicklungsstand und Perspektiven von CCS-Technologien in Deutschland. Gemeinsamer Bericht für die Bundesregierung. Berlin

BMWI/BMU (2007) Bericht zur Umsetzung der in der Kabinettsklausur beschlossenen Eckpunkte. Berlin, 5.12.2007, http://www.bmu.de/files/pdfs/allgemein/application/pdf/gesamtbericht_iekp.pdf

Bulling-Schröter et al (2011) Gesetzentwurf CO_2-Speicher-Verbotsgesetz. Berlin, http://dip21.bundestag.de/dip21/btd/17/052/1705232.pdf

BUND (2011) Bundesregierung beschließt riskantes CCS-Gesetz. Bund für Umwelt und Naturschutz Deutschland, Pressedienst, 13.04.2011, http://www.schattenblick.de/infopool/umwelt/meinunge/umst0024.html

BUND (2012) Schlechtes Gesetz für riskante Technologie. Pressemeldung v. 28.6.2012

Bundeskabinett (2007) Eckpunkte für ein integriertes Energie- und Klimaprogramm. Berlin, http://www.bmu.de/klimaschutz/downloads/doc/39875.php

Bundesrat (2009) Stellungnahme des Bundesrates: Entwurf eines Gesetzes zur Regelung von Abscheidung, Transport und duaerhafter Speicherung von Kohlendioxid, Drucksache 282/09. Berlin.

Bundesrat (2011) Antrag der Länder Brandenburg, Sachsen. Bundesrat Drucksache 214/2/11 vom 24.5.2011

Bundesrat (2012) Stenografischer Bericht 898. Sitzung. Berlin, BR, 29.6.1012

Bundesregierung (2007) Das Integrierte Energie- und Klimaschutzprogramm (IEKP). BMU, Berlin

Bundesregierung (2010) Energiekonzept für eine umweltschonende, zuverlässige und bezahlbare Energieversorgung. BMU, Berlin

Bundesregierung (2011a) Eckpunkte "Der Weg in die Energie der Zukunft". Berlin, http://www.bundesregierung.de/Content/DE/Artikel/2011/06/2011-06-06-energiewende-kabinettsbeschluss-doorpage-energiekonzept.html

Bundesregierung (2011b) Gesetzentwurf Gesetz zur Demonstration und Anwendung von Technologien zur Abscheidung, zum Transport und zur dauerhaften Speicherung von Kohlendioxid. Berlin, 14.2.2011, http://www.greenpeace.de/fileadmin/gpd/user_upload/themen/klima/Gesetzentwurf_-_CCS.PDF

Bundesverband Windenergie (2010) Stellungnahme anlässlich der öffentlichen Anhörung. Berlin, DBT, Ausschuss für Wirtschaft und Technologie, Ausschussdrucksache 17(9):265, 19.10.2010

CDU Schleswig-Holstein/FDP Schleswig Holstein (2009) Koalitionsvertrag vom 17.Oktober 2009 Aufbruch in eine bessere Zukunft. Kiel

Chiavari J (2010) The legal framework for carbon capture and storage in the EU (Directive 2009/31/EC). In: Oberthür S, Pallemaerts M (eds) The new climate policies of the European Union: internal legislation and climate diplomacy. VUB Press, Brussels, pp 151–177

CO_2 Handel (2011) Bund schafft sich Schlupfloch im CO_2-Speichergesetz. CO_2-Handel.de, 28.04.2011, http://www.co2-handel.de/article186_16262.html

Deutsche Bischofskonferenz (2007) Der Klimawandel: Brennpunkt globaler, intergenerationeller und ökologischer Gerechtigkeit. Bonn, http://www.dbk.de

Deutscher Bundestag (2009a) Antwort der Bundesregierung auf die Kleine Anfrage – Drucksache 16/12616 -, Drucksache 16/1333. Berlin, http://dipbt.bundestag.de/dip21/btd/16/133/1613333.pdf

Deutscher Bundestag (2009b) Gesetzentwurf der Bundesregierung: Entwurf eines Gesetzes zur Regelung der Abscheidung, Transport und dauerhafter Speicherung von Kohlendioxid. Berlin, Drucksache 16/12782 vom 27.4.2009, http://dip21.bundestag.de/dip21/btd/16/127/1612782.pdf

Deutscher Bundestag (2009c) Stenografischer Bericht, 219. Sitzung, Plenarprotokoll 16/219. Berlin, http://dipbt.bundestag.de/dip21/btp/16/16219.pdf

Deutscher Bundestag (2011a) Ausschuss für Umwelt, Naturschutz und Reaktorsicherheit – 44. Sitzung (6. Juni 2011), Öffentliche Anhörung Berlin, Deutscher Bundestag, http://www.bundestag.de/bundestag/ausschuesse17/a16/Oeffentliche_Anhoerungen/index.html

Deutscher Bundestag (2011b) Gesetzentwurf der Bundesregierung: Entwurf eines Gesetzes zur Demonstration und Anwendung von Technologien zur Abscheidung, zum Transport und zur dauerhaften Speicherung von Kohlendioxid. Berlin, Drucksache 17/5750, 9.5.11, http://dipbt.bundestag.de/dip21/btd/17/057/1705750.pdf

Deutscher Bundestag (2011c) Stenografischer Bericht 120. Sitzung. Berlin, Plenarprotokoll 17/120 vom 7.7.2011, http://dipbt.bundestag.de

Deutscher Bundestag (2011d) Stenografischer Bericht, 92. Sitzung, 23.2.2011, Plenarprotokoll 17/92. Berlin, http://dipbt.bundestag.de

Deutscher Bundestag (2011e) Stenografischer Bericht, 108. Sitzung, 12.5.2011, Plenarprotokoll 17/108. Berlin, http://dipbt.bundestag.de

Deutscher Bundestag (2012a) Beschlussempfehlung des Vermittlungsausschusses. Drucksache 17/10101, 17. Wahlperiode 27.06.2012

Deutscher Bundestag (2012b) Stenograpfischer Bericht 187. Sitzung. Berlin, DBT, 28.6.2012

Deutscher Städte- Und Gemeindebund (2010) DStGB-Stellungnahme zum Kohlendioxid-Speicherungsgesetz-Entwurf. Berlin, 28.09.2010 http://www.dstgb.de/dstgb/Schwerpunkte/

DGB (2010) DGB-Stellungnahme zum Gesetzentwurf zur Kohlendioxidspeicherung. Berlin, Deutscher Gewerschaftsbund, 10.9.201, http://www.dgb.de/

Dietrich L (2009) Entwurf zum CCS-Gesetz lässt noch Fragen offen und bedarf der Konkretisierung. Dow Jounes Energy Weekly 13. März 2009, No. 11: 8–9

Dietrich L (2011) Durch Länderklausel bleibt der Erfolg des CCS-Gesetzes offen. Dow Jones Energy Weekly 26. April 2011, Nr. 16: S. 8–9

Dimas S (2006) Speech 06/492: General Assembly of the Technology Platform on Zero Emission Fossil Fuel Power Plants. Brussels, September 2012, http://europa.eu/rapid/pressReleasesAction.do?reference=SPEECH/06/492

DNR (2010) DNR-Forderungspapier für den Umbau der Energiepolitik in Deutschland. Bonn, 29.6.2010, http://www.dnr.de/downloads/dnr-forderungspapier-energie-29.6.10.pdf

DUH (2010) Stellungnahme zum Entwurf. Berlin, 27.7.2010, http://www.duh.de/uploads/media/DUH_Stellungnahme_CCS_Gesetzentwurf_07-2010.pdf

DUH (2011) Stellungnahme zum Entwurf der Bundesregierung. Deutscher Bundestag, Anhörung des Ausschusses Umwelt, Naturschutz & Reaktorsicherheit, Berlin, 6.6.2011, Drucksache 17 (16)265-B, http://www.bundestag.de/dokumente/textarchiv/2011/34591171_kw23_pa_ccs/index.html

Eckardt F, Van Riesten H, Henning B (2011) CCS als Governance- und Rechtsproblem. Zeitschrift für Umweltpolitik & Umweltrecht 34(4):409–436

EKD (2009) Nachhaltige Entwicklung im Zeichen des Klimawandels. Hannover, Denkschrift des Rates der EKD, http://www.ekd.de/EKD-Texte/68911.html

Ethik-Kommission (2011) Deutschlands Energiewende – Ein Gemeinschaftwerk für die Zukunft. Berlin, 30.5.2011, http://www.bundesregierung.de/Content/DE/__Anlagen/2011/05/2011-05-30-abschlussbericht-ethikkommission,property=publicationFile.pdf

Europäische Kommission (2008) Vorschlag für eine Richtlinie des Europäischen Parlaments und des Rates über die geologische Speicherung von Kohlendioxid. Brüssel, 23.1.2008, KOM (2008) 18 endg., http://eur-lex.europa.eu/LexUriServ/LexUriServ.do?uri=COM:2008:0018:FIN:DE:PDF

Europäische Kommission (2011) Beschluss der Kommission zur Einführung eines Fragebogens für den ersten Bericht über die Durchführung der Richtlinie 2009/31/EG. Amtsblatt der Europäischen Union, L 37/19 vom 11.2.2011, http://eur-lex.europa.eu/LexUriServ/LexUriServ.do?uri=OJ:L:2011:037:0019:0024:DE:PDF

European Commission (2007) Sustainable power generation from fossil fuels: aiuming for near-zero emissione from coal after 2020. Brussels, COM(2006) 843 final, http://eur-lex.europa.eu/LexUriServ/LexUriServ.do?uri=COM:2006:0843:FIN:EN:PDF

European Commission (2009) Directive 2009/31/EC of the European Parliament and of the Council of 23 April 2009 on the geological storage of carbon dioxide. http://eur-lex.europa.eu/LexUriServ/LexUriServ.do?uri=OJ:L:2009:140:0114:0135:EN:PDF

European Commission (2011a) Energy Roadmap 2050. European Commission, Brussels, COM (2011) 885 final, http://eur-lex.europa.eu/LexUriServ/LexUriServ.do?uri=COM:2011:0885:FIN:EN:PDF

European Commission (2011b) A Roadmap for moving to a competitive low carbon economy in 2050. Brussels, Communication from the Commission, COM/2011/0112 final, http://eur-lex.europa.eu/LexUriServ/LexUriServ.do?uri=CELEX:52011DC0112:EN:NOT

European Union (2009) Directive 2009/31/EC of the European Parliament and of the Council on the geological storage of carbon dioxide. Brussels, 5.6.2009, http://eur-lex.europa.eu/LexUriServ/LexUriServ.do?uri=OJ:L:2009:140:0114:0135:EN:PDF

E-World (2012) Merkel hält CCS in Deutschland für chancenlos. E-world, 16.7.12; http://www.e-world-2013.com

FAZ (2011) Streit über CO_2-Speicher. Frankfurter Allgemeine Sonntagszeitung vom 03.07.2011

FDP (2009) KOPP/KAUCH: Klimaschutz und Energiesicherheit brauchen ein CCS-Gesetz. Berlin, Pressemitteilung 661, 17.6.2011, http://16wp.fdp-fraktion.de/files/541/661-Kopp-Kauch-Klimaschutz.pdf

FDP (2011) Beschlüsse zum 62. Bundesparteitag. Rostock, 13.-15.5.2011, http://www.fdp.de/files/1208/BPT-Sicher_bezahlbar_und_umweltfreundlich.pdf

Fischedick M et al (2008) Gesellschaftliche Akzeptanz von CO_2-Abscheidung und -Speicherung in Deutschland. Energiewirtschaftliche Tagesfragen 58(11):20–23

Fischer W, Hake J-F (2012) CCS in Deutschland - gescheitert? In: Schumann D, Pietzner K (eds) Akzeptanzforschung zu CCS in Deutschland – Aktuelle Ergebnisse, Praxisrelevanz und Perspektiven. Oekom-Verlag, München

Fischer W, Hake J-F, Kuckshinrichs W, Schenk O (2010) Carbon capture and storage – Politische und gesellschaftliche Positionen in Deutschland. Technologiefolgenabschätzung – Theorie und Praxis 19(3):38–45

Fröhlich A (2011a) CCS: SPD gibt der Linken einen Korb Ness: Koalitionsvertrag bleibt, wie er ist. Potsdamer Neueste Nachrichten vom 19.5.2011, http://www.pnn.de/brandenburg-berlin/371216/

Fröhlich A (2011b) SPD lehnt Verhandlung über CCS ab. Der Tagesspiegel vom 3.2.2011, http://www.tagesspiegel.de/berlin/brandenburg/spd-lehnt-verhandlung-ueber-ccs-ab/3792126.html

Geden O (2012) Die Modifikation des 2-Grad-Ziels: Klimapolitische Zielmarken im Spannungsfeld von wissenschaftlicher Beratung, politischen Präferenzen und ansteigenden Emissionen. Berlin, SWP-Studien 2012/S12

Geisslinger E, Havlicek T (2011) Schleswig-Holstein bleibt sauber. TAZ, 15.04.2011

Germanwatch (2009) German-Watch Stellungnahme zum Entwurf. Bonn, 13.5.2009, http://www.germanwatch.org/klima/ccsges09.pdf

GFZ (2011) Projekte CO_2-Speicherung. Potsdam, http://www.gfz-potsdam.de/portal/gfz/Struktur/GeoEngineering-Zentren/CO2-Speicherung/M2_Projekte

Götze S (2009) Ministerien einig über CO_2-Lagerung. TAZ v. 18.2.2009

Greenpeace (2010) Stellungnahme zum Referentenentwurf. Hamburg, 30.8.2010, http://www.greenpeace.de/fileadmin/gpd/user_upload/themen/klima/Greenpeace-Stellungnahme-CCS2010.pdf

Grefe C (2009) Grün, aber gefährlich. DIE ZEIT, 20.05.2009 Nr. 22.

Hebel C (2011) Kraft entdeckt die Industrie. WDR, 3.6.2011, http://www.wdr.de/themen/politik/nrw04/atomkraft/110603.jhtml?stdComments=1

Hellriegel M (2010) Der neue Gesetzentwurf zu Carbon Capture and Storage. Neue Zeitschrift für Verwaltungsrecht 29(24):1530–1534

Iz Klima (2009) Scheitern des CCS-Gesetzes verhindert Investitionen. IZ Klima, Newsletter 06/2009, Berlin

Iz Klima (2010) CCS-Gesetz: Verbände-Anhörung im Wirtschaftsministerium. IZ Klima, Newsletter 08/2010, Berlin

Iz Klima (2011) Anhörung zum CCS-Gesetz im Bundestags-Umweltausschuss: Rechtsprobleme mit der Länderklausel. IZ Klima, Newsletter 06/2011, Berlin.

Knopf S, May F, Müller C, Gerling JP (2010) Neuberechnung möglicher Kapazitäten zur CO_2-Speicherung in tiefen Aquifer-Strukturen. Energiewirtschaftliche Tagesfragen 60(4):76–80

Koalitionsvertrag (2009a) Der Koalitionsvertrag zwischen CDU CSU und FDP: Wachstum. Bildung. Zusammenhalt. Berlin, 17.10.2009 www.cdu.de/doc/pdfc/091026-koalitionsvertrag-cducsu-fdp.pdf

Koalitionsvertrag (2009b) Gemeinsinn und Erneuerung: Ein Brandenburg für Alle. Koalitionsvertrag SPD/Die Linke, Potsdam, http://www.brandenburg.de/media/lbm1.a.4868.de/koalitionsvertrag.pdf

Krischer (2011) CCS-Gesetz: Doch keine Länderklausel? Berlin, 2.5.2011, http://oliver-krischer.eu/detail/nachricht/ccs-gesetz-doch-keine-laenderklausel.html

Kuckshinrichs W, Markewitz P, Linssen J, Zapp P, Peters M, Köhler B, Müller TE, Leitner W (2010) Weltweite Innovationen bei der Entwicklung von CCS-Technologien und Möglichkeiten der Nutzung und des Recyclings von CO_2. Forschungszentrum Jülich, Jülich, Berichte des Forschungszentrums Jülich, Reihe Umwelt, Bd. 60

Lakoff G, Wehling E (2008) Auf leisen Sohlen ins Gehirn. Politische Sprache und ihre heimliche Macht. Carl-Auer, Heidelberg

Land Brandenburg (2011) Brandenburg lehnt CCS-Gesetzentwurf ab. Energieland Brandenburg – Newsletter 4/2011, http://www.energie.brandenburg.de/media/bb1.a.2865.de/Energie_Newsletter_4_2011.pdf

Land Brandenburg (2012) Energiestrategie 2030. Potsdam, Ministerium für Wirtschaft und Europaangelegenheiten, 21.2.2012.

Land Sachsen-Anhalt (2010) Landesentwicklungsplan 2010. Magdeburg, http://www.sachsen-anhalt.de/index.php?id=855

Landesregierung Brandenburg (2011) Pressemitteilung vom 5.12.2011, http://www.brandenburg.de

Landtag Sachsen-Anhalt (2011) Plenarprotokoll 6/3. Magdeburg, 13.5.2011, http://www.landtag.sachsen-anhalt.de/intra/landtag3/ltpapier/plenum/6/003stzg_6.pdf

Landtag (2012) Kohlendioxid-Speicherungsausschlussgesetz Mecklenburg-Vorpommern. Schwerin, 30.6.12, GVOBl. M-V 2012, S. 142

Léger C (2011) Analysis of the German technological innovation system of carbon capture and storage (CCS) and the role of policies. Karlsruhe, Fraunhofer Institut für System- und Innovationsforschung (ISI), Diplomarbeit, Karlsruhe

Liebing I (2010) Auf der Suche nach Verbündeten. Schleswig-Holsteinischer Zeitungsverlag, 28.8.2010.

Marchetti C (1989) How to solve the CO_2 problem without tears. Int J Hydrog Energy 14 (8):493–506

Markusson N, Shackley S, Evar B (eds) (2012) The social dynamics of carbon capture and storage: understanding representation, governance and innovation. Routledge, London/New York

Matern M, Fröhlich A (2011) Schwarz-Gelb droht Schlappe bei CCS-Gesetz. Potsdamer Neueste Nachrichten v. 14.9.11; http://www.pnn.de/brandenburg-berlin/576905/

MAZ (2012a) Das verspätete Gesetz. Märkische Allgemeine, 29.6.2012

MAZ (2012b) Kohlendioxid-Gesetz: Vattenfall winkt ab. Märkische Allgemeine, 29.6.2012

Meierhofer H (2011) Persönliche Erklärung. Berlin, 8.7.2011, http://www.horst-meierhofer.de/

Mikulcak K (2012) German government backtracks on the Energy Transition. European Energy Review, 12 July 2012, http://www.europeanenergyreview.eu/site/pagina.php?id=3799#artikel_3799

NABU (2011) Stellungnahme. Berlin, 14.2.2011, http://www.nabu.de/themen/klimaschutz/nationalerklimaschutz/10833.html

Nestle I (2011) Schriftliche Fragen zu CCS (mit Antworten). Berlin, http://ingrid-nestle.de/energiewirtschaft/energieueberblick/kohle-und-ccs/detail/nachricht/schriftliche-frage-zu-ccs-und-klimazielen.html

OZ (2012) Schleswig-Holstein will CO_2-Speicherung gesetzlich verbieten. Ostseezeitung v. 28.6.2012

Referentenentwurf (2010) Gesetz zur Demonstration und Anwendung von Technologien zur Abscheidung, zum Transport und zur dauerhaften Speicherung von Kohlendioxid. Berlin, 14.7.2010, http://oliver-krischer.eu/fileadmin/user_upload/gruene_btf_krischer/2010/Referentenentwurf_CCS_Gesetz.pdf

Reichert G (2009) Klimaschutz in der Europäischen Union. Centrum für Europäische Politik, Freiburg

Rempe U (2012) Braunkohle-Gipfel tagte in Leuna. Leuna_Echo v. 06.03.2012

Rief J (2011) Pressemitteilung. Berlin, 08.07.2011, http://www.josef-rief.de/

RNE (2008) Position des Nachhaltigkeitsrates zu aktuellen Fragen der Energie- und Klimapolitik. Berlin, 27.10.2008

Robbe P (2011) Zur Reichweite des § 2 Abs. 5 des Entwurfs eines Kohlendioxid-Speichergesetzes (BT-Drs. 175750). Wissenschaftlicher Dienst des Deutschen Bundestages, 12.5.2011, WD 3-3000-166/11, Berlin

Sachverständigenrat Für Umweltfragen (2011) Wege zur 100% erneuerbaren Stromversorgung. SRU, Sondergutachten, Berlin

Sattler KO (2011) Brisanz in der Tiefe. Das Parlament. Nr. 12/21.3.2011, http://www.das-parlament.de/2011/12/WirtschaftFinanzen/33821072.html

Schleswig-Holsteinischer Landtag (2009) Plenarprotokoll 17/3, 3. Sitzung am 19.11.2009 Kiel, http://www.landtag.ltsh.de/infothek/wahl17/plenum/plenprot/2009/17-003_11-09.pdf

Schleswig-Holsteinischer Landtag (2011) Penarprotokoll 41. Sitzung, 23.2.2011, Plenarprotokoll 17/41, Kiel, http://www.landtag.ltsh.de/plenumonline/februar2011/texte/01_ccs_akt_std.htm

Schramm S (2010) Das zweite Wendland. Die Zeit, 18.11.2010, Nr. 47, http://www.zeit.de/2010/47/U-CSS-Widerstand

Schumann D, Pietzner K (eds) (2012) Akzeptanzforschung zu CCS in Deutschland – Aktuelle Ergebnisse, Praxisrelevanz und Perspektiven. Oekom-Verlag, München

Schumann D, Pietzner K, Esken A (2010) Umwelt, Energiequellen und CCS: Regionale Unterschiede und Veränderungen von Einstellungen der deutschen Bevölkerung. Energiewirtschaftliche Tagesfragen 60(5):52–56
Schumann D, Fischer W, Hake J-F (2012) Akzeptanz der Transformation des Energiesystems in der Bevölkerung. Energiewirtschaftliche Tagesfragen 62(6):29–33
Seidler C (2010) Minister stellen Gesetz zu CO_2-Speichern vor. Spiegel online, 14.7.10
Skrylnikow I (2010) CCS: Carbon Dioxide Capture and Storage – Technologische Risiken und regulatorische Herausforderungen. Natur und Recht 32(8):543–550
Skudelny J (2011) Rundschreiben zu den Auswirkungen der Ereignisse in Japan auf die deutsche Energieversorgung. Berlin, 22.3.2011, http://www.judith-skudelny.de
SPD/GRÜNE (2010) Koalitionsvertrag: Gemeinsam neue Wege gehen. NRW-SPD, Bündnis 90/die Grünen, Düsseldorf, http://static.rp-online.de/layout/fotos/HB8TULLE.pdf
SPD/GRÜNE (2012) Koalitionsvertrag 2012–2017. (Düsseldorf, Juni 2012) http://www.nrwspd.de/meldungen/1/116611/Koalitionsvertrag-2012-2017.html
SPD/GRÜNE/SSW (2012) Bündenis für den Norden. Koalitionsvertrag 2012 bis 2017. Kiel, http://www.sh.gruene.de/cms/default/dokbin/411/411582.koalitionsvertrag_spd_buendnis90die_grue.pdf
Spd-Parteivorstand (2011) Aktuell: Neue Energie. Berlin
SRU (2009) Abscheidung, Transport und Speicherung von Kohlendioxid. Berlin, Stellungnahme Nr. 13, April 2009, http://www.umweltrat.de
Stoll P-T, Lehmann F (2008) Die Speicherung von CO_2 im Meeresuntergrund – die völkerrechtliche Sicht. Zeitschrift für Umweltrecht 19(6):281–286
TAZ (2012) “Keine anlasslosen Vorratsdaten”. TAZ. Die Tageszeitung, 30.12.2011
Totz S (2011) CO_2-Endlager: Greenpeace veröffentlicht Gesetzentwurf. Hamburg, Greenpeace, 20.02.2011, http://www.greenpeace.de/themen/klima/nachrichten/artikel/co2_endlager_unter_berlin_und_hamburg/
Treber M (2011) Neuer Schwung für die Energiewende. KlimaKompakt Nr. 71, Mai 2011, http://www.germanwatch.org/
Treblin J (2010) Landtag in Sachsen-Anhalt streicht CCS-Passus. www.klimaretter.info
Vattenfall (2011a) Pressemitteilung vom 5.12.2011, http://www.vattenfall.de
Vattenfall (2011b) Pressemitteilung vom 8.7.2011. http://www.vattenfall.de
Viebahn P, Vallentin D, Höller S, Fischedick M (2012) Integrated assessment of CCS in the German power plant sector with special emphasis on the competition with renewable energy technologies. Mitig Adapt Strateg Glob Chang 17(6):707–730
VIK (2010) Statistik der Energiewirtschaft. Verlag Energieberatung, Essen
Von Der Marwitz (2012) Aktuelle Pressemitteilung, 28.6.2012
Von Der Marwitz HG (2010) Positionspapier zu Carbon Capture and Storage (CCS). Berlin, August 2010, www.von-der-marwitz-mdb.de
Von Hirschhausen C, Herold J, Oei PY, Haftendorn C (2012) CCS-Technologie ein Fehlschlag – Umdenken in der Energiewende notwendig. DIW Wochenbericht Nr. 6. 2012, S. 3–9
Wasserzeitung (2009) Wasserwirtschaft einig: CCS bringt große Risiken für Lebensmittel Nr. 1. Wasserzeitung 2/2009, http://www.wv-nord.de/fileadmin/documents/WASSERZEITUNG_OKT_09.pdf
WBGU (2006) Die Zukunft der Meere – zu warm, zu hoch, zu sauer. Wissenschaftlicher Beirat der Bundesregierung Globale Umweltveränderungen, Sondergutachten 2006, Berlin
WBGU (2011) Welt im Wandel: Gesellschaftsvertrag für eine Große Transformation. Wissenschaftlicher Beirat der Bundesregierung Globale Umweltveränderungen, Berlin
Wendler S (2011) Schulterschluss in Schwarze Pumpe. Lausitzer Rundschau vom 27.1.2011
Werdermann F (2011) Die CCS-Taktierer. Der Freitag, 8.7.2011, http://www.freitag.de/politik/1127-die-ccs-taktierer
Wiese C (2012) Energiekommissar Öttinger erwägt CCS-Pflicht. DOW JONES NEWSLETTER, 19. März 2012, http://www.finanznachrichten.de

WWF (2010) WWF Stellungnahme zum CCS Gesetzesentwurf. Berlin, 27.8.2010, http://www.wwf.de/presse/details/news/wwf_stellungnahme_zum_ccs_gesetzesentwurf/

WWF (2011) Stellungnahme zum Gesetzentwurf. DBT, Anhörung des Ausschusses Umwelt, Naturschutz & Reaktorsicherheit, Berlin, 6.6.2011, Drucksache 17(16)265-E

Zabka M (2011) McAllister ist gegen CO_2-Lager. General-Anzeiger, 15.2.2011

Chapter 13
CCS Policy in the EU: Will It Pay Off or Do We Have to Go Back to Square One?

Olga Schenk and Jürgen-Friedrich Hake

Abstract The paper outlines the CCS policy in the EU and discusses its implementation. The European institutions established CCS as a key element in the package of policies aiming to realize climate change mitigation targets set by the EU. While the EU has developed a range of policies to support the development and deployment of CCS, their implementation in the member states faces obstacles. Some European countries question whether CCS should be deployed as a climate change mitigation option. For those member states which endorse CCS, the key difficulty that is yet to be overcome on the way to the commercial deployment is a development of the long-term incentives that would create a business case for CCS.

Keywords European Union • CCS • Energy policy • NER300

13.1 Introduction – Why Does the EU Need CCS?

Climate change mitigation is a field of work where the supranational organs of the European Union (EU) have actively brought their legislative competence and political influence to bear since the early 1990s. A common European objective involves reducing greenhouse gas (GHG) emissions by cutting emissions by 80–95 % before 2050 compared to the reference year 1990 (European Council 2011). At the level of the individual sectors, the European Commission's proposals assume that GHG emissions can be almost completely eliminated in the area of power generation (European Commission 2011a). The most important measures that will make a reduction of this magnitude possible not only include expanding the capacity of renewables and funding energy savings measures but also the implementation of CCS technology. The use of CCS in fossil-fired power plants in

---

O. Schenk (✉)
School of Public and Environmental Affairs, Indiana University,
1315 East Tenth Street, Bloomington, IN 47405, USA
e-mail: oschenk@indiana.edu

J.-F. Hake
Institute of Energy and Climate Research – Systems Analysis and Technology Evaluation (IEK-STE), Forschungszentrum Jülich GmbH, D-52425 Jülich, Germany
e-mail: j.-f.hake@fz-juelich.de

W. Kuckshinrichs, J.-F. Hake (eds.), *Carbon Capture, Storage and Use*,
DOI 10.1007/978-3-319-11943-4_13

Europe aims mainly to facilitate the climate-friendly use of coal, and thus help to consolidate the pillars of EU energy policy: security of supply, competitiveness, and sustainable development.

The European Commission's Green Paper on energy published in 2000 is quite pessimistic with regard to the future of coal in the European energy mix. The reasons for this are the competitive disadvantages of domestic hard coal production, competition with other energy carriers, particularly with natural gas, and the environmental impacts along the entire process chain from coal mining to coal combustion (European Commission 2000b). The accession of new countries to the EU in 2004 with considerable lignite and hard coal reserves, rising prices for natural gas, and a steadily increasing demand for fossil fuels from the newly industrialised countries altered the strategic importance of coal in the EU's energy mix (Mills 2004, 2010). The relatively low costs of coal and the stability of costs, as well as the contribution to the diversification of the European energy mix have become more important.

The implementation of CCS made it possible to unite the strategy of cutting GHG emissions with the use of coal. In the middle of the last decade, CCS was defined as one method of combating climate change by the EU in its political strategic documents. Since then, the EU institutions have developed a series of measures in an attempt to ease the commercial application of CCS from 2020 onwards. They are oriented towards the central challenges associated with CCS. The implementation of CCS necessitates the development of a political framework for the long-term competitiveness of the technology, large-scale demonstration in the power plant sector, and investments in research and development (R&D). These challenges are not country-specific. They take account of the characteristics of CCS as a new, risky and capital-intensive technology, which solely serves climate policy objectives.

The structure of this chapter reflects the development of European policy, which is dedicated to the above-mentioned challenges. Based on the analysis of key European legislation on CCS, the chapter will first outline the strategic anchoring of CCS in the energy and climate policy of the EU (Sect. 13.2.1). Then, the political measures supporting CCS in the long term will be detailed (Sect. 13.2.2). Key provisions within the European CCS Directive will be explained because of their central significance for the implementation of CCS. This section will close with an overview of the stages of implementation of the CCS Directive in the member states. Corresponding to the challenges connected with the novelty of CCS, the chapter lists the instruments and priorities in the area of R&D support (Sect. 13.3). In addition, the central funding instruments for the demonstration of CCS (EEPR, NER300) and their implementation to date will be reviewed (Sect. 13.4). The investigation will conclude with an evaluation of the measures in place and an outlook for the remaining challenges for CCS in the EU.

13.2 CCS – A Cornerstone of the EU's Integrated Climate and Energy Policy

CCS as a method of mitigating climate change is mentioned for the first time in the communication by the European Commission that sparked off a discussion on strategies for the reduction of emissions following the first commitment period of the Kyoto Protocol (2008–2012) (European Commission 2005). The paper refers to the findings of a study by Pacala and Sokolow, which determined the 15 most promising strategies – mainly technological options – for reducing GHG emissions in the long term (Pacala and Sokolow 2004).

While the early identification of CCS as a future option for climate change mitigation in the EU assumes a broad area of application for carbon capture – from base-load power plants to hydrogen units and synfuel production facilities – the subsequent Green Paper entitled 'A European Strategy for Sustainable, Competitive and Secure Energy' focuses on coal-fired power plants as one of the main areas for the application of CCS in Europe (European Commission 2006). In addition to the increased use of renewables and the promotion of energy efficiency, the Green Paper defines the implementation of CCS as the third option for reducing emissions in the long term – a concept that is reflected in the subsequent European thematic strategies for climate and energy policy. Figure 13.1 provides an overview of the most important European documents dealing with strategies and regulations supporting the development and implementation of CCS.

13.2.1 Integrated Energy and Climate Change Package in 2007 – Determination of Strategic Orientation for CCS

In preparation for the European Council meeting in spring 2007, the European Commission publishes an integrated energy and climate change package (Europe 2007). This is centred on the objective of limiting the increase in the average global temperature to not more than 2 °C above pre-industrial levels. Emissions are to be reduced predominantly via energy-policy measures. The Commission proposes cutting GHG emissions by at least 20 % by 2020[1] compared to levels in 1990 (European Commission 2007b). This requirement is supported by a series of measures, which envisage the need to revise the EU emissions trading system, to curb emissions from the traffic and transport sector, and to review European energy policy. Against the background of the three challenges facing energy policy – environmental compatibility, security of supply, and cost efficiency – the

[1] If international agreement is reached on climate change mitigation after 2012, the target will be increased to 30 % (European Commission 2007a).

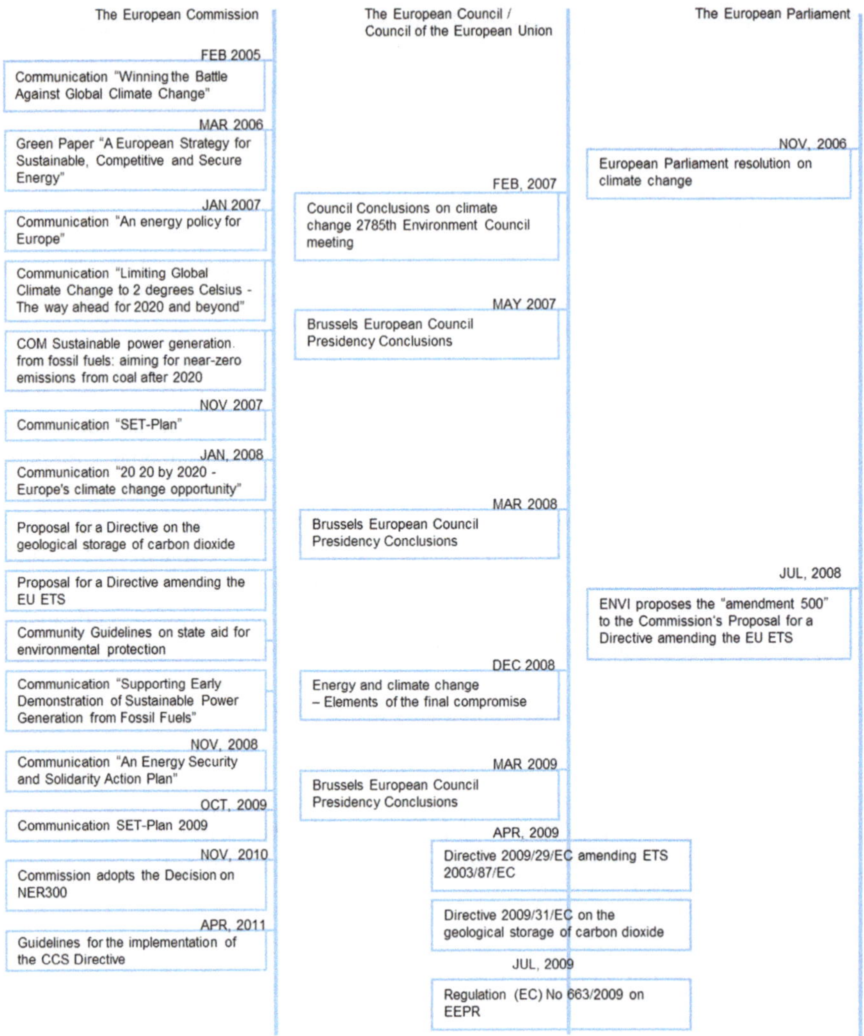

Fig. 13.1 Main points of EU policy and legislation in the area of CCS (as of June 2011) (Source: Schenk 2013)

European Commission develops an action plan of energy policy measures. The central points of the action plan are shown in Table 13.1.

The proposed action plan and accompanying measures (see Fig. 13.1) identify CCS as a key element in the package of measures aiming to realise the targets set for 2020. The Commission explains the high priority of CCS by referring to the importance of coal for both the European and the global energy mix. The Commission identifies four main areas of action that must be mobilised if the commercial application of CCS is to become a reality in 2020:

Table 13.1 Objectives and measures of the EU action plan for energy policy

Objective	Measures
Establishment of an internal energy market for the EU	Ownership unbundling of the networks
	Harmonisation of the European energy standards
	Increased transparency of energy markets
	Expansion of the energy infrastructure in Europe
	Introduction of binding network security standards
	Investments to ensure adequate electricity generation and gas supply capacities
	Strengthening the rights of energy consumers
More solidarity between EU member states in the event of an energy crisis	Development of measures
Long-term reduction of GHG emissions	EU emissions trading
Increased energy efficiency with the key goal of reducing global primary energy use by 20 % by 2020	
Increasing the level of renewable energy in the EU's overall energy mix to 20 % by 2020	
Transition to a low-carbon energy system	Development of a European Strategic Energy Technology Plan (European Commission 2007b)
Commercial use of CCS from 2020 onwards	Construction and operation of 10–12 large-scale demonstration plants by 2015
Nuclear safety and security	Development of additional European legislation
Development of shared external energy policy for Europe	
Monitoring of the development of a shared European energy policy	Establishment of an observatory

Source: European Commission (2007b)

- incentives for the large-scale demonstration and coordination of demonstration projects,
- development of the regulatory framework for carbon storage as well as the legal basis for retrofitting newly constructed power plants in the future,
- development of an incentive system for the long-term economic use of CCS, and
- recognition of CCS as an option for combating climate change in international climate change agreements,[2] as well as support for demonstration projects performed in cooperation with newly industrialised countries (European Commission 2007b).

[2] Cf. Chap. 14 by *Hake and Schenk* on international collaboration.

At the spring European Council in 2007, the heads of state and heads of government agreed on the strategic features of the integrated energy and climate package in the form of a three-year action plan (2007–2009), which is based mainly on the Commission's proposals. With respect to CCS, the European Council's action plan supports the series of measures aiming to make CCS commercially viable in 2020 as a method of mitigating climate change. The need to develop funding instruments for the construction and operation of up to 12 large-scale integrated demonstration power plants by 2015 is central in this context.

At the end of 2007, the Commission proposed the European Strategic Energy Technology Plan (SET-Plan). The SET-Plan's strategy is twofold: the funding of R&D, and the creation of an incentive system for the market introduction of strategic energy technologies. The early demonstration of the application of CCS in generating power from fossil fuels is defined by the SET-Plan as an area where Europe should invest resources as a priority.

13.2.2 Climate and Energy Package 2008 – Definition of Long-Term Prospects for CCS

While the strategic framework and the definition of important milestones on the way towards the commercial implementation of CCS represent the priorities of the policy-making process in 2007, the EU climate and energy package[3] proposed by the European Commission in January 2008 comprised concrete legislative proposals relevant to CCS (see Fig. 13.1). The package contained a proposal to reform the EU emissions trading system[4] by including CCS, new guidelines for state aid for environmental protection consistent with the position of the Commission towards state aid for CCS, as well as a proposed EU directive for carbon storage.

In the accompanying communication supporting early demonstration of CCS for power generation, the Commission forecasts that € 1 billion of public funds will be needed for an 800 MW block of a CCS demonstration power plant in order to offset the capital and operating costs associated with the application of CCS (European Commission 2008a). Despite the expected decrease in costs as a result of learning effects, the Commission assumes that the introduction of CCS will lead to higher fixed and variable costs for power generation. First of all, the investment costs for the construction of a facility equipped with CCS are higher than for conventional plants, no matter what CO_2 capture technology is applied. Secondly, the costs of

[3] The climate and energy package of the European Commission comprises the following documents: (i) proposal for updating the emissions trading system, (ii) decision on 'burden sharing' for those sectors that do not fall under the European emissions trading system, (iii) a proposal for a directive to increase the share of renewable energy, (iv) a proposal for the directive governing the geological storage of carbon dioxide (European Commission 2008b).

[4] European Union Emissions Trading System – EU ETS.

operating a plant equipped with CCS are higher than those for a conventional power plant due to efficiency losses as are the costs for carbon capture and transportation. The Commission therefore emphasises that the market introduction of CCS does not just require investments in R&D and demonstration. Specifically, it also needs long-term incentives to make the application of CCS more attractive than conventional power production from fossil fuels. Hence, the proposal to amend the EU ETS directive and the new guidelines for state aid for environmental protection contribute to ascertaining the long-term economics of CCS and the adequacy of the public funding of demonstration projects.

13.2.2.1 EU Directive on Emissions Trading and EU Guidelines on State Aid for Environmental Protection

The amendment of the EU ETS Directive applies to the third trading period 2013–2020, which stipulates the auctioning of all allowances for the electricity sector. Even in its original version, the EU ETS Directive allowed CCS installations to participate in emissions allowance trading via 'Procedures for unilateral inclusion of additional activities and gases' (Art. 24) (European Parliament and the Council 2003). The Commission's proposal therefore aimed to harmonise the integration of CCS into the EU system by explicitly including installations that capture, transport or store CO_2 emissions in the Directive's categories of activities (European Commission 2008c). With the aim of creating incentives for CCS, the current version of the Directive does not consider stored emissions as having been emitted. This releases operators from the obligation of surrendering allowances for emissions stored. The Directive, however, excludes the possibility of free allocation of CO_2 allowances for CCS installations.

In addition, on the recommendation of the European Parliament Committee for Environment, Public Health and Food Safety (ENVI), the amended EU ETS Directive makes allowances available from the EU ETS new entrants' reserve[5] to finance CCS demonstration programmes.[6] Up to 300 million allowances will be used to support the construction and operation of up to 12 commercial demonstration projects until 2015.

The new guidelines on state aid for environmental protection recognise the strategic significance of CCS for lowering GHG emissions in the long term. The guidelines demonstrate that the Commission considers state aid positive in the case of strategic energy technologies such as CCS. The guidelines do not define provisions for approving such aid due to a lack of previous experience.

---

[5] The new entrants' reserve (NER) is a small proportion of emission allowances in every national allocation plan (NAP) reserved specifically for new market participants.

[6] For a detailed description of the negotiation processes in connection with NER300 in the European Parliament and in the EU Council, see Claes and Frisvold (2009).

Another important aspect that contributes to the planning certainty of CCS is the development of a legal framework for the long-term storage of CO_2, which has not yet been practised at the level of the EU.

13.2.2.2 European Legal Framework for Carbon Storage

The supranational organs of the EU successfully created one of the world's first legal frameworks for the long-term operation of CO_2 storage sites both onshore and offshore. The Commission explained the need for a common European approach in the form of a European Directive regulating carbon storage by ensuring uniform environmental and safety standards (European Commission 2008d). The thematic priorities of the Directive include:

- modifying existing environmental legislation to meet the demands of carbon storage,
- the drafting of general framework conditions, standards, and criteria to ensure the safety of operating storage facilities, and
- the definition of the principles for rights of access for third parties to the transport network and to the storage sites.

The CCS Directive regulates carbon capture mainly by adapting existing legislation. The environmental risks of capturing CO_2 streams are accounted for by an amendment to Directive 2008/1/EC concerning integrated pollution prevention and control (IPPC). This subjects carbon capture installations to a licensing procedure. The criteria for granting a permit include the requirement to identify and use the best available technologies with respect to the capture techniques applied and the composition of the CO_2 stream. Furthermore, the amendment to Directive 85/337/EEC stipulates that an environmental impact assessment is a condition for displaying an IPPC permit. This demands that the operator describes and evaluates potential direct and indirect impacts of the project on the environment.

As the future use of CCS technologies is still open, and carbon capture technology is not yet state of the art in new-build power plants, the introduction of a 'capture-ready' standard has been discussed as one option for enabling the retrofitting of new-build, highly efficient power plants at a later point in time. At the moment, there are a few approaches for defining the capture readiness of power plants (IEAGHG 2007; TÜV NORD CERT). The CCS Directive supplements Directive 2001/80/EC on the limitation of emissions of certain pollutants into the air from large combustion plants with the introduction of the capture-ready standard for new-build power plants with a rated electrical output of 300 MW or more (European Parliament and the Council 2009a). The EU's capture-ready approach stipulates that sufficient space be provided for carbon capture and compression at the plant site provided that the following criteria have been met: accessible suitable storage sites, as well as the technical and economic feasibility of the transport facilities and of retrofitting for carbon capture.

During the consultation process on the proposal for the CCS Directive, the option was discussed of making the use of CCS obligatory for fossil-fired power plants from a set cut-off point. Although this regulation attracted particular interest from non-governmental organisations, the Commission decided against incorporating this option into the proposed directive based on the impact assessment of the economic, social, and ecological consequences of the mandatory implementation of CCS (Claes and Frisvold 2009).

As is the case with carbon capture, the CCS Directive mainly regulates the transportation of CO_2 via the application of existing legislation. An environmental impact assessment is also applied to the pipeline transport of the CO_2 stream. Of the three components that make up a CCS process chain – capture, transportation, and storage – CO_2 transportation is the one element that is mainly regulated at a national level. The amendments to Directive 2006/12/EC on waste and Regulation No. 1013/2006 on shipments of waste exclude the transport of CO_2 for the purpose of geological storage from the scope of both legislative acts. With regard to competition law, the CCS Directive demands transparent and non-discriminatory access for third parties to the CO_2 transport network.

The provisions of the CCS Directive are valid for the commercial application of carbon storage in onshore and offshore installations. The amendment to Directive 2000/60/EC establishing a framework for Community action in the field of water policy allows the CCS Directive to permit carbon storage in deep saline aquifers, which are considered to have the greatest storage potential in Europe.

As carbon storage in international waters does not fall within the legislative powers of the EU, the international instruments for the protection of the marine environment must be changed. In 2006, the Convention on the Prevention of Marine Pollution by Dumping of Wastes and Other Matter adopted in 1972 (commonly referred to as the London Convention) and its London Protocol from 1996 were amended. The amendment that took effect in 2007 explicitly allows carbon storage in sub-seabed geological formations. The decision to amend the Convention for the protection of the marine environment of the North-East Atlantic (also known as the OSPAR Convention) was made by the OSPAR Commission in 2007. While the decision of the OSPAR Commission allows CO_2 storage in sub-seabed geological formations, it also legally rules out the placement of CO_2 into the water-column of the sea and on the seabed (OSPAR Decision 2007/1, 2007).

The CCS Directive means that EU member states retain the right to decide on the location of storage sites, or to completely prohibit CO_2 storage within their territories (European Parliament and the Council 2009a). Those member states who wish to store CO_2 within their sovereign territory are obliged by the Directive to undertake an assessment of the storage capacity available within their territory. To this end, the Directive contains criteria that provide a basis for assessing the suitability of a geological formation for use as a storage site. The member states decide whether wells/boreholes or injection tests (exploration) are required to characterise a geological formation as a storage site. In such a case, the operator requires an exploration permit, which is granted by the competent authority of the member state. The procedure for granting the exploration permit, its validity, and its volume area is determined by the competent authority of the member state.

Just as the exploration of a storage site requires an exploration permit, CO_2 storage necessitates a storage permit, which is granted by the competent authority of the member state. The CCS Directive specifies the information to be included in applications for storage permits, the conditions for issuing storage permits, and the contents of storage permits. Although the implementation of the Directive lies in the hands of the member states, the Commission retains certain monitoring functions in an effort to increase public confidence in the new technology. The member states are obliged to make all storage permit applications received available to the Commission together with the accompanying documentation. The Commission can then issue a non-binding opinion on the applications. The competent authority is obliged to inform the Commission of the final decision on issuing the storage permit and to justify any departure from the Commission's opinion.

A key approach, for which the Commission has developed a uniform basis, is the monitoring of the storage complex (storage site and surrounding environment) after the injection of CO_2 into the geological formation. The central elements of monitoring are regular reporting by the operator to the competent authority and a system of inspections of the storage complexes by the competent authority of a member state. The Directive also lays down criteria for designing and updating a monitoring plan, which provides the basis for the regular reports submitted by the operator to the competent authority.

Uniform environmental standards are also in force for CO_2 storage. The criteria for granting a permit for IPPC, for example, require that the operator conducts an environmental impact assessment of CO_2 storage. Environmental damage caused by the operation of storage sites is regulated by the amended Directive 2004/35/EC on environmental liability with regard to the prevention and remedying of environmental damage. This Directive is based on the 'polluter pays' principle, and stipulates that the operator whose actions have caused the environmental damage is liable. Furthermore, the Directive specifies the measures that must be taken in case of CO_2 leakages (Art. 16). The operator is obliged to take the necessary corrective measures, and to notify the competent authority. If the operator fails to take the necessary corrective measures, the Directive transfers responsibility for so doing to the competent authority who shall subsequently recover the costs incurred for corrective measures from the operator. The incorporation of storage sites into the areas covered by Directive 2003/87/EC establishing a scheme for GHG emissions allowance trading within the EU also obliges operators to surrender emissions trading allowances for any leaked emissions.

The Directive also stipulates that the CO_2 stream should 'consist overwhelmingly' of carbon dioxide (European Parliament and the Council 2009a). The introduction of control mechanisms to ensure the purity of the CO_2 stream is the responsibility of the member states.

With respect to the long-term nature of CO_2 storage, the Directive specifies criteria and describes the procedure for the closure of storage sites. It stipulates the operator's post-closure obligations, which apply for a minimum period before responsibility can be transferred to the competent authority of the member state. During this minimum period determined by the competent authority, which may be no shorter than 20 years according to the Directive, the operator is responsible for sealing the storage site, removing the injection facilities, monitoring the storage site, submitting reports to the competent authority, implementing corrective measures, and surrendering allowances in the case of leakages. The competent authority performs its inspection obligations during this period. After the minimum period has elapsed, and 'when all available evidence indicates that the stored CO_2 will be completely and permanently contained', the operator, with the consent of the competent authority, can transfer responsibility for the storage complex and thus all obligations that were in force during the minimum period to the competent authority (European Parliament and the Council 2009a). When preparing the decision on transferring responsibility to the competent authority, the member state is obliged to provide the Commission with all relevant information. As is the case with the granting of storage permits, this procedure allows the Commission to issue a non-binding 'opinion', following which the competent authority must inform the Commission of its final decision and its reasons for this decision.

The operation and closure of the storage site, corrective measures in the case of a leakage, and the fulfilment of post-closure obligations are connected with financial contributions. The Directive therefore specifies that the member states must ensure that the operators provide proof of sufficient financial resources to meet their obligations as part of the application for a storage permit. In addition, the operator must provide the competent authority with a financial contribution to cover the anticipated costs of monitoring after the transfer of responsibility for a period of 30 years.

In analogy to the regulation of access to the CO_2 transport network, the CCS Directive also demands transparent and non-discriminatory access for third parties to CO_2 storage sites, while the specific legislation is subject to national regulations.

In late 2008, the European Parliament passed with an overwhelming majority the components of the energy and climate change package, including the proposal for a Directive on storing CO_2 (European Parliament 2008). After the agreement of heads of state and heads of government, the climate change package entered into force in spring 2009. The CCS Directive was to be transposed into national law by EU member states by 25 June 2011.

Spain was the only EU member state to implement the CCS Directive in its national legislation before the deadline (Armeni 2012). Romania managed to communicate its measures for national implementation on time, which meant that the Commission initiated infringement procedures against 25 EU member states. By the end of 2012, in addition to Spain and Romania, another 23 member states, including Germany, communicated their national implementing measures to the Commission (Europe 2012). Hungary and Poland have not yet reported their national implementing measures, and infringement procedures against these states are still ongoing. The Commission also assesses whether the national legal framework developed by the member states for CCS meets the requirements laid down in the CCS Directive (Armeni 2012).

13.3 Funding of Research and Development

Until the oil crises in the 1970s, coal research was funded exclusively by the European Coal and Steel Community (ECSC) (Ishihara et al. 1999). While European coal research previously concentrated mainly on mining technology, the oil crises in the 1970s and the drop in domestic coal production led to a realignment of research towards the areas of coal usage. The funding of coal research at a European level peaked in the early 1980s, but funding volumes have sunk continuously since then.[7]

EU support for R&D on individual technologies in the CCS process chain and clean coal technologies dates back to the EU's Third Framework Programme for Research and Technological Development[8] (FP) (1990–1994) (European Commission 2011b). Since then, EU expenditure on support for CCS research has increased. EU funding for CCS within FP5 (1998–2002) totalled € 16 million, and the total budget for CCS projects was € 32 million (European Commission 2004). Research in FP5 focused on CO_2 storage and monitoring.

The research and development budget within the framework of FP6 (2002–2006) was € 150 million with around € 70 million public funding (European Commission 2007b). FP6 concentrated mainly on researching carbon capture technologies. Another important priority of FP6 comprised activities aiming to develop a European Research Area (ERA). The foremost objective of the ERA is to develop a common European research policy to consolidate and improve the structure of European research (European Commission 2000a). A number of fields, instruments, collaborations and projects within the scope of ERA contribute to the coordination and harmonisation of European research. FENCO[9] ERA-NET is a network that funds the sustainable use of fossil energy carriers with a focus on CCS. FENCO ERA-NET provides a framework for representatives of ministries and national funding institutions to discuss national R&D programmes, share knowledge, and cooperate on the management of research projects.

Another important development connected with the implementation of FP6 is the creation of the technology platform known as the Zero Emissions Platform (ZEP) as one of the European stakeholder initiatives supporting the deployment of CCS technologies (ZEP 2006). ZEP advises the European Commission on setting R&D priorities and on funding measures in the field of CCS. On the basis of ZEP's recommendation, for example, the Commission adopted the goal of constructing up to 12 CCS demonstration plants in Europe by 2015.[10]

---

[7] For a list of European R&D and demonstration programmes in the area of coal from the early 1970s to the late 1990s, see Ishihara et al. (1999).

[8] Framework Programme for Research and Technological Development.

[9] FENCO – Fossil Energy Coalition; see www.fenco-era.net

[10] For more information on how interests are mediated between the Commission and the technology platform ZEP, see Schenk (2013).

To date, calls for project proposals within FP7 have led to 24 projects being funded within the area of CCS with a total of approx. € 90 million of EU funding (European Commission Research). The increase in funding volume from the EU's fifth to the ongoing seventh framework programme clearly shows that the framework programmes are the most important instrument for funding R&D in the area of CCS at the EU level.

A point of criticism with respect to public funding of CCS research was its focus on carbon capture technologies (ZEP 2010c). Against the background of a lack of public acceptance for CO_2 storage projects in certain European countries (see Chap. 11 by *Schumann*), the lack of early European storage projects was criticised because this made it impossible to describe positive experience in the public discussion on the hazards associated with CO_2 storage. The Commission reacted to this by refocusing the research priorities towards carbon storage in the later part of FP7. It is anticipated that a large number of carbon storage projects in Europe will allow more knowledge to be obtained and may even lead to a higher acceptance of CO_2 storage among the general public.

13.4 Support for the Demonstration of CCS: Instruments and Their Implementation

A discussion on appropriate instruments for supporting CCS demonstration projects focused initially on available mechanisms at the national level as well as revenues from the EU ETS and financial resources from the structural funds and cohesion funds (European Commission 2008a). At the end of 2008, the supranational organs of the EU announced the introduction of funding mechanisms with a very distinctive European component. The European Energy Programme for Recovery (EEPR)[11] and the allocation of allowances from the reserve for new entrants into the EU ETS (NER300) are the most important instruments for funding demonstration projects at the EU level. They are based on the same basic principle: both allow a financial contribution towards private and national investments in demonstration projects. Both instruments have a specific time frame. This public funding should thus provide an incentive for the speedy implementation of projects in order to ensure the competitive availability of CCS from 2020 onwards.

Although EEPR and the first NER300 call have similar selection criteria for projects, they are two different funding instruments. EEPR funds come from the Commission's budget and provide projects with short-term stimuli by subsidising investment costs. NER300 funding is generated by auctioning emissions allowances and it aims to subsidise the initial operating costs of the projects. With respect to efforts aiming to coordinate the resources of different actors, the SET-Plan 2009 seeks to implement a common concept in the mid to long term for EU funding of the

[11] European Energy Programme for Recovery.

development of low-carbon technologies (European Commission 2009). The first attempts at coordinating efforts in the area of sharing knowledge and experience were made within the framework of a demonstration network[12] set up by the European Commission. This network brings together those responsible for the EEPR projects and Statoil's Sleipner project.[13]

While the Commission can only issue recommendations with respect to national funding strategies for low-carbon technologies, as was the case with the proposal to invest 50 % of the revenues generated within the EU ETS in activities to combat climate change (European Parliament and the Council 2009b), funding at an EU level grants the Commission a say in selecting projects and in formulating criteria for the dissemination of knowledge and information gained from the projects.

13.4.1 The European Energy Programme for Recovery

The EEPR is part of the EU's Economic Recovery Plan launched in 2008 and also supports action in selected energy sectors. Funding within this special programme offers immediate financial support. The envisaged package of some € 3.98 billion aims to provide decisive impetus for the accelerated implementation of projects. Gas and electricity infrastructures, offshore wind energy, and CCS have been identified as strategic sectors in the energy industry. The implementation of projects in these sectors should considerably help to increase supply security and to reduce GHG emissions.

Interruption to the gas supplies from Russia during the winters of 2006 and 2009 led to the incorporation of infrastructural programmes as security of supply became even more important. This area also received the largest sum corresponding to € 2.365 billion, which is approx. 60 % of the total budget. Projects in the area of offshore wind energy received € 0.565 billion (14 % of the total budget). Projects in the area of CCS received funding worth € 1.050 billion – the remaining 26 % of the total budget.

Funding of CCS projects is limited to a maximum of one project per member state (European Parliament and the Council 2009c). The criteria for the eligibility of projects for funding are shown in Table 13.2.

The projects are granted funds within the framework of the authorized budget after they submit bills for completed work to the Commission. Funds that have already been received do not have to be paid back if the deadline for putting the facility into operation is pushed back. However, if the project is delayed considerably or discontinued, funding can be suspended. Although member states did not play a role in selecting projects to be funded within the framework of EEPR, they

[12] For more information, see www.ccsnetwork.eu

[13] Membership of the network is open to European demonstration projects that fulfil the criteria of a CCS demonstration project and that are committed to sharing knowledge.

Table 13.2 EEPR funding criteria

Technical characteristics of the facilities	Electrical output of at least 250 MW
	Ability to capture, transport, and geologically store at least 80 % of the CO_2
Knowledge exchange	A binding commitment from the project managers that they will share 'general knowledge' gained during operation of the demonstration facility with industry at large and with the Commission
Start-up	Expected by 31 December 2015

Source: European Parliament and the Council (2009c)

Table 13.3 EEPR-funded projects

Project name, location	Developer	Carbon capture	Carbon transportation	Carbon storage
Porto Tolle, Italy	Enel	Post-combustion	Pipeline	Deep saline formations, offshore
Jänschwalde, Germany	Vattenfall	Oxyfuel, post-combustion	Tanker	Exhausted natural gas field, deep saline formations, onshore
Bełchatów, Poland	PGE EBSA	Post-combustion	Pipeline	Deep saline formations, onshore
Don Valley Power Project, United Kingdom	2Co Energy	IGCC	Pipeline	EOR or deep saline formations, offshore
Rotterdam (ROAD), Netherlands	Electrabel, E.ON	Post-combustion	Pipeline	Exhausted natural gas field, offshore
OXYCFB 300 Compostilla Project, Spain	Endesa (PPP)	Oxyfuel	Pipeline	Deep saline formations

Source: Europe (2009) and GCCSI (2012)

support the implementation of the programme by technically monitoring and financially controlling the projects. The final selection of projects was made by the Commission together with experts from the European Investment Bank and was announced in December 2009. The following six projects were originally funded (Table 13.3).

In December 2011, the list of EEPR-funded projects lost one project. Vattenfall announced its decision to discontinue the Jänschwalde project due to the lack of a legal framework for carbon storage in Germany (Vattenfall 2011). In March 2012, Enel announced that the Porto Tolle project would have to be postponed due to delays in granting the administrative permit (Trabattoni 2012). In May 2012, Poland's largest energy provider PGE declared that it was putting a hold on further developments within the Bełchatów project (Bakewell 2012; Cienski 2012) The reason for this was the low price of CO_2 allowances, which would make the construction and operation of a CCS project uneconomic without further state aid. Experts believe that the final decision on discontinuing the Bełchatów project will depend on whether or not it is accepted into the NER300 funding instrument (Hallerman 2012). The Hatfield project, which was renamed the Don Valley Power Project after the operator changed, is awaiting a final investment decision, which depends on the level of public funding should it be accepted into the NER financing instrument (ZERO 2012). The final investment decisions regarding the ROAD project in the Netherlands and the Compostilla project in Spain have yet to be made.

13.4.2 NER300

NER300 is the most comprehensive mechanism worldwide providing public funding for CCS and innovative renewable energy technologies (EIB 2010). Funding within NER300 comprises the financing of commercial-scale demonstration projects in the areas of CCS and innovative renewable energy technologies. The grants paid are limited to 50 % of the relevant costs,[14] and individual projects receive no more than 15 % of the total budget for the call (European Commission 2010). The revenues generated by selling 300 million carbon allowances are used to fund projects within the framework of two calls for proposals. The selection procedure for the first call is currently under way. It aims to fund at least 8 CCS projects and 34 projects in the area of renewable energy (European Commission 2010). In order to benefit from funding in the first call, the projects must meet the funding criteria listed in Table 13.4.

The most important actors in the selection procedure, which occurs in a four-stage process, are the member states, the European Investment Bank and the

Table 13.4 NER300 funding criteria

Technical characteristics of the facilities	Electrical output of at least 250 MW
	Thermal integration during carbon capture
	Ability to capture at least 85 % of the CO_2
	Implementation of an independent research block to ensure the safety of carbon storage and improve monitoring techniques
Knowledge exchange	A binding commitment from the project managers that they will share 'general knowledge' gained during operation of the demonstration facility with industry at large and with the Commission
Start-up	Expected by 31 December 2015[a]

Source: European Commission (2010)
[a]Or at the latest 4 years after the decision to provide financial aid (European Commission 2010)

[14] The relevant costs are investment costs incurred because of the application of CCS. For a definition, see European Commission (2010).

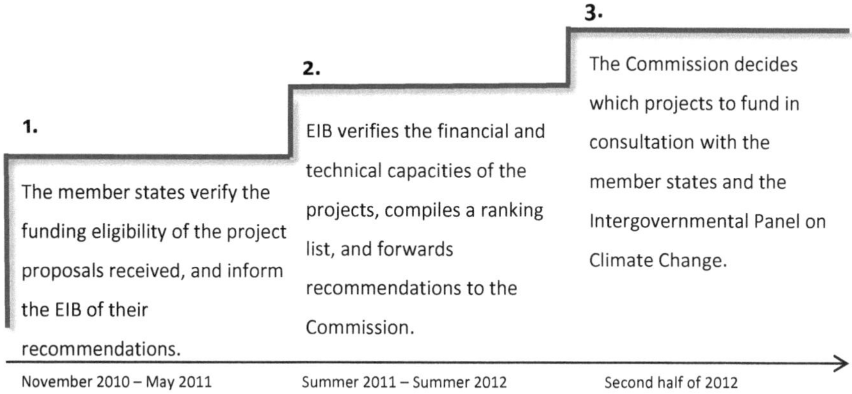

Fig. 13.2 ER300 selection procedure (Source: European Commission 2010)

European Commission. Figure 13.1 provides an overview of the selection procedure for projects.

The Commission transfers 300 million emissions allowances on behalf of the member states to the EIB, which then sells the allowances and manages the money. The EIB transfers the revenues to the member states accordingly. These then pay the projects selected for funding a certain sum depending on the amounts of CO_2 stored (European Commission 2010). CCS demonstration projects are paid grants once a year over a period of 10 years from the time they are first put into operation. If a member state guarantees that surplus funds will be paid back, the entire funding amount or part of it can be paid before the facility is put into operation (Fig. 13.2).

In trading the first lot of 200 million emissions allowances, the EIB sold some 120 million allowances between December 2011 and May 2012 for an average price of € 8.12 each (EIB 2012). If we use this price to calculate the revenues for 300 million allowances, the total amount of funding available through NER300 would be around € 2.5 billion. In early 2011, when the project applications were submitted and were being evaluated by the member states, the average price for an emissions allowance was approx. € 15, and the total funding volume was therefore expected to be € 4.5 billion (Limousin 2011; ZEP 2010a).

Within the first call for the submission of projects, 22 proposals in the area of CCS were evaluated by the member states. Of these, 13 CCS project proposals were forwarded to the EIB for further evaluation. Half of these proposals came from the United Kingdom. The decision on the allocation of funding for the first call will be made in the second half of 2012.[15]

[15] At the time when this chapter was written, no decision had yet been made. For more on the decision at the end of 2012, see *Kuckshinrichs and Markewitz*, Chap. 15.

13.5 CCS in the EU – An Initial Assessment

While climate change mitigation has become pivotal in European policy, political action in the area of energy policy is mainly influenced by the nation-states. On the one hand, the number of areas and the extent to which they are covered by energy policy legislation in Europe has grown continuously since the oil crisis in 1973/74. The implementation of the internal energy market considerably extended the political influence of the European institutions. On the other hand, the EU had no independent legislative powers in the area of energy, which means that legislation was introduced in connection with other areas such as environment or competition. Only with the Lisbon Treaty the European Union was granted legislative competence in energy matters. Although the principles of European energy policy continued to hold, whereby each member state decides on its own energy mix, the EU provides incentives for selected energy carriers – first and foremost for renewables – and thus influences the energy mix in the member states via legislation aiming to protect the environment. Environmental protection, which has already emerged as the 'motor of European integration' (Oberthür and Kelly 2008; Schreurs and Tiberghien 2007), has led to a steady increase in EU legislation in the area of energy. The policy field connected with the introduction of CCS represents part of this development.

The policy-making process, which anchors CCS in the EU as an option for combating climate change, was executed at a remarkable speed. Faced with two options – the mandatory introduction of CCS and the development of an economic framework for the voluntary implementation of CCS – the EU institutions decided in favour of the second option. In the absence of an international climate regime, the mandatory introduction of CCS could not prevail against the expected competitive disadvantages caused by high electricity costs for European companies. Between 2005 and 2009, European institutions successfully established CCS as a cornerstone of the EU's integrated energy and climate policy, developed a legal framework for geological storage, incorporated CCS into the European emissions trading system, and elaborated funding instruments for CCS. With respect to the extent of the policy development supporting CCS, the EU is one of the pioneers internationally. While the EU has been remarkably successful in formulating CCS policy, positive feedback with regard to the implementation of CCS policy is rarer.

The implementation of the CCS Directive clearly shows that there is no consensus in some European countries on whether CCS should be deployed as a method of mitigating climate change. Today, some European countries, such as Austria, are completely against carbon storage. The termination of the Jänschwalde project in Germany by Vattenfall illustrates that the delayed implementation of the CCS Directive in member states can impact on the European CCS demonstration programme.

Against the background of the hesitant development of EEPR projects and the volatility of EU ETS prices for allowances, a discussion has emerged as to whether CCS support via the EU ETS and subsidies within the demonstration programme

(EEPR, NER300) are enough to achieve the target of making CCS commercially available by 2020. Experts speak of altered framework conditions, which have important impacts on the implementation of CCS.[16] Topics and measures dealing with climate change mitigation are fading into the background because of the enduring global recession (ZEP 2011a). In the current situation, companies are reverting back to their core area of business and are cautious with respect to investments in climate protection measures. The unsuccessful Post-Kyoto Protocol negotiations and the low level of importance accorded to climate protection policy in the USA and in China lead to uncertainty concerning whether climate change mitigation can maintain its driving force for innovations and investments in the long term.

The role of NER300 financing as one of the main instruments supporting CCS demonstration projects in Europe is being increasingly questioned in this context (ZEP 2011b). The instrument had attracted criticism from the very beginning because of the uncertainty regarding the price level of allowances, which ultimately determines the volume of funding available to finance projects. Pessimistic expectations were confirmed by initial experiences trading the allowances. The low price for emissions allowances does not just cast doubt on the incentive system for demonstration programmes in the EU. It has also fuelled discussions on the competitiveness of CCS technology after the demonstration phase. Even if support in the form of public funding could aid the successful demonstration of CCS technologies, an answer is still required to the question of what mechanisms would allow the pan-European implementation of CCS after 2020 (ZEP 2010c). Long-term incentives are decisive for a stable development of low-carbon technologies. Some of these mechanisms currently being discussed are (ZEP 2010b):

- a lower EU ETS cap in order to bring the price of emissions allowances to a level that would make CCS application competitive[17];
- a CO_2 tax tied to the allowance price;
- a bonus-malus system, which would pay subsidies to CO_2 emitters that keep below a set quota of CO_2 emissions per kilowatt hour, and enforce a levy on the emitters that exceed this quota (Birkeland et al. 2010).

The decision for or against further funding of CCS revolves around the relationship between the European and national levels of the policy-making and policy-implementation processes in the EU. The lack of a European public awareness enabled CCS legislation to be rapidly introduced at an EU level. This policy-making process resulted in CCS assuming a key role in achieving the European energy and climate policy objectives, and in R&D and demonstration of the technologies being supported in the form of subsidies at the EU level. Difficulties implementing CCS in the member states give rise to the question of whether CCS

[16] Cf. video footage of the ETP ZEP General Assembly 2011 (ZEP 2011d).

[17] The European technology platform ZEP predicts that an EU ETS allowance price of € 35 would make CCS application competitive for coal-fired power plants (ZEP 2011c).

requires further political action at the EU level or whether the implementation of CCS now solely depends on support from the member states. Another question that has yet to be answered is what impacts it will have on European energy and climate policy, if CCS technology fails to have the desired effect as a method of mitigating climate change in the EU due to a lack of social consensus within the member states.

References

Armeni C (2012) An update on the state of CCS regulation from Europe [Online]. GCCSI. Available: http://www.globalccsinstitute.com/community/blogs/authors/chiara-armeni/2012/02/13/update-state-ccs-regulation-europe. Accessed 18 June 2012

Bakewell S (2012) PGE says Belchatow carbon capture project needs state support [Online]. Bloomberg Business Week. Available: http://www.businessweek.com/news/2012-05-30/pge-says-belchatow-carbon-capture-project-needs-state-support. Accessed 12 June 2012

Birkeland L, Tjetland G, Hoff E, Eickhoff C, Domitrovic D, Bernstone C, Pages A (2010) Improving the regulatory framework, optimizing organization of the CCS value chain and financial incentives for CO_2-EOR in Europe. Available: http://bellona.org/ccs/ccs-news-events/publications/article/improving-the-regulatory-framework-optimizing-organization-of-the-ccs-value-chain-and-financial-inc.html. Accessed 20 June 2012

Cienski J (2012) Poland's clean coal going up in smoke [Online]. Financial Times. Available: http://blogs.ft.com/beyond-brics/2012/05/25/polands-clean-coal-going-up-in-smoke/#axzz1xa1yvmM7. Accessed 12 June 2012

Claes DH, Frisvold P (2009) CCS and the European Union: magic bullet or pure magic? In: Meadowcroft J, Langhelle O (eds) Catching the carbon. The politics and policy of carbon capture and storage. Edward Elgar, Cheltenham

EIB (2010) Climate change – Commission launches major investment programme for innovative low-carbon technologies [Online]. European Investment Bank. Available: http://www.eib.org/about/press/2010/2010-193-climate-change-commission-launches-major-investment-programme-for-innovative-low-carbon-technologies.htm. Accessed 14 June 2012

European Investment Bank (2012) EIB – NER300 monetisation monthly report – May 2012. Available: http://www.eib.org/attachments/ner_monthly_report_may_2012.pdf

Europe (2007) Commission proposes an integrated energy and climate change package to cut emissions for the 21st century [Online]. IP-07-29. Available: http://europa.eu/rapid/pressReleasesAction.do?reference=IP/07/29. Accessed 6 Dec 2012

Europe (2009) List of 15 energy projects for European economic recovery [Online]. MEMO/09/542. Available: http://europa.eu/rapid/pressReleasesAction.do?reference=MEMO/09/542&format=HTML&aged=0&language=en&guiLanguage=en. Accessed 27 Oct 2010

Europe (2012) 72009L0031: National Execution Measures [Online]. Available: http://eur-lex.europa.eu/LexUriServ/LexUriServ.do?uri=CELEX:72009L0031:EN:NOT. Accessed 18 June 2012

European Commission (2000a) Communication from the Commission to the Council, the European Parliament, the Economic and Social Committee and the Committee of the Regions – Towards a European research area. Office for Official Publications of the European Communities, Brussels, http://eur-lex.europa.eu/legal-content/EN/TXT/PDF/?uri=CELEX:52000DC0006&from=EN

European Commission (2000b) Green Paper – Towards a European strategy for the security of energy supply. Office for Official Publications of the European Communities, Luxembourg

European Commission (2004) European CO_2 capture and storage projects Commission of the European Communities, Directorate-General for Research, Directorate J – Energy,

Unit J2 –Energy production and distribution systems, Brussels. Information and Communication Unit Directorate-General for Research European Commission, Brussels, http://ec.europa.eu/research/energy/pdf/co2capt_en.pdf
European Commission (2005) Communication from the Commission to the Council, the European Parliament, the European Economic and Social Committee and the Committee of the Regions – Winning the battle against global climate change Brussels
European Commission (2006) Green Paper – A European strategy for sustainable, competitive and secure energy, Brussels, http://europa.eu/documents/comm/green_papers/pdf/com2006_105_en.pdf
European Commission (2007a) Communication from the Commission to the Council, the European Parliament, the Economic and Social Committee and the Committee of the Regions – A European strategic energy technology plan (SET-PLAN) 'Towards a low carbon future' Brüssels. Luxembourg: Publications Office of the European Union
European Commission (2007b) Communication from the Commission to the Council, the European Parliament, the Economic and Social Committee and the Committee of the Regions – Limiting global climate change to 2 degrees Celsius. The way ahead for 2020 and beyond Brussels
European Commission (2008a) Communication from the Commission to the European Parliament, the Council, the European Economic and Social Committee and the Committee of the Regions supporting early demonstration of sustainable power generation from fossil fuels. Brussels. Available: http://eur-lex.europa.eu/legal-content/EN/TXT/PDF/?uri=CELEX:52008DC0013&from=EN
European Commission (2008b) Communication from the Commission to the European Parliament, the Council, the European Economic and Social Committee and the Committee of the Regions 20 20 by 2020 Europe's climate change opportunity. Brussels. Available: http://eur-lex.europa.eu/LexUriServ/LexUriServ.do?uri=COM:2008:0030:FIN:en:PDF. Accessed 14 Oct 2014
European Commission (2008c) Proposal for a Directive of the European Parliament and of the Council amending Directive 2003/87/EC so as to improve and extend the greenhouse gas emission allowance trading system of the community. Brussels. Available: http://eur-lex.europa.eu/LexUriServ/LexUriServ.do?uri=COM:2008:0016:FIN:EN:PDF
European Commission (2008d) Proposal for a Directive of the European Parliament and of the Council on the geological storage of carbon dioxide and amending Council Directives 85/337/EEC, 96/61/EC, Directives 2000/60/EC, 2001/80/EC, 2004/35/EC, 2006/12/EC and Regulation (EC) No 1013/2006. Brussels. 17 August 2010: http://eur-lex.europa.eu/LexUriServ/LexUriServ.do?uri=COM:2008:0018:FIN:EN:PDF
European Commission (2009) Communication from the Commission to the Council, the European Parliament, the Economic and Social Committee and the Committee of the Regions – investing in the development of low carbon technologies (SET-Plan) Brussels
European Commission (2010) Commission Decision of 3 November 2010 laying down criteria and measures for the financing of commercial demonstration projects that aim at the environmentally safe capture and geological storage of CO_2 as well as demonstration projects of innovative renewable energy technologies under the scheme for greenhouse gas emission allowance trading within the Community established by Directive 2003/87/EC of the European Parliament and of the Council. Official Journal of the European Union, Brussels
European Council (2011) European Council, 4 February 2011, conclusions. Available: http://www.consilium.europa.eu/uedocs/cms_data/docs/pressdata/en/ec/119175.pdf
European Commission (2011a) Communication from the Commission to the Council, the European Parliament, the Economic and Social Committee and the Committee of the Regions – A roadmap for moving to a competitive low carbon economy in 2050 Brussels
European Commission (2011b) EU support for CCS [Online]. Available: http://ec.europa.eu/research/energy/eu/research/ccs/support/index_en.htm. Accessed 22 Dec 2011
European Commission Research. EU energy research. projects [Online]. Available: http://ec.europa.eu/research/energy/eu/projects/index_en.cfm#results. Accessed 17 Aug 2010
European Council (2011) European Council, 4 February 2011, conclusions. Available: http://www.consilium.europa.eu/uedocs/cms_data/docs/pressdata/en/ec/119175.pdf

European Parliament (2008) EP seals climate change package [Online]. Brussels. Available: http://www.europarl.europa.eu/sides/getDoc.do?pubRef=-//EP//TEXT+IM-PRESS+20081208BKG44004+0+DOC+XML+V0//EN. Accessed 7 Apr 2011

European Parliament and the Council (2003) Directive 2003/87/EC of the European Parliament and of the Council of 13 October 2003 establishing a scheme for greenhouse gas emission allowance trading within the Community and amending Council Directive 96/61/EC. Official Journal of the European Union, Brussels

European Parliament and the Council (2009a) Directive 2009/29/EC of the European Parliament and of the Council of 23 April 2009 amending Directive 2003/87/EC so as to improve and extend the greenhouse gas emission allowance trading scheme of the Community. Official Journal of the European Union, Brussels

European Parliament and the Council (2009b) Directive 2009/31/EC of the European Parliament and of the Council of 23 April 2009 on the geological storage of carbon dioxide and amending Council Directive 85/337/EEC, European Parliament and Council Directives 2000/60/EC, 2001/80/EC, 2004/35/EC, 2006/12/EC, 2008/1/EC and Regulation (EC) No 1013/2006. Official Journal of the European Union, Brüssels

European Parliament and the Council (2009c) Regulation (EC) No 663/2009 of the European Parliament and of the Council of 13 July 2009 establishing a programme to aid economic recovery by granting Community financial assistance to projects in the field of energy. Official Journal of the European Union, Brussels

GCCSI (2012) Large-scale integrated projects database [Online]. GCCSI. Available: http://www.globalccsinstitute.com/projects/browse. Accessed 29 June 2012

Hallerman T (2012) Fate of Poland's only proposed CCS projects remains in question [Online]. GHG Reduction Technologies Monitor. Available: http://ghgnews.com/index.cfm/fate-of-polande28099s-only-proposed-ccs-project-remains-in-question/. Accessed 14 June 2012

IEAGHG (2007) CO_2 capture ready plants. Technical Study Report Number: 2007/4. IEA Greenhousegas Programme. Available: http://ieaghg.org/docs/General_Docs/Reports/2007-4%20Capture%20Ready.pdf

Ishihara Y, Benson S, Morrison G (1999) Coal research, development and demonstration funding. CCC/25, IEA Clean Coal Centre 1999

Limousin L (2011) Ner 300: Inventory of CCS applications to Member States [Online]. Carbon capture journal. Available: http://www.carboncapturejournal.com/displaynews.php?NewsID=750. Accessed 14 June 2012

Mills SJ (2004) Coal in an enlarged European Union. CCC/85. IEA Clean Coal Centre, London

Mills SJ (2010) Prospects for coal, CCTs and CCS in the European Union. CCC/173. IEA Clean Coal Centre, London

Oberthür S, Kelly CR (2008) EU leadership in international climate policy: achievements and challenges. Int Spect 43:35–50

OSPAR Decision 2007/1 2007. OSPAR Decision 2007/1 to prohibit the storage of carbon dioxide streams in the water column or on the sea-bed. OSPAR Convention for the Protection of the Marine Environment of the North-East Atlantic

Pacala S, Sokolow R (2004) Stabilization wedges: solving the climate problem for the next 50 years with current technologies. Science 305(5686):968–972

Schenk O (2013) Interest mediation and policy formulation in the European Union. Influence of transnational technology-oriented agreements on European policy in the field of carbon capture and storage, Energy & Environment 165, Advances in Systems Analysis 3. Jülich

Schreurs MA, Tiberghien Y (2007) Multi-level reinforcement: explaining European Union leadership in climate change mitigation. Glob Environ Politics 7(4):19–45

Trabattoni A (2012) Enel's Italian Porto Tolle coal conversion timeframe uncertain: CEO [Online]. Platts. Available: http://www.platts.com/RSSFeedDetailedNews/RSSFeed/Coal/8037757. Accessed 12 June 2012

TÜV NORD CERT. Carbon capture ready – die Lösung zur Prüfung neuer Kraftwerke hinsichtlich CCS [Online]. Available: http://www.tuev-nord.de/de/klimaschutz/kraftwerkspruefung-1639.htm. Accessed 12 Aug 2010

Vattenfall (2011) Hängepartie um CCS-Gesetz erzwingt Aus für Milliardeninvestition in der Lausitz [Online]. Available: http://www.vattenfall.de/de/pressemitteilungen-detailseite.htm?newsid=E3B4752A90F9447FA1B1D95CA6860B77. Accessed 12 June 2012

ZEP (2006) Strategic research agenda. The European Technology Platform for Zero Emission Fossil Fuel Power Plants (ZEP). Available: http://www.zero-emissionplatform.eu/website/docs/ETP%20ZEP/ZEP%20Concepts%20Final%20V2.pdf

ZEP (2010a) The CCS demonstration programme, Philippe Paelinck, Co-Chair Taskforce Demonstration & Implementation & Director, CO_2, Alstom Power [Online]. ZEP General Assembly 2010. Available: http://www.zeroemissionsplatform.eu/events/details/45-zep-general-assembly-2010.html. Accessed 2 July 2012

ZEP (2010b) Facilitating the widespread deployment of CCS, Eivind Hoff, Co-Chair, Taskforce Policy & Regulation, ZEP & Director, Bellona Europa [Online]. ZEP General Assembly 2010. Available: http://www.zeroemissionsplatform.eu/events/details/45-zep-general-assembly-2010.html. Accessed 29 June 2012

ZEP (2010c) Shared learnings and continued momentum, Niels Peter Christensen, Co-Chair, Taskforce Technology & Chief Geologist, Vattenfall [Online]. ZEP General Assembly 2010. Available: http://www.zeroemissionsplatform.eu/events/details/45-zep-general-assembly-2010.html. Accessed 1 July 2012

ZEP (2011a) CCS state of play, Dr. Graeme Sweeney, Chairman, ZEP & EVP, CO_2, Royal Dutch Shell [Online]. ZEP General Assembly 2011. Available: http://www.zeroemissionsplatform.eu/events/details/136-zepga2011.html. Accessed 29 June 2012

ZEP (2011b) The costs of CCS post-2020 & incentives, Dr. Lars Stromberg, Co-Chair Taskforce Technology, ZEP & Senior Advisor, Vattenfall AB [Online]. ZEP General Assembly 2011. Available: http://www.zeroemissionsplatform.eu/events/details/136-zepga2011.html. Accessed 29 June 2012

ZEP (2011c) The costs of CO_2 capture, transport and storage. Post-demonstration CCS in the EU. ETP ZEP. Available: http://www.zeroemissionsplatform.eu/library/publication/165-zep-cost-report-summary.html

ZEP (2011d) ZEP General Assembly 2011 [Online]. Available: http://www.zeroemissionsplatform.eu/events/details/136-zepga2011.html. Accessed 29 June 2012

ZERO (2012) Don Valley Power Project [Online]. ZERO. Available: http://www.zero.no/ccs/projects/powerfuel-hatfield-colliery-pre-combustion-project. Accessed 12 June 2012

Chapter 14
International Cooperation in Support of CCS

Jürgen-Friedrich Hake and Olga Schenk

Abstract At the international level, there are a number of international organizations and political initiatives supporting the development and use of CCS. Their role and their main activities can be assigned to the major challenges associated with the development and use of CCS: (1) The cooperation to include CCS in the Clean Development Mechanism (CDM), (2) developing incentives for the demonstration of CCS, and (3) pooling and dissemination of knowledge from R&D activities. The international organizations and initiatives considered comprise e.g. Intergovernmental Panel on Climate Change (IPCC), G8, International Energy Agency (IEA), Carbon Sequestration Leadership Forum (CSLF), and the Global Carbon Capture and Storage Institute (GCCSI).

Keywords International cooperation • IPCC • G8 • IEA • CSLF • GCCSI

14.1 Introduction

International cooperation plays a pivotal role in many respects for the development and deployment of CCS. As is the case for alternative climate protection measures, CCS as one option for combating climate change will only make a visible contribution to reducing greenhouse gas (GHG) emissions if it is used by the largest GHG emitters in the world. The implementation of CCS in only a few countries could create competitive disadvantages because of the additional costs of electricity generation, and it could jeopardise the objective of reducing global GHG emissions

J.-F. Hake (✉)
Institute of Energy and Climate Research – Systems Analysis and Technology Evaluation (IEK-STE), Forschungszentrum Jülich GmbH, D-52425 Jülich, Germany
e-mail: j.-f.hake@fz-juelich.de

O. Schenk
School of Public and Environmental Affairs, Indiana University,
1315 East Tenth Street, Bloomington, IN 47405, USA
e-mail: oschenk@indiana.edu

W. Kuckshinrichs, J.-F. Hake (eds.), *Carbon Capture, Storage and Use*,
DOI 10.1007/978-3-319-11943-4_14

that is the ultimate goal of CCS deployment. In this context, international cooperation is based on a twofold approach: it aims to implement CCS as early as possible to combat climate change in industrialised countries, and it uses important international activities to support CCS implementation in the developing and newly industrialised countries.

CCS is also a new, capital-intensive and risky technology (Rai et al. 2009) – characteristics that demand international cooperation in the area of research and development (R&D). Burden sharing should not only be understood as a defensive strategy with respect to government demands for new energy technologies. It also serves to ensure competitive advantages for companies in future markets. International cooperation in the area of R&D occurs within research networks and within the framework of R&D or pilot projects. Research networks usually involve cooperation between representatives from government and industry, and thus aim to encourage the exchange of information on technical, political, regulatory, economic, and sometimes social aspects associated with the implementation of CCS. In practical projects, representatives from industry and research institutions usually work together in an effort to achieve technological improvements.

This chapter is divided into two main parts. The first part outlines the most important international collaborations in the area of CCS, and provides an overview of the findings of research to date on the international aspects of funding CCS technologies. The second part discusses the importance of international cooperation for the implementation of CCS in Germany.

14.2 International Cooperation: Priorities and Discussion

At the international level, there are a number of international organisations and political initiatives supporting the development and use of CCS in different ways. The main activities of these international organisations can be assigned to the major challenges associated with the development and use of CCS:

- The cooperation to include CCS in the Clean Development Mechanism (CDM) aims to enable the long-term competitiveness of CCS.
- In addition, several international organisations focus on developing incentives for the demonstration of CCS technology.
- Finally, the pooling and dissemination of knowledge from R&D activities, and pilot and demonstration projects throughout the world is another important goal of international collaboration.

Among the international organisations concerned with CCS, the Intergovernmental Panel on Climate Change (IPCC) occupies an important position. The Special Report on CCS, which was compiled by IPCC at the invitation of the seventh session of the Conference of the Parties (COP) to the United Nations Framework Convention on Climate Change (UNFCCC), confirms the feasibility of the permanent[1] geological storage of CO_2 (IPCC 2005). This is considered a milestone in the history of CCS development (Meadowcroft and Langhelle 2009; Pena 2009). Since the publication of the special report, CCS is no longer considered an exotic technology, but rather has developed into a serious option for combating climate change on the political agenda of several OECD countries.

14.2.1 International Cooperation Supporting Competitiveness

The UNFCCC, which came into force in 1994 (United Nations 1992), aims to stabilise the concentration of GHG emissions in the atmosphere in order to avoid anthropogenic interference with the climate system. The subsequent Kyoto Protocol, which came into force in 2005, lays down binding targets and implementation mechanisms for reducing GHG emissions in industrialised countries. The integration of CCS as a climate protection option into the existing instruments of international climate change mitigation is considered one of the key prerequisites for the long-term competitiveness of CCS (Russial 2011; Kerr et al. 2009).

In line with the new appraisal of CCS as an important option for combating climate change, which began to take root after the publication of the IPCC Special Report on CCS in 2005, the IPCC updated its Guidelines for National Greenhouse Gas Inventories.[2] The IPCC guidelines now comprise statistical methods for registering emissions saved by the use of CCS in the inventories of Kyoto Parties (IPCC 2006). In the literature, this development is assessed as an important step towards including CCS in an international portfolio of climate change mitigation options and towards improving the competitiveness of CCS (de Coninck and Bäckstrand 2011; Kerr et al. 2009). However, the Kyoto Parties have not yet recognised the amended guidelines. Furthermore, the emissions reductions brought about by the use of CCS do not yet have major impact on the energy balances of individual nation-states. Specific statistical methods are therefore necessary in order to account for them. The new guidelines will therefore only come into force if a Post-Kyoto Protocol is agreed upon.

While the emissions reductions in developing and newly industrialised countries played a minor role in the Kyoto Protocol negotiations, the steadily rising demand for fossil energy carriers from non-OECD countries and the expected rise in this

---

[1] "Observations from engineered and natural analogues as well as models suggest that the fraction retained in appropriately selected and managed geological reservoirs is very likely [...] to exceed 99 % over 100 years and is likely [...] to exceed 99 % over 1,000 years". "Very likely" is a probability between 90 and 99 %, "Likely" is a probability between 66 and 90 % (IPCC 2005, p. 14)".

[2] The new IPCC Guidelines amend the version published in 1996.

demand in the long term[3] has changed how the role of the developing and newly industrialised countries is assessed in terms of international climate change mitigation. The inclusion of CCS in the CDM is an important thread in the discussion on the importance of different incentives for the reduction of GHG emissions in developing and newly industrialised countries (Philibert et al. 2007). This step should simplify technology transfer, help to cut emissions, and promote the demonstration of CCS in developing and newly industrialised countries.

CDM is an instrument for international climate change mitigation that allows industrialised countries as Parties to the Convention on Climate Change to fulfil their obligations in terms of reducing emissions in the form of project activities in developing countries (United Nations 1997). Research dedicated to examining the debate surrounding the inclusion of CCS in the CDM not only identifies the challenges in terms of the compatibility of CCS and CDM arising from the technical specificities of CCS and the organisational aspects of CDM, but it also lists the conditions under which CCS can be sustainably applied within the CDM (Watanabe et al. 2007; de Coninck 2008b; Bode 2009; Pollak and Wilson 2009).

The novelty of CCS as a climate change mitigation option and the lack of experience in the long-term geological storage of CO_2 are the most important challenges. They lead to uncertainties, not only with respect to the licensing and implementation of CCS CDM projects, particularly cross-border projects (Philibert et al. 2007), but also in dealing with the consequences of short-term and long-term leakages (Bode 2009; Pollak and Wilson 2009; de Coninck 2008b). The resulting implications for the CDM governance structures point towards the need to formulate more comprehensive regulations, tailored to the specificities of CCS, for the licensing and monitoring of CCS CDM projects (de Coninck 2008b; Bode 2009; Pollak and Wilson 2009). The advantages of central control mechanisms in dealing with CCS CDM projects include the consolidation of the governance structures in the CDM host countries,[4] as well as limiting the potential flooding[5] of the CDM market with a large number of CCS projects.

At its sixth session[6] in 2010, the Conference of the Parties to the Kyoto Protocol decided to include CCS in the CDM (UNFCCC 2010). It asked the Subsidiary Body for Scientific and Technological Advice (SBSTA) to draft the modalities and procedures for including CCS projects in the CDM. The modalities for implementing CCS activities within the framework of the CDM, as decided by

[3] See the forecasts in the 'World Energy Outlook' published annually by the International Energy Agency (IEA).

[4] For more on the risks associated with the management of CCS projects in CDM host countries, see Pollak and Wilson (2009).

[5] Certified emission reductions (CERs) allocated as a result of CDM projects are not limited by a quantified emissions budget. Instead, the total number of available CERs increases in the system as a whole. The flooding of the CDM market with CERs would endanger the integrity of the instrument in terms of the objective of a global emissions reduction, because the CDM aims to shift the reduction of emissions and not to reduce them (Bode 2009).

[6] The 6th Conference of the Parties serving as the meeting of the Parties to the Kyoto Protocol (CMP).

the Conference of the Parties at its seventh session, stipulate that the following CCS-specific requirements be fulfilled:

- a risk and safety assessment prior to project approval,
- a clear indication of liability obligations for each project phase,
- proof of sufficient financial resources,
- a detailed description of planned monitoring measures,
- the transfer of 5 % of the CERs issued to a reserve account (for surrender in the case of CO_2 leaks[7]), and
- a prohibition on cross-border storage sites.

The requirements adopted reflect the main points of the debate outlined earlier in that they stipulate a series of CCS-specific regulations for the implementation of CCS CDM projects, and assign a key role to the CDM Executive Board as one of the central control mechanisms. The modalities are subject to regular amendment in line with experience gained in the projects.

14.2.2 International Cooperation Supporting the Demonstration of CCS Technologies

The summary of the debate surrounding the inclusion of CCS in the CDM shows that the novelty of CCS is a key factor of uncertainty. Although CCS is not a new technology as such – carbon capture has been used in the USA since the 1970s as a method of enhanced oil recovery (US DOE 2009) – there are still no commercial-scale integrated plants that apply CCS for coal- or gas-fired power generation (cf. Table 14.1). At the international level, intergovernmental forums such as G8, and the international organisations IEA, CSLF, and GCCSI have introduced measures supporting the demonstration of CCS.

The political declaration adopted at the G8 summit in Gleneagles recognises the anthropogenic causes of climate change, and emphasises the importance of measures in the energy sector among strategies for reducing GHG emissions (G8 Gleneagles 2005). The action plan decided upon at Gleneagles comprises suggestions aiming to accelerate the development and implementation of CCS (cf. Table 14.2). In this context, the G8 announced their support for the international organisations IEA and CSLF, and requested that they identify short-term and long-term incentives for the implementation of CCS.

With this request, the G8 expanded the competencies of the International Energy Agency (IEA), whose initial focus after it was founded in 1974 was on managing the strategic oil reserves of OECD members (Lesage et al. 2009).[8] The IEA has

[7] Should the number of CERs not suffice to cover fugitive CO_2 emissions, other mechanisms and rules will apply.

[8] For an analysis of the driving factors behind the institutional transformation of the International Energy Agency, see Van de Graaf and Lesage (2009).

Table 14.1 Active CCS projects worldwide

Project name	Location	Project phase	Capture	Transportation	Storage
Enhance Energy EOR Project	Canada, Alberta	Under construction	Pre-combustion	Pipeline	EOR
Great Plains Synfuel Plant and Weyburn-Midale Project	Canada, Saskatchewan	In operation	Pre-combustion	Pipeline	EOR
Boundary Dam Integrated Carbon Capture and Sequestration Demonstration Project	Canada, Saskatchewan	Under construction	Post-combustion	Pipeline	EOR
Shute Creek Natural Gas Processing	USA, Wyoming	In operation	Pre-combustion	Pipeline	EOR
Lost Cabin Gas Plant	USA, Wyoming	Under construction	Pre-combustion	Pipeline	EOR
Occidental Gas Processing Plant	USA, Texas	In operation	Pre-combustion	Pipeline	EOR
Enid Fertilizer Plant	USA, Oklahoma	In operation	Pre-combustion	Pipeline	EOR
Val Verde Natural Gas Plants	USA, Texas	In operation	Pre-combustion	Pipeline	EOR
Kemper County IGCC Project	USA, Mississippi	Under construction	IGCC	Pipeline	EOR
ADM Illinois Industrial Carbon Capture and Sequestration Project	USA, Illinois	Under construction	Carbon capture in industry	Pipeline	Deep saline formations
In Salah CO_2 Injection	Algeria, Ouargla Wilaya	In operation	Pre-combustion	Pipeline	Deep saline formations
Sleipner CO_2 Injection	Norway, North Sea	In operation	Pre-combustion	On-site	Deep saline formations
Snøhvit CO_2 Injection	Norway, Barents Sea	In operation	Pre-combustion	Pipeline	Deep saline formations
Gorgon Carbon Dioxide Injection Project	Australia, Western Australia	Under construction	Pre-combustion	Pipeline	Deep saline formations

Overview: GCCSI (2011)

Table 14.2 Work packages of international organisations in response to calls from the G8 in the area of CCS

G8 summit	Key points of G8 resolutions on CCS	IEA CSLF work packages in response to G8 requirements
2005 Gleneagles, United Kingdom	Accelerated development and marketing of CCS by supporting international cooperation (IEA, CSLF) in the:	
	analysis of obstacles to the acceptance of CCS technology among the general public;	
	evaluation of short-term options for CCS in the area of fossil fuels;	
	examination of the definition, costs, areas of application, and incentives for plants with direct carbon capture ('capture-ready' plants);	
	exploration of geological storage options in developing countries (G8 Gleneagles 2005).	
2006 St Petersburg, Russia	Support for CSLF activities in relation to the preparation and implementation of CCS demonstration projects (cf G8 Saint-Petersburg 2006).	
2007 Heiligendamm, Germany	Objective of accelerating CCS development by:	
	supporting national R&D activities and international collaboration between industrialised countries and developing countries, as well as between IEA and CSLF;	
	supporting national activities concerned with the safety of CO_2 storage sites, and in developing the legal framework for CO_2 storage;	
	developing mechanisms for industrial-scale CCS;	
	encouraging industry to consider the 'capture-ready' concept for new-build fossil-fired power plants (G8 Heiligendamm 2007).	
2008 Hokkaido, Japan	Support for IEA in developing roadmaps for CCS and other innovative energy technologies.	Organisation of three workshops on the short-term options for CCS by IEA and CSLF:
	Support for the initiation of 20 large-scale CCS demonstration projects worldwide by 2010 in order to enable the commercial implementation of CCS by 2020 (G8 Hokkaido 2008).	2006 in San Francisco (IEA/CSLF 2006); 2007 in Oslo (IEA/CSLF 2007a); 2007 in Calgary (IEA/CSLF 2007b). The results of the workshops were incorporated into the IEA report on the G8 summit in Hokkaido (2008) (IEA 2008).

(continued)

Table 14.2 (continued)

G8 summit	Key points of G8 resolutions on CCS	IEA CSLF work packages in response to G8 requirements
2009 L'Aquila, Italy	Confirmation of the objective of initiating 20 large-scale CCS demonstration projects worldwide by 2010 by:	
	accelerating the development of political and regulatory framework conditions;	
	identifying sources of funding for CCS demonstration projects;	
	supporting the work of IEA, CSLF, and GCCSI;	
	identifying potential partnerships with multilateral financial institutions (G8 L'Aquila 2009).	
2010 Muskoka, Canada	Confirmation of the objective to initiate 20 large-scale CCS demonstration projects worldwide by 2010.	A joint IEA-CSLF report on progress on the way towards the G8 objectives of Hokkaido (IEA/CSLF 2010).
	Some G8 members pledge to speed up the development of CCS projects in order to put them into operation by 2015 (G8 Muskoka 2010).	

Source: Schenk (2013)

since developed into an important advocate of CCS. The World Energy Outlook, published annually, emphasises the importance of coal in the energy mix of OECD members and particularly of non-members. With respect to the impacts of coal combustion on the climate, the IEA warns of the need for corrective action in the form of climate policy and energy policy measures. In this context, the IEA has an important function as a communicator to national governments. The IEA derives policy recommendations on the basis of statistical and technical information, and disseminates these in IEA member countries as well as in non-member countries, such as China and Russia – via its increasing outreach activities. The significance of CCS as a climate change mitigation measure within the IEA was made clear by the creation of a special Carbon Capture and Storage Technology Unit with analytical and advisory functions. While CCS is only one area of the many IEA activities, the IEA works with international organisations who are exclusively dedicated to supporting the demonstration and implementation of CCS technologies.

The Carbon Sequestration Leadership Forum (CSLF) is an international cooperation that was established by the US Department of Energy in 2003 with the aim of supporting CCS technologies. Reducing emissions by deploying innovative GHG reduction technologies provides an alternative to the obligation to reduce emissions as formulated in the Kyoto Protocol. The international cooperation, which is aimed at technology development and deployment as a strategy for cutting

GHG emissions, received considerable support in the USA,[9] Australia,[10] and Canada.[11]

CSLF and similar international collaborations are based on voluntary participation and the voluntary use of resources. Collaboration within CSLF involves regular meetings at the ministerial level and the level of government representatives. As CSLF does not have its own budget, the respective activities are financed by the members themselves. The hallmark of CSLF is the 'endorsement' of selected CCS projects with CSLF recognition. This is intended to enhance the prestige of the projects by internationally recognising them, and thus help them to acquire better funding and facilitate cooperation. To date, 34 international projects have been endorsed.[12] CSLF-recognised projects were initially research and pilot projects dealing with individual elements in the CCS process chain. Recently, some European CCS demonstration projects[13] have also been endorsed by CSLF. The interaction between the projects and knowledge sharing in the area of carbon storage were both intensified with the first CSLF Storage and Monitoring Projects Interactive Workshop, which took place in Saudi Arabia in 2010. Other results of cooperation within the framework of CSLF include studies, reports, discussion papers, and workshops on CCS-relevant topics.

Following the IEA and CSLF recommendations, the G8 heads of state and heads of government in the Hokkaido declaration (G8 Hokkaido 2008) called for the introduction of 20 industrial-scale CCS demonstration projects worldwide by 2010 in order to enable the commercial maturity of CCS by 2020 (cf. Table 14.2).

In the same year, the Australian government announced a decision to set up an international organisation to provide financial incentives for CCS demonstration projects worldwide. Annual funding of some A$ 100 million was pledged to the Global Carbon Capture and Storage Institute (GCCSI) for 4 years. In October 2010, GCCSI announced support for six international projects to the tune of A$ 18 million. The funding was primarily for feasibility studies, engineering and design. After the resignation of Australia's Prime Minister and GCCSI founder Kevin Rudd in June 2010, speculation emerged about cuts in GCCSI funding. In January 2011, the new Australian government announced cuts to the GCCSI budget in connection with recovery measures after the flood catastrophe in Queensland.

The status report on the G8 summit in Muskoka, Canada, highlighted the progress made internationally in developing CCS (IEA/CSLF 2010). At the same time, the momentum of international support for the market introduction of CCS,

[9] The USA signed the Kyoto Protocol in 1998 but has yet to ratify it.

[10] Australia signed the Kyoto Protocol in 1998 but only ratified it in 2007. Australia's obligation comprises an increase of 8 % in GHG emissions compared to the reference year 1990.

[11] Canada signed the Kyoto Protocol in 1998, ratified it in 2002, and withdrew from the Kyoto Protocol in 2011. Canada's obligation to reduce emissions was 6 % compared to the reference year 1990.

[12] As of May 2012.

[13] Cf. Schenk and Hake, Chap. 13.

which had found its expression in the Gleneagles action plan, was lost. The fate of CCS will be decided in each of the nation-states. In the literature, a similar development was observed for the G8 area of energy as a whole. The attempt to coordinate the energy policies of the G8 members in the Gleneagles action plan faded gradually into the background because of divergences in some of the countries (Lesage et al. 2009, 2010).

The literature describing the essential prerequisites for successful commercial introduction of CCS emphasises the importance of international cooperation for the implementation of demonstration projects, as well as for technology development (Gibbins and Chalmers 2008; de Coninck et al. 2009; Hagemann et al. 2011). The recognised advantages of international cooperation on new, capital-intensive technologies like CCS[14] include knowledge sharing and the sharing of resources, support for economic, political, and technological goals, as well as support for the competitiveness of national industries (Justus and Philibert 2005).

A feature of international cooperation in the area of CCS is its institutional fragmentation. The studies that analyse international cooperation on initiating and implementing demonstration projects emphasise the diversity of active organisations and the loose coordination of activities (de Coninck et al. 2009; de Coninck and Bäckstrand 2011). A concerted approach in terms of an international portfolio of CCS demonstration projects is considered essential by some authors in order to maximise learning effects, consolidate social awareness of CCS, and to facilitate cost-efficient and safe CO_2 storage (de Coninck et al. 2009; Hagemann et al. 2011).

Another feature that characterises the demonstration of CCS technologies worldwide is the discontinuation of several projects considered important for the significance of CCS. The multiple restructuring of the American FutureGen project, the discontinuation of project plans for the RWE IGCC project in Hürth and Vattenfall's EU-funded demonstration project at Jänschwalde in Germany, as well as delays in the construction of the ENEL Porto Tolle project in Italy, are only a few examples of the current status[15] of CCS demonstration projects.

Back in 2005, the IPCC Special Report drew attention to the adequate technical maturity of the individual components of the CCS process chains, which would allow CCS to be demonstrated in an integrated project. GCCSI lists a total of 74 large-scale integrated CCS projects worldwide.[16] Of these 74 projects, 14 are 'active'. Most of these projects use carbon capture for natural gas production, and inject the CO_2 into geological formations in order to improve oil recovery (enhanced oil recovery, EOR). Since last year, this list now includes two integrated large-scale projects that deploy CCS in electricity generation for the first time worldwide.

SaskPower's Boundary Dam Integrated Carbon Capture and Sequestration Demonstration Project involves retrofitting one block of the existing coal-fired power

[14] For details on the characteristics of CCS and analogous technologies, see (Rai et al. 2009).

[15] As of May 2012.

[16] As of May 2012.

plant in the Canadian province of Saskatchewan. The 100 MW block is due to go into operation in 2014. The project is supported by the Canadian government to the tune of C$ 240 million. In addition, the project benefits from revenues from the enhanced oil recovery made possible by using the captured carbon.

Mississippi Power's Kemper County IGCC Project aims to construct a 582 MW IGCC power plant that will be operated with lignite and natural gas from the Gulf of Mexico. Public funding for the project totals US$270 million in the form of subsidies from the US Department of Energy, tax relief worth US$412 million granted within the framework of the National Energy Policy Act of 2005, and loan guarantees from the US federal government (Mississippi Power 2011). The captured CO_2 will be sold for use in enhanced oil recovery. The power plant is due to go into operation in 2014.

The studies analysing the failure to create an international portfolio of demonstration projects in the electricity sector do not believe that this negative outcome is due to technical deficits but rather to economic challenges (Hawkins et al. 2009). These are the result of a reticence on the part of industry to invest in a new technology when the economic viability of its application in the electricity generation sector depends on the uncertain development of the CO_2 price. CCS application for enhanced oil and gas recovery generates revenues for companies, but because of the amounts injected, it falls short of the requirements that would make CCS a climate change mitigation option that would cut CO_2 emissions in the medium term. Support from national actors who could help CCS to advance from the R&D phase to the demonstration phase developed from initial euphoria after the publication of the IPCC Special Report on CCS to cautious restraint. This restraint can be partially explained by the lack of social consensus on the risks of CCS and the role of coal in the energy mix in the individual countries. It also reflects the much higher costs of CCS demonstration plants, which deviate considerably from initial expectations.

14.2.3 International Cooperation and Knowledge Sharing

In addition to using international cooperation as an instrument to accelerate CCS demonstration, international cooperation has also established itself as a platform for knowledge sharing. Between strategic ministerial meetings in the task forces, CSLF's regular work concentrates on the exchange of information between government representatives of OECD member and non-member countries. With its Implementing Agreements (IAs), IEA has also developed a form of cooperation that facilitates collaboration in the area of energy technologies between governments, industry representatives, and other stakeholders in OECD member and non-member countries (IEA 2010).

In the area of CCS technologies, the IEA Greenhouse Gas R&D Programme (IEAGHG) and the IEA Clean Coal Centre are the most important networks whose main aim is knowledge sharing. Neither the IEAGHG nor the Clean Coal Centre pursue R&D themselves. Instead, they provide coordinated information in the form of reports and expert networks, offer assistance for R&D and demonstration projects, and organise specialist conferences and workshops. The Clean Coal Centre is concerned mainly with topics related to the area of coal-fired power generation. The IEAGHG, in contrast, concentrates on issues concerning the entire CCS process chain.

The studies published by the Clean Coal Centre and IEAGHG are different to the reports published by international organisations such as the IEA itself in that they detail the technological aspects of CCS technology but do not contain any policy-related information.

Voluntary international collaboration for technology development is regarded in the literature as an alternative to international cooperation based on international obligations. Research has concentrated on classifying the wide range of collaborations and on evaluating their effectiveness for technology development and GHG emissions reductions (de Coninck et al. 2008). De Coninck et al. differentiate between four types of technology-oriented collaborations on the basis of the aim of the cooperation. According to them, international cooperation focuses on:

- knowledge sharing and coordination,
- RD&D,
- technology transfer,
- technology mandates and incentives (de Coninck et al. 2008).

According to this classification, CSLF collaborations in the area of CCS are assigned to the knowledge sharing and coordination category, and the IEA Implementation Agreements to the RD&D category. Due to the limited budget, CSLF is considered to indirectly affect the development of CCS technology – by influencing national R&D funding. In another paper, de Coninck refers to CSLF's contribution to the development of CCS technology as 'disappointing' (de Coninck 2008a, p. 2). The member structure of CSLF, which comprises policy makers with a limited mandate, and the weak leadership of the USA, which is mainly due to domestic difficulties associated with the use of CCS in electricity generation, led to de Coninck concluding that CSLF's contribution to coordinating the CCS demonstration projects worldwide is insufficient. The impact of CSLF activities on GHG emissions reductions is considered difficult to measure but it is assumed to be small (de Coninck et al. 2008; Bäckstrand 2008).

R&D cooperation is considered to contribute indirectly to technology development, as organisations such as the IEAGHG and Clean Coal Centre do not conduct R&D themselves but rather pool and disseminate existing knowledge. Reducing GHG emissions is not a primary objective of R&D cooperation. In addition, the impact of collaborations on the development of environmental technologies, without implementing incentives for these technologies, is unclear.

Cooperation in the area of technology transfer, such as the multilateral funds for the implementation of the Montreal Protocol and the Global Environment Facility, is regarded as being environmentally effective, provided that the collaborations have access to sufficient financial resources and the issue of intellectual property rights has been clarified. Collaborations aiming to develop incentives by introducing quotas and technology standards could also be environmentally effective according to the authors if the international community were to enforce mandatory regulations.

Both categories (collaborations for technology transfer and for the development of technology mandates and incentives) – which the authors consider to have the biggest impact on technology development – pose challenges for international CCS cooperation. International cooperation is characterised by institutional fragmentation in the areas of technology transfer and capacity building. In addition to multilateral initiatives, such as the provision of funds by the GCCSI to CSLF (A$ 1.2 million) and to the World Bank's CCS Trust Fund (A$ 2.4 million) for the capacity building programme, several bilateral initiatives are also active (Hagemann et al. 2011). This development is symptomatic of international cooperation in the area of CCS as a whole, which is characterised by institutional fragmentation and little sharing of responsibilities.

Furthermore, the implementation of international technology standards has not proven successful in the area of CCS. The capture-ready standard developed as a result of international cooperation within the IEAGHG Implementation Agreement has been implemented within the framework of individual CCS regulations, e.g. in the EU, but not within the framework of larger international cooperation as the standard for new-build fossil-fired power plants.

14.3 Germany's Role in International Collaboration

Germany's involvement in international cooperation in the area of CCS is characterised by little political engagement. The countries that play a leading role in supporting CCS include the USA, Australia, and the UK. They have established international organisations and have provided impetus in existing political initiatives like the G8 processes. In addition, these countries have made efforts to consolidate their international leadership with an ambitious, state-subsidised national programme for the demonstration of CCS. Internationally, Germany has concentrated less on political cooperation and more on the exchange of ideas in the area of technology development. It is expected that political engagement at the international level will decline further because of difficulties establishing CCS as a climate option in Germany.

Within the framework of the G8 initiative, the German presidency of the G8 in 2007 was dominated by global economic issues and the form of the G8 partnership with Africa. In the period after the Gleneagles G8 summit in 2005 until the summit in Hokkaido in 2008, no new initiatives were introduced in support of CCS by the G8 presidents (cf. Table 14.2). Germany is due to hold the presidency of the G8 summit again in 2015. As no significant progress is expected in CCS demonstration in Germany by 2015, CCS will likely be a marginal topic during the German G8 presidency.

In 2003, Germany joined the CSLF and has been represented by the Federal Ministry of Economics and Technology (BMWi) since then. In 2005, CSLF members were invited by BMWi to Germany for their bi-annual meeting, and were welcomed by then Federal Minister for Economics and Technology Wolfgang Clement. Of the 34 CSLF projects, one project is currently located in Germany. This is the carbon storage project CO2SINK, funded by the EU within its Sixth Framework Programme. Another project that was proposed for recognition by CSLF was the Vattenfall CCS demonstration facility in Jänschwalde, which was funded within the European Economic Recovery Plan. As Vattenfall ceased planning for the demonstration plant at Jänschwalde, the CSLF recognition process was not completed for the project.

As a member of the international networks IEA Clean Coal Centre and IEAGHG, BMWi set up the COORETEC initiative in 2003 in Germany with the aim of following international R&D developments and presenting the results of German R&D programmes. As this collaboration is rooted in exchanging ideas in the area of technology development, and provides important insights into the status of international technology development, it will also remain important within the scope of the sixth Energy Research Programme after COORETEC has been restructured (BMWi 2011).

14.4 Summary and Outlook

To summarise, it can be said that attitudes towards CCS as a climate change mitigation option have undergone rapid development over the last 10 years from the initial euphoria to cautious restraint. The development of international cooperation reflects these changes. The recognition of CCS as a potential method for combating climate change associated with the publication of the IPCC Special Report was accompanied by the establishment of several international organisations and new priorities in existing collaborations. The G8, IEA and CSLF aim to realise the early commercialisation and demonstration of CCS. In this context, the IEA Implementing Agreements also ensure high visibility by offering platforms for a more intensive exchange of information between representatives from governments and industry. The period between 2008 and 2010 is characterised by the termination of several CCS projects that were important for the international portfolio, and meant that the G8 target of initiating 20 integrated demonstration projects worldwide by 2010, particularly in the electricity generation sector, was not achieved. Despite approaches, such as the recognition of CCS demonstration projects by the CSLF, no international cooperation succeeded in furthering the demonstration of CCS. A lack of economic resources was considered the reason for setbacks in creating an international portfolio of projects. The GCCSI, which was set up in 2009 and provided with a budget by the Australian government, represents an attempt to close the gap in funding for demonstration projects. It is too early to tell whether the GCCSI activities will be successful in their overarching aim of accelerating the commercialisation of CCS via demonstration programmes. However, in considering the incentives aiming to support the different phases of technology development – from R&D and demonstration right up to commercialisation – it becomes apparent that one decisive prerequisite must be fulfilled so that CCS can have the desired effect as a method of combating climate change throughout the world. This involves setting a CO_2 price in the top carbon emitter countries that is high enough to make the implementation of CCS in the area of electricity generation attractive. Successful international cooperation plays a crucial role in establishing this framework condition, which ultimately amounts to an international agreement with binding emissions reduction targets. If this condition is implemented, different approaches and forms of international cooperation would foster a cost-efficient global demonstration programme, an optimal exchange of information, technology transfer, and capacity building measures. As long as the top carbon emitters do not implement binding CO_2 reduction targets, funding CCS as a climate change mitigation option does not make sense. After all, even a successful, state-funded demonstration of the technology will not lead to the global commercial implementation of CCS.

References

Bäckstrand K (2008) Accountability of networked climate governance: the rise of transnational climate partnerships. Glob Environ Polit 8(3):74–102

BMWI (2011) Forschung für eine umweltschonende, zuverlässige und bezahlbare Energieversorgung Das 6. Energieforschungsprogramm der Bundesregierung, Bundesministerium für Wirtschaft und Technologie, Berlin

Bode S (2009) CO_2-Abscheidung und -Lagerung im Rahmen des Clean Development Mechanism: Chancen und Risiken. Gaia-Ecol Perspect Sci Soc 4:300–306

de Coninck H (2008a) The international race for CO_2 capture and storage: and the winner is...? FACET [Online]. Available: http://www.aicgs.org/documents/facet/coninck.facet12.pdf. Accessed 14 July 2009

de Coninck H (2008b) Trojan horse or horn of plenty? Reflections on allowing CCS in the CDM. Energy Policy 36:929–936

de Coninck H, Bäckstrand K (2011) An international relations perspective on the global politics of carbon dioxide capture and storage. Glob Environ Chang 21:368–378

de Coninck H, Fischer C, Newell RG, Ueno T (2008) International technology-oriented agreements to address climate change. Energy Policy 36:335–356

de Coninck H, Stephens JC, Metz B (2009) Global learning on carbon capture and storage: a call for strong international cooperation on CCS demonstration. Energy Policy 37(6):2161–2165

G8 Gleneagles (2005) Aktionsplan von Gleneagles – Klimawandel, saubere Energie und nachhaltige Entwicklung. G8-Weltwirtschaftsgipfel 2005 in Gleneagles, BMWA [Online]. Available: http://www.bmwi.de/BMWi/Redaktion/PDF/Publikationen/g8-wwg-gleneagles-2005,property=pdf,bereich=bmwi,sprache=de,rwb=true.pdf

G8 Heiligendamm (2007) Growth and responsibility in the world economy. Summit declaration. [Online]. Available: http://portal3.sre.gob.mx/groupfive/images/Heiligendamm/CumbreG8Heiligendamm.pdf. Accessed 14 Dec 2010

G8 Hokkaido (2008) G8 Hokkaido Toyako summit leaders declaration [Online]. Available: http://www.mofa.go.jp/policy/economy/summit/2008/doc/doc080714__en.html. Accessed 16 Aug 2010

G8 L'AQUILA (2009) Responsible leadership for a sustainable future [Online]. Available: http://www.g8italia2009.it/static/G8_Allegato/G8_Declaration_08_07_09_final,0.pdf. Accessed 20 Dec 2010

G8 Muskoka (2010) G8 Muskoka declaration – Recovery and new beginnings [Online]. Available: http://www.bundesregierung.de/nsc_true/Content/DE/StatischeSeiten/Breg/G8G20/Anlagen/G8-erklaerung-muskoka-en,property=publicationFile.pdf/G8-erklaerung-muskoka-en. Accessed 20 Dec 2010

G8 Saint-Petersburg (2006) St Petersburg plan of action on global energy security [Online]. Available: http://en.g8russia.ru/docs/11.html. Accessed 14 Dec 2010

GCCSI (2011) The global status of CCS: 2011. Global CCS Institute, Canberra

Gibbins J, Chalmers H (2008) Preparing for global rollout: a 'developed country first' demonstration programme for rapid CCS deployment. Energy Policy 36:501–507

Hagemann M, Moltmann S, Palenberg A, de Visser E, Höhne N, Jung M, Bakker S (2011) CATO-2 Deliverable WP 2.3-D03, Background paper on "Role of CCS in the international climate regime". CATO2-WP2.3-D03, Program Office CATO-2, Utrecht

Hawkins D, Peridas G, Steelman J (2009) Twelve years after Sleipner: moving CCS from hype to pipe. Energy Procedia 1(1):4403–4410

IEA (2008) Towards a sustainable energy future – IEA programme of work on climate change, clean energy and sustainable development. In support of the G8 plan of action, OECD/IEA

IEA/CSLF (2006) Near-term opportunities for carbon dioxide capture and storage. Issues identification workshop. In support of the G8 plan of action, Paris

IEA/CSLF (2007a) Near-term opportunities for carbon dioxide capture and storage. Global assessments workshop. In support of the G8 plan of action, Paris

IEA/CSLF (2007b) Results from the Calgary workshop, November 27 & 28 2007. 3rd workshop on near-term opportunities for carbon capture and storage. Recommendations on near-term opportunities for carbon dioxide capture and storage to the G8 summit Hokkaido, Paris

IEA/CSLF (2010) IEA/CSLF report to the Muskoka 2010 G8 summit – Carbon capture and storage – Progress and next steps. OECD/IEA, CSLF, prepared with the co-operation of the Global CCS Institute

IPCC (2005) IPCC special report on carbon dioxide capture and storage. Intergovernmental Panel on Climate Change, Cambridge University Press, New York

IPCC (2006) 2006 IPCC guidelines for national greenhouse gas inventories. In: Eggleston HS, Buendia L, Miwa K, Ngara T, Tanabe K (eds) Prepared by the National Greenhouse Gas Inventories Programme. IGES, Hayama, Japan

Justus D, Philibert C (2005) International energy technology collaboration and climate change mitigation. Synthesis report. OECD/IEA, Paris

Kerr T, Havercroft I, Dixon T (2009) Legal and regulatory developments associated with carbon dioxide capture and storage: a global update. Energy Procedia 1:4395–4402

Lesage D, van de Graaf T, Westphal K (2009) The G8's role in global energy governance since the 2005 Gleneagles Summit. Glob Gov 15:259–277

Lesage D, Graaf TVD, Westphal K (2010) Global energy governance in a multipolar world. Ashgate, Farnham/Burlington

Meadowcroft J, Langhelle O (2009) The politics and policy of carbon capture and storage. In: Meadowcroft J, Langhelle O (eds) Catching the carbon. The politics and policy of carbon capture and storage. Edward Elgar, Cheltenham

Mississippi Power (2011) Questions & Answers [Online]. Available: http://www.mississippipower.com/kemper/facts-and-faqs.asp. Accessed 30 May 2012

Pena N (2009) CCS in the United States 1998–2008: from resistance to support. CCT2009. IEA CCC, Dresden

Philibert C, Ellis J, Podkanski J (2007) Carbon capture and storage in the CDM. COM/ENV/EPOC/IEA/SLT(2007)10. OECD/IEA

Pollak M, Wilson EJ (2009) Risk governance for geological storage of CO(2) under the Clean Development Mechanism. Clim Policy 9:71–87

Rai V, Victor DG, Thurber MC (2009) Carbon capture and storage at scale: lessons from the growth of analogous energy technologies. PESD Stanford, Freeman Spogli Institute for International Studies, Stanford

Russial TJ (2011) Carbon capture and storage – legal and regulatory framework. CCC/179
Schenk O (2013) Interest mediation and policy formulation in the European Union. Influence of transnational technology-oriented agreements on European policy in the field of carbon capture and storage, Energy & Environment 165, Advances in Systems Analysis 3. Jülich
UNFCCC (2010) Carbon dioxide capture and storage in geological formations as clean development mechanism project activities. In: FCCC/KP/CMP/2010/12/ADD.2 (ed)
UNITED NATIONS (1992) United Nations framework convention on climate change
US DOE (2009) Enhanced Oil Recovery / Co_2 Injection [Online]. Washington D.C.: U.S. Department of Energy. Available: http://fossil.energy.gov/programs/oilgas/eor/. Accessed 2 Mar 2009
United Nations (1997) Kyoto Protocol to the United Nations Framework Convention on Climate Change, 11 December, Kyoto
van de Graaf T, Lesage D (2009) The International Energy Agency after 35 years: reform needs and institutional adaptability. Rev Int Organ 4:293–317
Watanabe R, Duckat R, Sterk W (2007) Carbon capture and storage under the Clean Development Mechanism. Impact on the long-term climate goal, energy supply planning, and development paths. JIKO Policy Paper 4/2007. Wuppertal Institute for Climate, Environment, and Energy, Wuppertal

Part IV
Conclusion

Chapter 15
Evaluation Index of Carbon Capture and Utilization: A German Perspective and Beyond

Wilhelm Kuckshinrichs and Peter Markewitz

Abstract While there is no lack of technical options for CCS and storage capacities are available, the question arises as to whether and under what conditions CCS could become a key element within the framework of implementing climate protection strategies. To answer this question, an integrated technology evaluation is required covering technical, economic, environmental, and social considerations. In order to play a decisive role in climate protection strategies, five key challenges are identified, that must be overcome: (1) 'demonstration of an industrial scale and commercial availability', (2) 'environmental and safety requirements', (3) 'cost efficiency and economic viability', (4) 'coordination of energy and climate policy', and (5) 'public acceptance'. Given the different quantitative and qualitative analyzes of the five key challenges and the results of subsequent expert interviews, the OECD approach of a composite index for evaluation is used. The experts surveyed criterion (2) 'environmental and safety requirements' as being most fulfilled (rating of 3.00), although 3.00 does not even come close to achieving the maximum (5.00), and criterion (5) 'public acceptance' (rating of 1.32) as being least fulfilled. Criterion (1) 'demonstration of an industrial scale and commercial availability' was also evaluated relatively positively (rating of 2.84), while (3) 'cost efficiency and economic viability' fared poorly with a rating of 1.86.

Keywords Integrated technology assessment • Evaluation criteria • Demonstration • Commercial availability • Environmental impacts • Safety requirements • Cost efficiency • Economic viability • Energy and climate policy • Public acceptance • Composite index

W. Kuckshinrichs (✉) • P. Markewitz
Institute of Energy and Climate Research – Systems Analysis and Technology Evaluation (IEK-STE), Forschungszentrum Jülich GmbH, D-52425 Jülich, Germany
e-mail: w.kuckshinrichs@fz-juelich.de; p.markewitz@fz-juelich.de

W. Kuckshinrichs, J.-F. Hake (eds.), *Carbon Capture, Storage and Use*,
DOI 10.1007/978-3-319-11943-4_15

15.1 Introduction and Motivation

In order to limit the temperature rise expected because of the greenhouse effect, carbon dioxide emissions must be drastically reduced. Numerous analyses of and projections for the global energy system emphasize the importance of CCS in strategies aiming to reduce greenhouse gases (e.g. the Stern Report, Energy Technology Perspectives, and the World Energy Outlook (IEA 2009, 2010, 2011, 2012a, b; Stern 2006). In this context, capturing carbon dioxide is an important mitigation measure for CO_2 point sources in the energy conversion sector and in industry, and it is the focus of numerous research and development projects throughout the world.

At present, three technology lines are favoured for carbon capture: post-combustion, oxyfuel, and pre-combustion. Although the post-combustion and oxyfuel processes are being tested in smaller test facilities, practical demonstration is still required before first-generation technologies can be implemented on an economic and industrial scale. In the long term, interesting options could replace the currently favoured physical and chemical scrubbing using membranes, as well as carbonate looping, which count as second-generation technologies. For the oxyfuel process, the cryogenic air separation process could be improved (three-column process) and a transition to other oxygen production processes (membranes, chemical looping) is also possible.

For the storage of CO_2, a range of options are being discussed both at the national and European level. These include unused deep underground rock formations containing highly saline fluids (on-shore and under the seabed), depleted natural gas and crude oil fields (enhanced gas and oil recovery, EGR/EOR), and coal seams (enhanced coal-bed methane, ECBM). In national and international research projects, potential storage capacities are being analysed and concepts developed for the long-term and safe trapping of CO_2. With respect to the acceptance of CO_2 storage, strong reservations abound in Germany, as illustrated by the planned on-shore storage facility in Schleswig-Holstein and by the discussion in Lower Saxony. At the moment, neither the general public nor politicians in the north and north-west of Germany appear to be willing to accept potential CO_2 storage sites.

While there is no lack of technical options for CCS and storage capacities are available, the question arises as to whether and under what conditions CCS could become a key element within the framework of implementing climate protection strategies. To answer this question, an integrated technology evaluation is required that goes beyond a purely technical evaluation. This chapter therefore looks at possible implications that the technology evaluation of carbon capture and utilization could have for energy, climate, and industrial policy. First, the preceding chapters will be used as a basis to outline and classify the most important conclusions regarding the main challenges facing technology development and implementation (technical, economic, environmental, and social aspects). In a second step, an overall assessment will be made using a methodology based on the process proposed by the OECD for developing indices to evaluate technology. The method accounts for different weightings of the individual aspects. To conclude, possible implications will be analysed for the situation of CCS in Europe.

15.2 Key Conclusions of the Integrated Technology Evaluation

Before CCS can play a decisive role in the process of implementing strategies to mitigate climate change, there are a number of key challenges that must be overcome. Simply fulfilling these requirements may not necessarily be enough to guarantee the success of CCS because of the possible development of competing technologies aiming to reduce CO_2 emissions (e.g. renewables, energy efficiency).

The challenges affect all areas of an integrated technology evaluation from the technical, economic, and environmental aspects right up to the social aspects. They comprise:

- demonstration on an industrial scale and commercial availability
- environmental and safety requirements
- cost efficiency and economic viability
- coordination of energy and climate policy
- public acceptance

The main arguments concerning these points in the preceding chapters are summarized here and presented in a graph in Fig. 15.1.

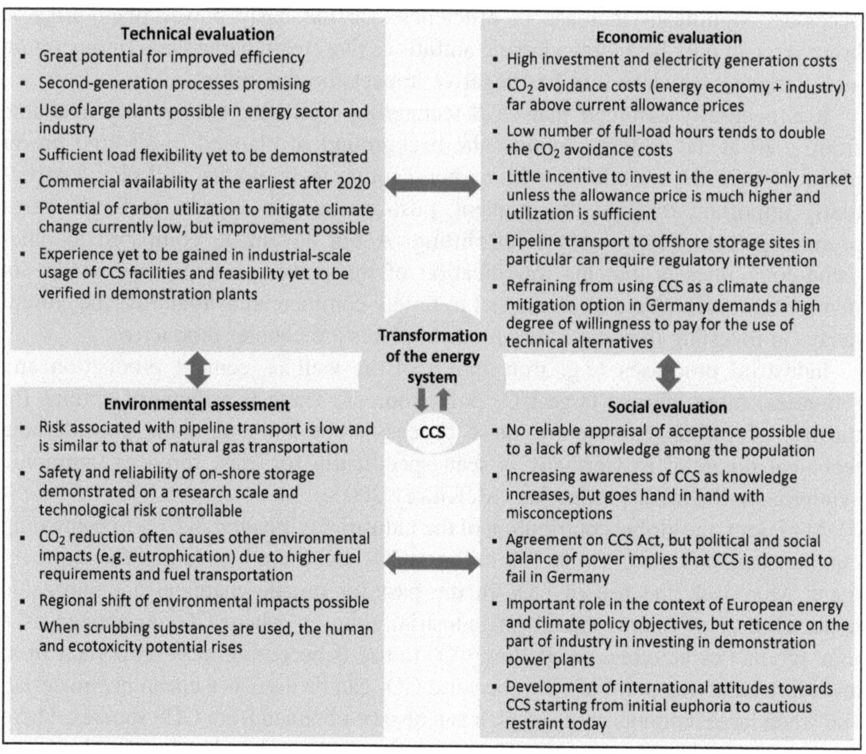

Fig. 15.1 Key conclusions of the integrated evaluation of CCS

15.2.1 *Challenges for Technology and Actors*

15.2.1.1 Demonstration on an Industrial Scale and Commercial Availability

According to *Markewitz and Bongartz* (Chap. 2), all three technology lines have great potential to improve efficiency depending on the processes involved, although the energy penalties remain considerable. In all cases, the thermodynamic integration of the carbon capture process is particularly challenging. Interesting options exist in the long term for replacing the currently favoured physical or chemical scrubbing. Alternatives here include the use of membranes as well as carbonate looping. For the oxyfuel process, the cryogenic air separation process could be improved (three-column process), and the transition to other oxygen production processes (use of membranes, chemical looping) is also possible.

Increasing the flexibility of coal-fired power plants with and without carbon capture is one of the main challenges from a technical perspective, because an increasingly volatile feed-in of electricity into the grid will place much greater demands on the flexibility and mode of operation of power plants (e.g. higher load ramps, greater load ranges, more start-up and shut-down cycles). How well CCS power plants will be able to meet these demands is a question that cannot be answered at the moment. From a technical point of view, a basic power plant process with the highest possible efficiency is generally considered essential. The necessary significant increase in efficiencies in the basic power plant process, however, can only be achieved using ambitious live steam parameters (temperature and pressure), which in turn has negative impacts on flexibility.

It is generally assumed that CCS technology will be commercially available from 2020 at the earliest. Against the background of planned fossil-fired power plants worldwide, retrofitting with carbon capture technologies will play a particularly important role. At the moment, post-combustion appears to be the most promising technology line for retrofitting. A big advantage compared to other technology lines is that the modification of the power plant process would not involve too much effort. With respect to timely commercial availability, the current delays in investing in demonstration power plants are counter-productive.

Industrial processes (e.g. iron and steel, as well as cement production and refineries) often involve large CO_2 point sources. There is a range of options for the use of carbon capture for these processes. In the long term, considerable technical potential in Germany is seen specifically for blast furnaces, ammonia synthesis, and clinker production (McKinsey 2007).

At present, the global contribution of the industrial utilization of CO_2 to combating climate change is quite low at 130 million tCO_2, but there is potential for improvement. Moreover, the use of CO_2 in the past for organo-chemical and inorganic applications was mainly based on industrial sources, where CO_2 is created as a joint product or an emission. Putting CO_2 to use is becoming more important from an industrial policy point of view, because CO_2 can be used as a cheap raw material, and when large amounts are needed, it can also be obtained from CCS sources. There are many possible ways of using CO_2, which should be analysed in detail with respect to their climatic relevance and their value-added potential. As global carbon emissions are increasing and will continue to do so in future, it can be assumed that the utilization of CO_2 will not replace carbon storage but will supplement it.

The relevance of utilizing CO_2 motivated by industrial policy for climate change mitigation not only depends on the amount itself, but also on the duration of CO_2 fixation (*Müller et al.*, Chap. 4). The fixation potential varies widely depending on the use of CO_2 and is calculated based on the combination of small to large quantities and short to long durations of fixation. At the same time, attention should be paid to whether the activation or utilization of CO_2 requires the use of other resources or energy that would interfere with the balance of CO_2. In addition, there is a need to clarify whether the use of CO_2 from CCS sources substitutes another source that would not require geological storage. The best method for analysing the entire energy and CO_2 balance is the life cycle assessment – an established approach for evaluating the environmental impacts of processes and products. However, in practice, conclusions can only be drawn separately for each use of CO_2.

15.2.1.2 Environmental and Safety Requirements

Carbon capture technologies often lead to amplification of other environmental effects (*Schreiber et al.*, Chap. 5). The rise in other environmental effects is usually triggered by the decline in net efficiency, and the related additional requirements for fuels and chemicals (e.g. scrubbing substances), as well as increased volumes of waste. A detailed analysis of the reasons shows that optimizing the reduction of power plant emissions is in itself not enough to prevent this increase. In particular, the provision of fuel often involves a high proportion of different environmental impacts. If scrubbing substances are additionally used, the human toxicity and ecotoxicity potential rises mainly because of emissions during production. Heavy metal emissions during the dumping of hazardous waste and ash also contribute to increased toxicity. A comparison of the studies shows that the processes of the upstream and downstream chains are often not represented in the same detail as the electricity generation and subsequent carbon capture processes. These processes should therefore be investigated in more detail.

A consideration of the entire life cycle also shows that there may be local or regional environmental effects upstream. While acidification and eutrophication are reduced at power plant sites, they increase in regions where the fuel is extracted and along transportation paths.

Furthermore, a comparison with the overall effects of a region helps to relate different impacts to each other. The desired effect of reducing greenhouse gas emissions is obvious. However, more detailed consideration must be given to emissions promoting acidification and human toxicity, especially for post-combustion plants. The most important method of reducing the majority of environmental impacts is reducing efficiency losses. New technological developments, such as membranes, are promising. Nevertheless, further analyses with a detailed description of the system boundaries and the parameters are required in order to provide robust information on the respective environmental impacts of the different technologies.

Essential safety requirements concern transportation and storage activities. Pipelines are particularly interesting for transporting large amounts of CO_2 over long distances. At present, CO_2 pipelines throughout the world have a total length of more than 4,000 km. The transportation of carbon dioxide is state of the art.

The release of large amounts of CO_2 can pose local risks to humans and the environment. As CO_2 is heavier than air under ambient conditions, it can collect in sinks for example, and at very high concentrations (7–10 vol.%) it can pose a life-threatening danger. Comparisons of natural gas and CO_2 pipelines show that the frequency of failures is similar. The purity of the CO_2 stream is particularly relevant for protection against corrosion. Experience with the standards in the USA can only be transferred to the European situation to a limited extent. With respect to impurities, the captured CO_2 stream in power plants is very different to the volumes of CO_2 currently transported in the USA.

Bongartz et al. (Chap. 3) summarized risk assessments using probabilistic approaches. Frequencies of occurrence were assumed for the different scenarios and used as a basis for determining the ranges of critical CO_2 concentrations. The available studies were used to qualitatively evaluate the categorized transportation risks (e.g. valve leakage, leak, rupture) in terms of frequency and range of critical CO_2 concentrations with the aid of a risk matrix (frequency classes, hazard classes). The findings show that the majority of risks associated with transportation are either insignificant or very small.

Reservoir rocks with the potential for geological storage are mainly sandstones, as they are characterized by sufficient porosities and permeabilities, allowing CO_2 to be injected efficiently into these formations. Overall, four retention mechanisms in the layers of the storage formation facilitate permanent and safe storage: (i) structural retention below an impermeable caprock, (ii) immobilization via capillary binding in pore space, (iii) dissolution of CO_2 in the formation water, and (iv) mineral binding via carbonization.

Near the town of Ketzin on the Havel in Brandenburg, the first continental European field laboratory for CO_2 storage was set up and put into operation as a pilot site in 2004. The pilot site in Ketzin is thus the first and to date the only active CO_2 storage project in Germany. The injection of CO_2 is accompanied by one of the most extensive scientific research and development programmes in the world. The findings on a research scale (*Kühn et al.*, Chap. 6) show that: (i) the geological storage of CO_2 at the pilot site in Ketzin is safe and reliable, and poses no danger to humans or the environment, (ii) a well-thought-out combination of different geochemical and geophysical monitoring methods can detect minute amounts of CO_2 and image its spatial distribution, (iii) the interactions between fluid and rock induced by CO_2 injection at the pilot site in Ketzin have no significant impacts and do not influence the integrity of the reservoir rock or the caprock, and (iv) numerical simulations can depict the temporal and spatial behaviour of injected CO_2.

15.2.1.3 Cost Efficiency and Economic Viability

Martinsen et al. (Chap. 10) use the energy system model IKARUS to estimate the value of CCS technologies in Germany within the framework of greenhouse gas reduction scenarios ('system value' in the following). This value is determined here by the additional avoidance costs that would be incurred if climate change mitigation targets were to be achieved without CCS technologies. It is therefore an implicit measure of the level of willingness of society to pay for refraining from the use of CCS technologies.

The actual present value of the costs avoided by deploying CCS technologies for the period 2005–2050 is approx. $€_{2010}$ 100 billion. The value is calculated by balancing across all sectors (end-use, conversion, primary energy incl. imports). In the end-use sectors (industry, households, traffic and transport, commerce, trade and services), relatively expensive savings measures can be avoided if CCS is implemented in the conversion sector. In the same way, the primary energy sector including imports also plays a role, where most of the additional costs associated with the import of biomass products (e.g. bioethanol) are avoided when CCS is implemented, but additional costs are incurred for fossil fuels, which predominate until 2035. Despite the costs caused by CCS technologies, the conversion sector also contributes to the system value because an additional increase in renewable energy capacity is avoided. Overall, this applies to all sectors but the extent is very different.

The construction of CCS facilities represents an investment with long-term and high capital tie-up. In addition, the projections of the plant costs for CCS power plants still involve uncertainties, despite the continuing development of demonstration facilities. Increased knowledge and ongoing technological development lead to the investment costs of the first commercial CCS plants being predicted as 70–90 % higher than those of conventional plants. The costs for the transportation and storage of CO_2 depend on the quantities to be transported, the transport distance, and the type and location of the geological storage facility, and they vary considerably. In all cases, the costs of capturing CO_2 dominate.

Even high plant utilization gives rise to much higher electricity generation costs, particularly for coal-based CCS plants (lignite: +80 %) (*Kuckshinrichs and Vögele*, Chap. 7). The CO_2 avoidance costs are € 34–38/tCO_2 (lignite), € 41–48/tCO_2 (hard coal), and approx. € 67/tCO_2 for natural gas plants. Only if the price of allowances rises to the same level will the use of CCS power plants during normal operation be cost-effective.

A very low number of full-load hours tends to cause the CO_2 avoidance costs to double. As a result, a relatively high CO_2 price would be necessary to justify operation with a low number of full-load hours.

CCS power plants must be refinanced through the electricity market. Furthermore, the use of CCS power plants can have an effect on the price of electricity on the wholesale market under certain conditions. The price on the electricity market is determined by the costs of the last power plant used, whereby the power plants are used in order of their marginal costs (merit order) and the costs for electricity imports must be considered.

In general, the question arises as to the degree to which potentially higher revenues due to merit order effects cover the additional investment costs for CCS power plants. Owing to the high uncertainties with respect to the additional investment costs, it can be assumed that CCS plants will only become interesting to investors when the allowance price is at least € 40/tCO_2. Development in the area of renewable energy must also be considered. The increased use of renewables will lead to a decrease in the average annual price on the electricity market as long as sufficient 'cheap' back-up capacities are available, i.e. power plants with low operating costs. In addition, merit-order effects arise where the use of CCS also dampens the price of electricity. Merit-order effects also tend to boost the level of domestic production. It should be noted that this could cause reciprocal effects. Price effects caused by the increased use of renewable energy will make it more difficult to refinance CCS power plants, and the electricity price effects of CCS power plants will reduce the revenues for renewable energy (which in turn impacts on the level of Renewable Energy Act (EEG) surcharges).

If renewable energy is further integrated into the electricity system, with the current market design ('energy only') there is a danger that the power plant capacities of an existing fleet will be potentially underused. In addition to the generation cost effect caused by a low number of full-load hours, the drop in residual demand would lead to a merit order effect. As a result, there would be a short-term cost recovery problem for fossil plants in the installed power plant fleet. Regardless of the possible concrete design of capacity markets, the comparatively high refinancing needs compared to conventional power plants will prevail if capacity revenues are incorporated.

The use of CCS as a CO_2 mitigation measure for industrial plants is technically feasible in principle, but neither demonstration nor commercial CCS plants are currently in operation on the industrial scale. As a result, estimates for plant costs continue to be associated with great uncertainties.

A cost analysis of a cement plant with a capacity of one million tonnes of cement per year shows an increase of 32 % in production costs when oxyfuel technology is used (*Fleer and Kuckshinrichs*, Chap. 8). In the case of carbon capture with post-combustion technology, production costs increase by about 100 %. Retrofitting an oil refinery with a capacity of ten million tonnes of crude oil per year with oxyfuel technology leads to an increase of roughly 15 % in processing costs. This results in CO_2 avoidance costs of about € 55/tCO_2 for the oxyfuel cement plant, and about € 62/tCO_2 for the oxyfuel refinery. Avoidance costs are much higher for the cement plant with post-combustion capture (€ 143/tCO_2).

15.2.1.4 Coordination of Energy and Climate Policy

The policy-making process anchoring CCS as a climate change mitigation option in the EU was executed at a remarkable speed (*Schenk and Hake*, Chap. 13). Faced with two options – the mandatory introduction of CCS or the development of a framework for the industrial implementation of CCS – the EU institutions decided in favour of the second option. Between 2005 and 2009, European institutions successfully established CCS as a cornerstone of the EU's integrated energy and climate policy, developed a legal framework for geological storage, incorporated CCS into the European emissions trading system, and elaborated funding instruments for CCS. This dynamic momentum makes the EU one of the pioneers internationally. Despite this, feedback with respect to the implementation of CCS policy is less positive. The implementation of the CCS Directive reveals the lack of consensus on whether CCS should be used as an option for combating climate change. Today, some European countries, such as Austria, are completely against carbon storage.

The emissions trading system (EU ETS) and the demonstration programme (EEPR, NER300) are instruments that support CCS. However, companies are hesitant to invest in demonstration projects at the moment. The role of NER300 financing as one of the main instruments supporting CCS demonstration projects in Europe is being increasingly questioned in this context. The instrument attracted criticism from the very beginning because of the uncertainty regarding the price level of allowances. Pessimistic expectations were confirmed by initial experiences trading the allowances. The low price for emissions allowances also ignited discussions on the competitiveness of CCS technology after the demonstration phase. Long-term incentives are decisive for a stable development of low-carbon technologies. Within the scope of these instruments, an EU ETS cap, a carbon tax, and a bonus-malus system are being discussed as part of a carbon standard.

Attitudes towards CCS as a climate change mitigation option have undergone rapid development over the last 10 years from initial euphoria to cautious restraint (*Hake and Schenk*, Chap. 14). The development of international cooperation reflects these changes. The recognition of CCS as a potential method for combating climate change with the publication of the IPCC Special Report was accompanied by the establishment of several international organizations and new priorities in existing collaborations. The G8, IEA, and Carbon Sequestration Forum (CSLF) aim to facilitate the timely commercialization and demonstration of CCS.

However, the period between 2008 and 2010 is characterized by the discontinuation of several internationally important CCS projects. This means that the G8's aspiration of initiating 20 integrated demonstration projects worldwide by 2010 was not met. Despite approaches, such as the recognition of CCS demonstration projects by the CSLF or the creation of the Global Carbon Capture and Storage Institute (GCCSI) in Australia, no international cooperation has succeeded in furthering the demonstration of CCS. A decisive prerequisite has yet to be implemented – the introduction of a sufficiently high CO_2 price in the main emitter countries.

Germany's CCS Act, for example, shows that there is neither consensus on a national prohibition nor on the demonstration of the commercial application of CCS. The answer to the question of whether CCS is an option for Germany for reducing CO_2 emissions has been pushed into an uncertain future by the Act (*Fischer*, Chap. 12).

Compared to the first bill in 2009, the adopted CCS Act has shrunk to a research law with a theoretical potential for smaller demonstration projects which will probably not be exploited in Germany. If the potential of CCS should be demonstrated for large power plants or for industrial plants, the Act would have to be amended with respect to the storage amounts. In 2017, the CCS Act will be evaluated, and the discussion on CCS could become heated once again.

A minority of explicit political advocates of CCS hope that suitable energy economy and European framework conditions will emerge in future. The advocates include the state governments of Brandenburg, Saxony, and Saxony-Anhalt. In North Rhine-Westphalia, where in 2010 around 54 % of German lignite was mined, the last word has yet to be spoken on lignite policy and thus indirectly on the implementation of CCS (*Fischer*, Chap. 12).

A clearer picture will emerge over the next few years as to whether the targets for expanding the capacity of renewables will be achieved and how smooth the transformation of the energy system will be (Mikulcak 2012), what role coal in general and lignite-fired base-load power plants could play, and how emissions trading and its CO_2 prices (EU ETS) will develop. Should it emerge that the EU and its member states are not in a position to implement their extremely ambitious action plan for CO_2 mitigation politically or economically, e.g. because international climate change mitigation goes along with the less demanding willingness of states to act to mitigate CO_2 (bottom-up), the perspectives for CCS could become more clouded. The EU's CCS policy could also have an impact, e.g. possible additional CCS regulations, making CCS mandatory for new as well as for old power plants. CCS plays a key role in the EU plan for a low-carbon economy in 2050 (European Commission 2011b) and the more specific 2050 Energy Roadmap (European Commission 2011a), even though commercial application is not expected until after 2030 and CCS projects are not progressing well in the EU member states.

Another obstacle for CCS is the lack of acceptance for the solution of the 'back-end' CCS problem, namely storage. The northern federal states will pull out all the stops to block even potential storage projects. This problem might be less virulent if CO_2 were to be stored below the seabed (off-shore), particularly within the context of enhanced oil/gas recovery. Research work is being conducted on storage in deep ocean sediments, the safety of such sites, and on the consequences of leaks for the marine environment, which also includes regions off the German coast. It remains to be seen whether a 'loophole' (CO_2 Handel 2011) will emerge for federal storage projects below the seabed. That CCS opponents take this option seriously is reflected in the coalition agreement of the new state government in Schleswig-Holstein: it wants to 'preclude' CCS 'for the whole of Germany – particularly in the exclusive economic zone.' (SPD/Grüne/SSW 2012). For this reason, should a European CO_2 transportation infrastructure be created, CO_2 could be exported to onshore storage sites in other countries or injected into their deep ocean sediments. The statements issued in response to the legislative compromise indicate an interest in this option in sections of the political and industrial arenas. Assuming that there is interest in carbon capture in Germany and that the respective transportation infrastructure existed, then the acceptance of CO_2 pipelines through the federal states would also have to be ensured – considering the massive opposition to the planned Hürth pipeline in 2008, this represents a huge challenge for politics and society. It may be mastered if CCS were to be considered independently of lignite and if it were to become an integral part of a comprehensive strategy for a low-carbon society that would bring advantages with it for citizens, the economy, and the environment. Within the framework of the Rotterdam Climate Initiative,[1] an attempt is being made to implement this strategy. Germany can learn from this social experiment.

15.2.1.5 Public Acceptance

The acceptance of technologies cannot yet be reliably measured because the population still knows too little about CCS technologies (*Schumann*, Chap. 11). CCS acceptance research therefore focuses on investigating awareness and knowledge of CCS as well as spontaneous attitudes towards it among the general public. Such studies also concentrate on identifying factors that have an impact on spontaneous attitudes towards the technologies as well as on analysing the impact of information and methods of communication on changes in and the stability of spontaneous attitudes.

With respect to how well known CCS is among the general public, the findings of international and national studies confirm that at least awareness of the concept of ‘carbon capture and storage’ has increased considerably over the course of time. The increasing awareness of the concept of ‘carbon capture and storage’, however, is not accompanied by an increase in knowledge of the technologies. As the findings of international and national studies show, misconceptions about CCS (still) abound among the general public. This can be explained by the fact that lay people often find it difficult to distinguish between environmental problems, such as ozone depletion, global warming, acid rain or smog.

In addition, information on CCS and the communication of CCS should consider the fact that citizens have spontaneous attitudes towards the technologies even though they know little or nothing about CCS. In Germany, these spontaneous attitudes towards CCS are on average (still) mainly neutral, although women are more sceptical of the technologies than men.

The regional differences in spontaneous attitudes before and after the receipt of information demonstrate that citizens of Schleswig-Holstein do not only have more negative attitudes towards CCS than citizens of the region along the Rhine or of the ‘rest’ of Germany, but that the debate surrounding carbon storage in Schleswig-Holstein (cf. *Fischer*, Chap. 12) has already led to the emergence of negative attitudes towards CCS in this region which are not necessarily spontaneous attitudes any more but rather stable opinions. As the present findings also suggest, these negative attitudes in Schleswig-Holstein are mainly related to the fact that citizens here consider the personal risks associated with carbon storage to be much greater than citizens of the other regions.

However, the results also illustrate that spontaneous attitudes towards CCS in all regions are most heavily influenced by the perception of the social benefits of the

[1] http://www.rotterdamclimateinitiative.nl

technologies and that this influence is positive: the greater the social benefits of CCS, the more positive the spontaneous attitudes towards the technologies.

How stable these perceptions of the benefits of CCS are or how easily they can be changed by new information cannot be conclusively analysed using the present findings as a basis. The influence of information on the perception of the benefits of CCS, as well as the influence of the perception of benefits on the stability of attitudes towards CCS, must therefore be systematically investigated in future studies in order to assess the importance of the perception of benefits as an indicator for evaluating the future public acceptance of CCS in Germany.

At the moment, the lack of acceptance for the solution to the 'back-end' problem associated with CCS, namely storage, in the northern federal states is blocking all potential storage projects. Policies and legislation on CCS in Germany clearly reflect this negative stance, even if CCS is emphasized as a (necessary) option for energy-intensive and carbon-intensive industries.

15.2.2 The Big Picture: Where Do We Stand?

The preceding deliberations explain the challenges considered to be most important for technology evaluation. What big picture emerges? What is the success factor like as a whole? And how sensitive is the result with respect to the separate challenges?

The greatest success factor applies to the safety requirements for the transportation and storage of CO_2. The assessment of the environmental requirements from a life cycle perspective, however, is cause for concern in that the envisaged reduction in the global warming potential may lead to other environmental impacts, such as eutrophication, and thus induce regional shifts in environmental impacts. With a view to public acceptance, no conclusions can be drawn at this point because the public does not yet know enough about CCS technologies. Irrespective of this, the public still forms opinions on CCS technologies, which are characterized by the negative attitudes, e.g. in the northern federal states where potential storage sites are located. This negative attitude is also reflected in the national CCS Act, which does not permit commercial storage of large quantities of CO_2. Implementation on an economic and industrial scale has not yet been demonstrated in Europe despite EU financial incentive systems. As a result, commercial availability is not likely in the near future. Compared to the CO_2 avoidance costs of other large technical options, those of CCS technologies are average, but the electricity generation costs increase rapidly and incentives to invest in the technologies are not enough. The price of CO_2 is currently low and the number of full-load hours is dropping due to the increasing integration of renewable technologies, which means that refinancing the high investment costs is too uncertain in today's market design. With respect to political factors, the outcome of the overall evaluation is negative. The EU appears to be an institutional driving force for CCS technologies, promoting them in its energy, climate, and technology policy. However, this is not as successful as it may appear. Internationally, the euphoria surrounding CCS has dissipated, and CCS advocates in Germany (such as those in individual state governments) can only hope for improvements in future. Should climate change mitigation in Germany prove to be insufficient using the options currently preferred within the framework of the transformation of the energy sector, there may be a re-evaluation of CCS in Germany. Even though none of this can yet provide a decisive answer to the question of whether CCS has a future in Germany or not, the economic climate combined with the political and social balance of power imply that CCS is doomed to failure.

Table 15.1 'translates' the detailed explanations of the preceding section into a quantified technology evaluation.

Methodologically, an approach used by the OECD for a composite index and applied in its technology evaluations is taken here (OECD 2008); it combines individual indices to form an overall index.

$$I=\sum_{i=1}^{n} w_i \cdot x_i$$

where

$$1 \leq x_i \leq 5; \ \sum w_i = 1; \ 0 \leq w_i \leq 1$$

First, each criterion (=individual index) is assigned a success factor x on a scale of 1–5. The lower the scale value, the worse the technology assessment with respect to the criterion. Conversely, the higher the scale value, the better the assessment. In a second step, weightings w are introduced to account for the fact that the criteria could affect the overall assessment in different ways. The sum of the weighting factors is always 1.

For comparison, the case where the criteria are weighted equally is also analysed. It implicitly exists whenever – supposedly – weighting is not used. The chosen methodology means that the overall index can have values between 1 and 5. The results for the success factor and weighting are based on a survey of IEK-STE energy experts, and are shown in Table 15.1 (individual results: Tables 15.2 and 15.3).

The experts surveyed evaluated criterion (2) 'environmental and safety requirements' as being most fulfilled (rating of 3.00), although 3.00 does not even come close to achieving the maximum, and criterion (5) 'public acceptance' (rating of 1.32) as being least fulfilled. Criterion (1) 'demonstration, commercial availability' was also evaluated relatively positively (rating of 2.84), while (3) 'cost efficiency and economic viability' fared poorly with a rating of 1.86.

With respect to the weighting of the criteria, comparing the expert ratings with the case of equal weighting is relevant (Fig. 15.2). The three criteria'demonstration, commercial availability', 'environmental and safety requirements', and 'coordination of energy and climate policy' are assigned a lower weighting by the experts in the overall assessment, and the deviation in the case of the latter criterion is highest. The two criteria 'cost efficiency and economic viability' and

Table 15.1 Overall classification with varied weighting of the indicators

Criterion	Success factor (scale 1–5)	Weighting ($0 \leq x \leq 1$)	
	Expert rating	Equal weighting	Expert weighting
1. Demonstration, commercial availability	2.84	0.20	0.19
2. Environmental and safety requirements	3.00	0.20	0.18
3. Cost efficiency and economic viability	1.86	0.20	0.27
4. Coordination of energy and climate policy (EU, DE)	2.21	0.20	0.14
5. Public acceptance	1.32	0.20	0.22
Total	–	1	1

Source: Survey of experts in IEK-STE (The survey was designed for energy scientists in different disciplines (engineers, economists, sociologists, and political scientists) at IEK-STE who have been working on CCS technology for a number of years, and who have contributed to this book as authors)

'public acceptance' are weighted higher in the overall assessment, and the deviation in the case of the former criterion is higher. The deviation from equal weighting is clearest for this criterion overall. It should be noted that the two criteria with the lowest success factor – 'cost efficiency and economic viability' and 'public acceptance' – are incorporated into the overall assessment with the highest weighting factors assigned by the respondents.

The overall index comprising the evaluation and the weighting of criteria is slightly higher when all criteria are weighted equally than for the expert rating. However, at approx. 2.2, both results are far below the maximum possible value of the composite index (Fig. 15.2). The survey therefore supports the qualitative conclusions of the preceding chapters.

15.3 Possible Implications for Implementation in Europe

CCS is relevant as a measure for reducing energy-related and process-related CO_2 emissions. The importance of CCS as an element in a technology portfolio within the context of global CO_2 mitigation strategies was recently analysed in the studies *World Energy Outlook 2012* (IEA 2012b) and *Energy Technology Perspectives 2012* (IEA 2012a). Both studies consider CCS an extremely important part of ambitious scenarios (with 50 % likelihood that the increase in the average global temperature will be limited to 2° C).

The scenarios show that CCS in Europe is of crucial importance for ambitious climate mitigation policies, even though technical alternatives (energy efficiency, renewables) are also extremely important here. The likelihood that CCS will not be used in Germany does not in itself mean that the options open to the EU to reduce CO_2 accordingly are impossible, but it does give rise to justified doubts. Moreover, excluding CCS means that other measures, such as energy efficiency or renewables,

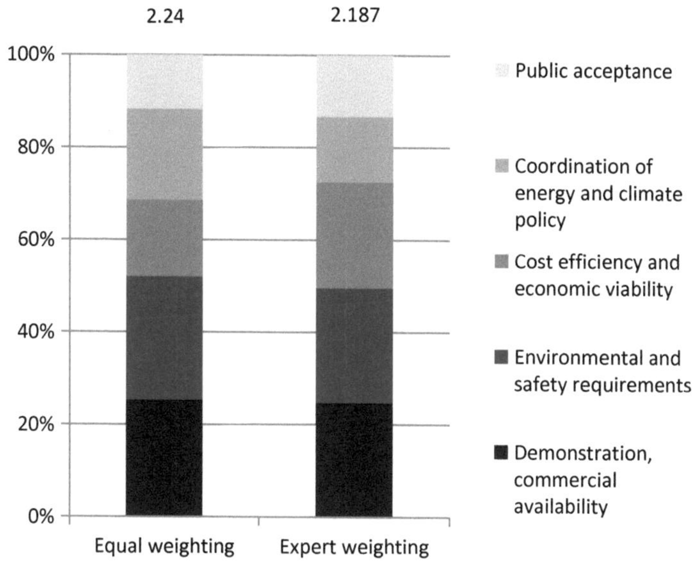

Fig. 15.2 Overall index with varied weighting of the indicators

must additionally achieve the CO_2 savings that CCS was originally intended to achieve. Within the framework of the ambitious reduction scenarios, this poses a particular challenge.

The danger that other European countries will also reject CCS cannot simply be dismissed, even though the waning support is not necessarily due to a lack of public acceptance but rather to other factors, such as high economic risks. Scepticism is warranted because the process for the financial support of demonstration power plants in the EU is currently floundering. The European Commission assigned all of the resources available within the *NER Funding Programme*, totalling some € 1.2 billion, to a series of projects concentrating on renewable energy. In the first round, € 275 million had been earmarked for CCS demonstration. The funds should be available for the second round (European Commission 2012a). As the construction of demonstration power plants, the test phase, and the subsequent tapping of new markets will take at least 15 years, there is a danger that if Europe continues to hesitate it will lose its envisaged leading position in the development and commercialization of CCS power plants. In this context, the reasons stated by the European Commission as to why CCS was not supported in the first round are worrying: 'Most CCS projects put forward were not confirmed by the Member States concerned, and could therefore not be considered for funding awards. Member States were unable to confirm the projects for various reasons: in some cases there were funding gaps, while in others the projects were not sufficiently advanced to allow for confirmation within the timeframe of the first call for proposals.' (European Commission 2012b). In other words, the current commitment of EU member states, the energy economy, and industry is simply not enough.

Appendix: Survey

Fifteen people were asked to participate. Fourteen people completed the survey.

Table 15.2 Individual ratings of the evaluation

	Participant														
	1	2	3	4	5	6	7	8	9	10	11	12	13	14	Ø
Demonstration, commercial availability	3.00	4.00	2.75	3.00	3.00	2.00	4.00	4.00	2.00	2.00	1.00	4.00	3.00	2.00	2.84
Environmental and safety requirements	2.50	3.00	4.00	3.00	3.00	3.00	2.00	5.00	2.50	2.00	4.00	3.00	3.00	2.00	3.00
Cost efficiency and economic viability	2.00	4.00	2.00	2.00	2.00	2.00	1.00	3.00	2.00	2.00	1.00	1.00	1.00	1.00	1.86
Coordination of energy and climate policy	1.50	3.00	2.00	2.00	1.00	2.00	3.00	3.00	1.50	1.00	4.00	2.00	2.00	3.00	2.21
Public acceptance	1.50	2.00	2.00	1.00	1.00	2.00	1.00	1.00	1.00	1.00	2.00	1.00	1.00	1.00	1.32

Table 15.3 Individual ratings of the weighting

	Participant														
	1	2	3	4	5	6	7	8	9	10	11	12	13	14	Ø
Demonstration, commercial availability	0.25	0.10	0.25	0.25	0.25	0.25	0.10	0.10	0.15	0.20	0.10	0.20	0.20	0.30	0.19
Environmental and safety requirements	0.05	0.20	0.15	0.25	0.25	0.15	0.20	0.09	0.20	0.20	0.20	0.20	0.20	0.20	0.18
Cost efficiency and economic viability	0.25	0.30	0.25	0.25	0.20	0.25	0.20	0.30	0.55	0.20	0.30	0.20	0.20	0.30	0.27
Coordination of energy and climate policy	0.25	0.10	0.20	0.20	0.15	0.15	0.25	0.01	0.05	0.20	0.10	0.10	0.20	0.00	0.14
Public acceptance	0.20	0.30	0.15	0.05	0.15	0.20	0.25	0.50	0.05	0.20	0.30	0.30	0.20	0.20	0.22
Total	1.00	1.00	1.00	1.00	1.00	1.00	1.00	1.00	1.00	1.00	1.00	1.00	1.00	1.00	1.00

References

CO_2 Handel (2011) Bund schafft sich Schlupfloch im CO_2-Speichergesetz [Online]. CO_2-Handel. de. http://www.co2-handel.de/article186_16262.html. 28 Apr 2011

European Commission (2011a) Energy roadmap 2050. European Commission, Brussels, COM (2011) 885 final. http://eur-lex.europa.eu/LexUriServ/LexUriServ.do?uri=COM:2011:0885:FIN:EN:PDF

European Commission (2011b) A roadmap for moving to a competitive low carbon economy in 2050. Brussels, Communication from the Commission, COM/2011/0112 final. http://eur-lex.europa.eu/LexUriServ/LexUriServ.do?uri=CELEX:52011DC0112:EN:NOT

European Commission (2012a) 23 innovative renewable energy demonstration projects receive €1.2 billion EU funding. Climate Action, [Online]. http://ec.europa.eu/clima/news/articles/news_2012121801_en.htm

European Commission (2012b) Questions and answers on the outcome of the first call for proposals under the NER300 programme. *MEMO* [Online], MEMO/12/999. http://europa.eu/rapid/press-release_MEMO-12-999_en.htm

IEA (2009) World energy outlook 2009. International Energy Agency OECD/IEA, Paris

IEA (2010) Energy technology perspectives – scenarios & strategies to 2050. International Energy Agency OECD/IEA, Paris

IEA (2011) World energy outlook 2011. International Energy Agency OECD/IEA, Paris

IEA (2012a) Energy technology perspectives 2012. International Energy Agency OECD/IEA, Paris

IEA (2012b) World energy outlook 2012. International Energy Agency OECD/IEA, Paris

McKinsey (2007) Kosten und Potenziale der Vermeidung von Treibhausgasemissionen in Deutschland – Sektorperspektive Industrie. McKinsey&Company, Inc.

Mikulcak K (2012) German government backtracks on the Energy Transition. European Energy Review, 12 July 2012. http://www.europeanenergyreview.eu/site/pagina.php?id=3799#artikel_3799

OECD (2008) Handbook on constructing composite indicators. OECD, Paris. http://www.oecd.org/std/leadingindicatorsandtendencysurveys/42495745.pdf

SPD/GRÜNE/SSW (2012) Bündnis für den Norden. Koalitionsvertrag 2012 bis 2017. Kiel. http://www.sh.gruene.de/cms/default/dokbin/411/411582.koalitionsvertrag_spd_buendnis90die_grue.pdf

Stern N (2006) Stern review: the economics of climate change. London. http://webarchive.nationalarchives.gov.uk/+/http://www.hm-treasury.gov.uk/stern_review_report.htm. 21 June 2012

References

MIX
Papier aus verantwortungsvollen Quellen
Paper from responsible sources
FSC® C105338
FSC
www.fsc.org

If you have any concerns about our products,
you can contact us on
ProductSafety@springernature.com

In case Publisher is established outside the EU,
the EU authorized representative is:
Springer Nature Customer Service Center GmbH
Europaplatz 3, 69115 Heidelberg, Germany

Printed by Libri Plureos GmbH
in Hamburg, Germany